TM 9-741

RESTRICTED

STAGHOUND

MEDIUM ARMORED CAR T17E1

TECHNICAL MANUAL

By **WAR DEPARTMENT**

DISCLAIMER:

This manual is sold for historic research purposes only, as an entertainment. It contains obsolete information and is not intended to be used as part of an actual operation or maintenance training program. No book can substitute for proper training by an authorized instructor.

This document reproduces the text of a manual first published by the Department of the Army, Washington DC. All source material contained in the reproduced document has been approved for public release and unlimited distribution by an agency of the U.S. Government. Any U.S. Government markings in this reproduction that indicate limited distribution or classified material have been superseded by downgrading instructions that were promulgated by an agency of the U.S. government after the original publication of the document. No U.S. government agency is associated with the publishing of this reproduction.

©2013 Periscope Film LLC
All Rights Reserved
ISBN#978-1-937684-40-2
www.PeriscopeFilm.com

TM 9-741

RESTRICTED

TECHNICAL MANUAL }
No. 9-741

WAR DEPARTMENT
Washington, December 15, 1942

MEDIUM ARMORED CAR T17E1

Prepared under the direction of the
Chief of Ordnance
(With the cooperation of the Chevrolet Motor Division,
General Motors Corporation)

CONTENTS

PART ONE — Operating Instructions

			Paragraphs	Pages
Section	I:	Introduction	1-5	3-10
	II:	Operation and controls	6-10	11-24
	III:	Armament and ammunition	11-14	25-33
	IV:	Preventive maintenance	15-21	34-39
	V:	Lubrication instructions	22-27	40-51
	VI:	Care and preservation	28-35	52-55
	VII:	Equipment and tools on vehicle	36-39	56-63
	VIII:	Operations under unusual conditions	40-41	64-67
	IX:	Materiel affected by gas	42-45	68-70

PART TWO — Organization Instructions

	X:	Organization maintenance	46-48	71-74
	XI:	Tools and equipment	49-50	75
	XII:	Organization spare parts and accessories	51-52	76
	XIII:	Engines	53-63	77-117
	XIV:	Engine ignition system	64-72	118-124
	XV:	Fuel system	73-80	125-134
	XVI:	Cooling system	81-90	135-149
	XVII:	Exhaust system	91-94	150
	XVIII:	Fluid coupling	95	151
	XIX:	Transmission	96-101	152-177
	XX:	Transmission gear reduction assembly	102-104	178

TM 9-741

MEDIUM ARMORED CAR T17E1

PART TWO — Organization Instructions (Cont'd)

		Paragraphs	Pages
SECTION XXI:	Propeller shafts and universal joints	105-113	179-187
XXII:	Transfer case	114-117	188-193
XXIII:	Front axle	118-123	194-199
XXIV:	Rear axle	124-126	200-203
XXV:	Brake system	127-138	204-224
XXVI:	Springs, radius rods, and shock absorbers	139-142	225-229
XXVII:	Steering gear	143-148	230-237
XXVIII:	Wheels, tires, wheel bearings, and tire pump	149-154	238-247
XXIX:	Batteries and starting system	155-161	248-251
XXX:	Generators and controls	162-166	252-256
XXXI:	Lighting system	167-177	257-263
XXXII:	Instruments and gages	178-184	264-271
XXXIII:	Electrical accessories	185-189	272-273
XXXIV:	Radio suppression	190-191	274-277
XXXV:	Electrical system wiring	192-195	278-306
XXXVI:	Turret and traversing system	196-203	307-313
XXXVII:	Hull	204-208	314-317
XXXVIII:	Gyrostabilizer	209-234	318-343
XXXIX:	Fire extinguishing system	235-238	344-348
XL:	Storage and shipment	239-240	349-352
XLI:	References	241-242	353-354
INDEX			355-368

TM 9-741
1-4

PART ONE — OPERATING INSTRUCTIONS

Section I

INTRODUCTION

	Paragraph
Purpose and scope	1
Content and arrangement of manual	2
References	3
Description	4
Data	5

1. PURPOSE AND SCOPE.

a. This technical manual is intended to serve temporarily (pending the publication of a more complete revision) to give information and guidance to personnel of the using arms charged with the operation, maintenance and minor repair of this materiel.

2. CONTENT AND ARRANGEMENT OF MANUAL.

a. Specific information for the guidance of operating personnel is contained in section I to section IX inclusive. Information chiefly for the guidance of organizational maintenance personnel is contained in section X to section XL inclusive.

3. REFERENCES.

a. All pertinent standard nomenclature lists, technical manuals, and other publications having reference to the materiel described herein are listed in section XLI.

4. DESCRIPTION.

a. The Medium Armored Car T17E1 is a four-wheeled, four-wheel-drive vehicle. The vehicle is operated and controlled in a manner similar to that of a four-wheel-drive truck. Various automatic features embodied in the design and the use of power mechanisms assist the driver in the control of the vehicle.

b. The vehicle is powered by two six-cylinder, valve-in-head, 97 horsepower water-cooled engines. The engines are located side by side in the rear of the hull providing unrestricted operation of the armament. The engines can be operated simultaneously or individually.

c. The power from each engine is transmitted through its four speed hydra-matic transmission and a gear reduction case to a single,

MEDIUM ARMORED CAR T17E1

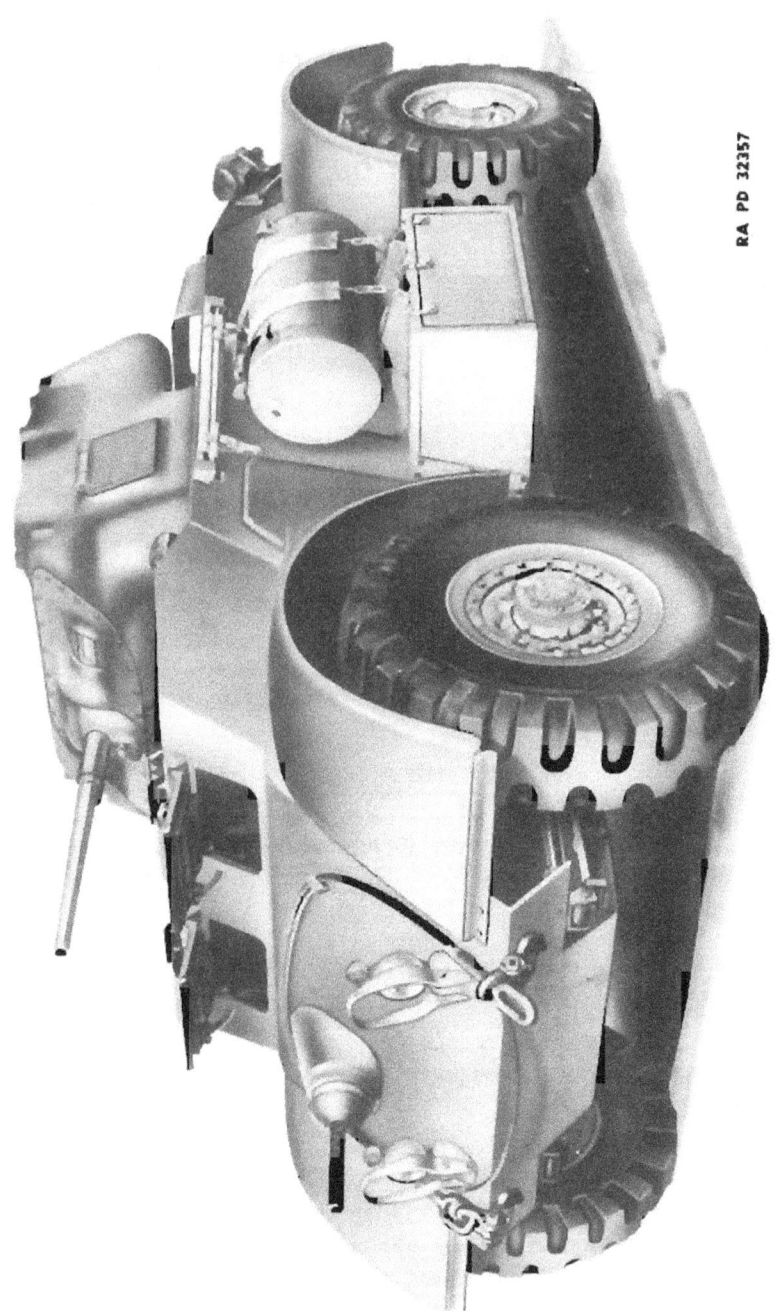

Figure 1 — Three-quarter Left Front — Medium Armored Car T17E1

INTRODUCTION

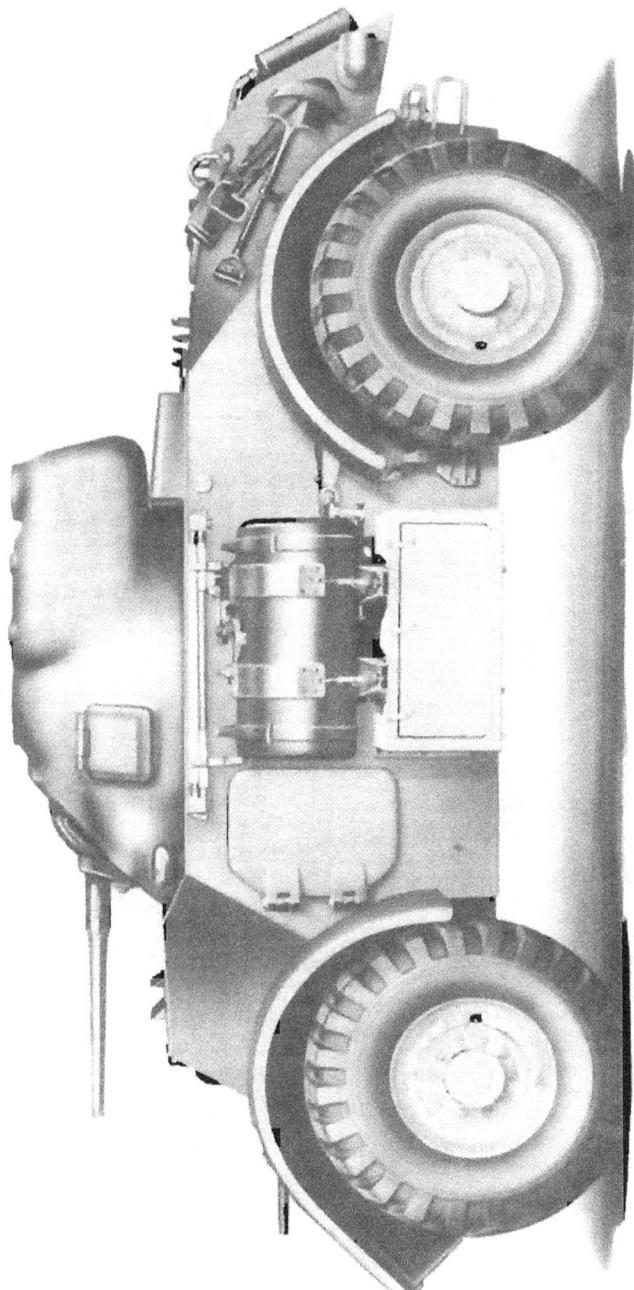

Figure 2 — Left Side — Medium Armored Car T17E1

MEDIUM ARMORED CAR T17E1

Figure 3 — Right Side — Medium Armored Car T17E1

INTRODUCTION

Figure 4 — Front — Medium Armored Car T17E1

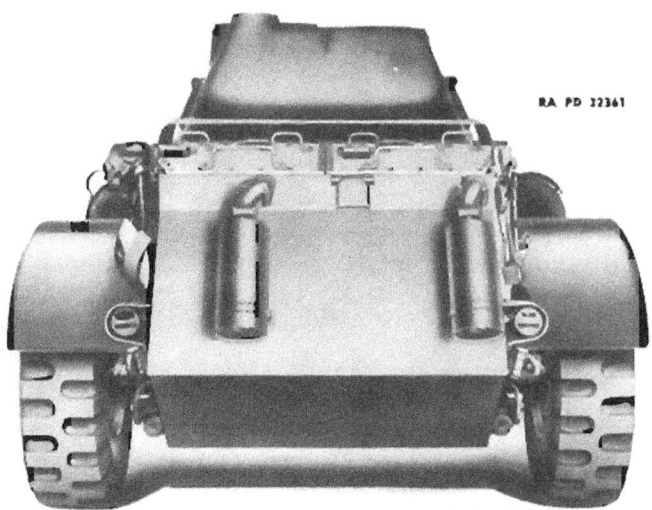

Figure 5 — Rear — Medium Armored Car T17E1

TM 9-741
4

MEDIUM ARMORED CAR T17E1

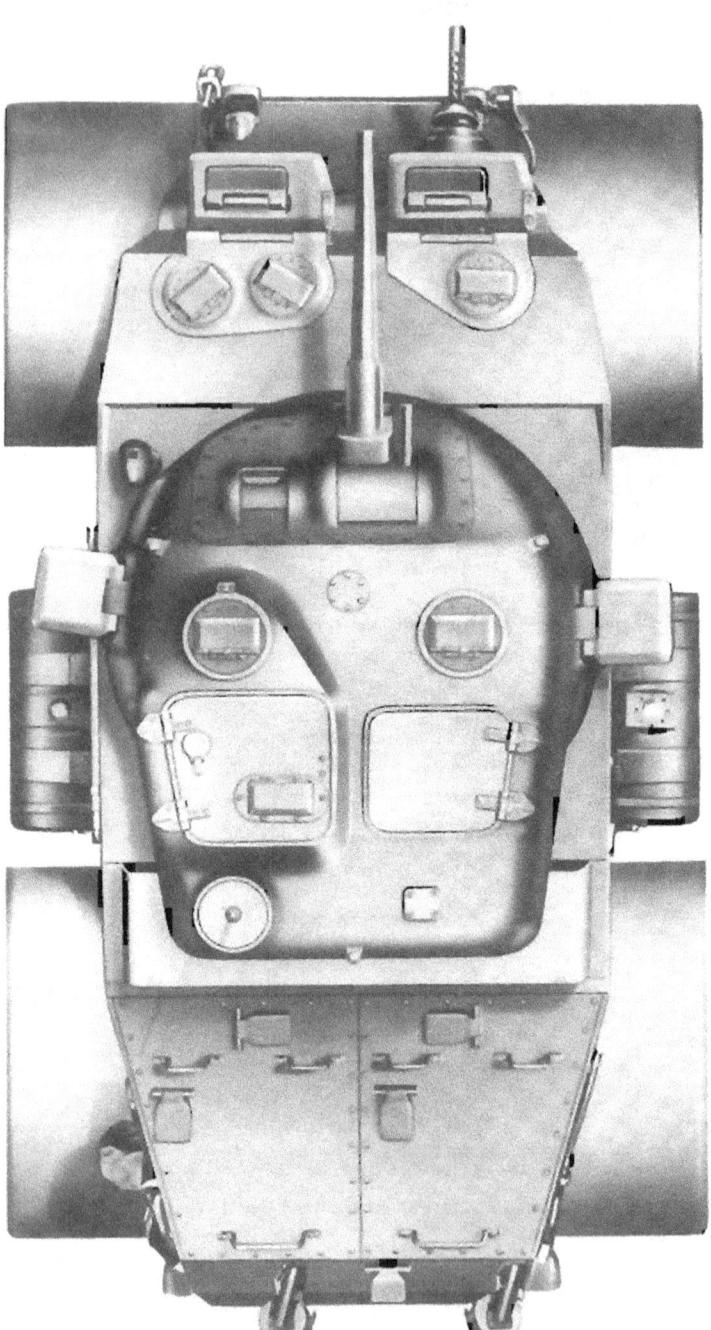

Figure 6 — Top — Medium Armored Car T17E1

INTRODUCTION

two-speed, transfer case. Drive shafts transmit the power from the transfer case to the front and rear axles, and thus, to the wheels.

d. The hull is so constructed that a frame is not required. The springs, steering gear, transfer case, and other units that would normally be attached to the frame, are attached directly to the hull.

e. The outside of the hull is constructed of armor plate and is fitted for mounting the turret.

f. This vehicle is designed for a five man crew. The driver and assistant driver are in the hull or front compartment; the gunner, loader, and radio operator are in the turret or rear compartment.

5. DATA.

a. General.

Wheelbase .... 120 in.
Weight with armament (without ammunition, radio,
  dunnage, fuel, and crew) .... 26,590 lb
Ground pressure, pounds per square inch
  0 inch depression 60.5
  4 inch depression 17.95
Ground clearance .... 13¼ in.
Tread (center to center of wheels) .... 89.1 in.
Over-all height .... 92 in.
Over-all length .... 211¼ in.
Over-all width .... 106 in.

b. Engine (two, valve-in-head).

Rated horsepower (each) at 3,000 rpm 97
Number of cylinders (each) 6
Weight of engine with accessories .... 830 lb

c. Armament.

In turret:
  37-mm gun 1
  .30 cal. machine gun (fixed) 1
In driver's compartment:
  .30 cal. machine gun (flexible) 1
  .45 cal. submachine gun 1
    (not mounted)

d. Ammunition.

37-mm .... 101 rounds
Cal. .30 .... 5000 rounds
Cal. .45 .... 450 rounds
4 boxes hand grenades (3 grenades each)

MEDIUM ARMORED CAR T17E1

e. **Protected Vision.** Protected vision is provided for the driver and assistant by indirect vision devices called periscopes, mounted above the windshield.

f. **Seats, Body Supports, and Safety Belts.** These are provided for each of the five members of the crew.

g. **Communication.**

Radio ....SCR506 or SCR508
Intra-car ....Telephone

h. **Fuel and Oil.**

Fuel capacity (including jettison tanks) ....112 gal
Number of miles without refueling (at 40 mph —
   hard surface level road, 72 F air temperature) ....500 miles
Octane rating of fuel 80
Oil consumption (approximate miles per qt) 700
Engine oil capacity (each engine) ....8 qt
Lubricants ....See lubrication guide

i. **Performance.**

Maximum sustained speed on hard road ....55 mph
Maximum allowable engine speed ....3,500 rpm
Minimum engine idling speed ....500 rpm
Maximum grade ascending ability ....65 percent
Maximum grade descending ability ....65 percent
Maximum fording depth (at slowest forward speed)...........32 in.

TM 9-741
6

Section II

OPERATION AND CONTROLS

Paragraph

General information on instruments and controls	6
Prestarting inspection	7
Starting the engines	8
Operating the vehicle	9
Cold weather precautions	10

6. GENERAL INFORMATION ON INSTRUMENTS AND CONTROLS.

a. It is of definite importance that everyone authorized to drive one of these vehicles be thoroughly familiar with the various controls and their proper use. Due to the location and action of the various controls, the size and weight of the vehicle, and the limited vision of the driver, it will require a little practice to become accustomed to the operation or "feel" of this vehicle.

b. The driver who obtains a thorough knowledge of the vehicle and adheres to good driving practices will obtain the maximum of economy and performance.

c. Instruments are provided which indicate the condition of such vital items as engine temperature, engine oil pressure, electrical charging rate, quantity of fuel, etc., all of which are of vital importance to continued operation. CAUTION: Watch these instruments on the panel in front of the driver closely. This is important.

d. Figures 7 to 18 inclusive illustrate the controls, instruments, and instruction plates which are referred to by the key numbers in the following text. The right- and left-hand sides of the vehicle are determined from the driver's position.

e. **Transmission Manual Control Lever** (1, fig. 7). This lever manually controls the automatic transmissions to select the neutral, drive, low, or reverse range. This lever has detents in the control linkage so that each position is positively selected (fig. 8).

f. **Transfer Case Shift Lever** (2, fig. 7). This lever selects the high, neutral, and low ratios in the transfer case assembly. The lever operates in a gate plate (3, fig. 7) that is notched for each position. High ratio is the bottom notch, neutral the center, and low the top notch. CAUTION: When shifting the transfer case shift lever, the transmission must be in neutral (fig. 9).

MEDIUM ARMORED CAR T17E1

1. Transmission Manual Control Lever
2. Transfer Case Shift Lever
3. Transfer Case Shift Lever Gate Plate
4. Engine Selector Lever
5. Carburetor Choke Levers—Left and Right Engines
6. Hand Throttles—Left and Right Engines
7. Hand Brake Lever
8. Transfer Case Front Axle Lever
10. Head Lamp Release Buttons
11. Brake Pedal
12. Accelerator Pedal
13. Steering Gear Motor Switch
14. Siren Foot Control
15. Hand Fire Extinguisher
16. Fire Extinguishing System Control Handles
17. Periscopes
18. Hand Brake Assist Pedal
19. Choke Levers Clamp Bolt
20. Throttle Levers Clamp Bolt
21. Compass

RA PD 56263

Figure 7 — Driver's Compartment

g. **Engine Selector Lever** (4, fig. 7). This lever engages the transmissions to the transfer case. To shift lever, pull up on the knob and move lever to left to disengage the left-hand engine; shift lever to the right to disengage right-hand engine. To engage both engines, place lever in center position. The transfer case shift lever must be in neutral when shifting engine selector lever (fig. 10).

h. **Carburetor Choke Levers** (5, fig. 7). The lever next to the hull controls the choke for the left-hand engine, the other lever, the choke

TM 9-741

OPERATION AND CONTROLS

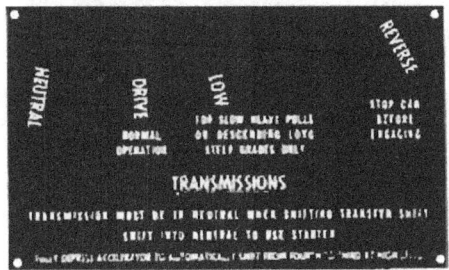

Figure 8 — Transmission Control Lever Instruction Plate

Figure 9 — Transfer Case Shift Lever Instruction Plate

for the right-hand engine. When levers are in forward position, the chokes are in the "OFF" position. When starting a cold engine, pull the lever back as this shuts off the air to the carburetor providing a rich

TM 9-741

MEDIUM ARMORED CAR T17E1

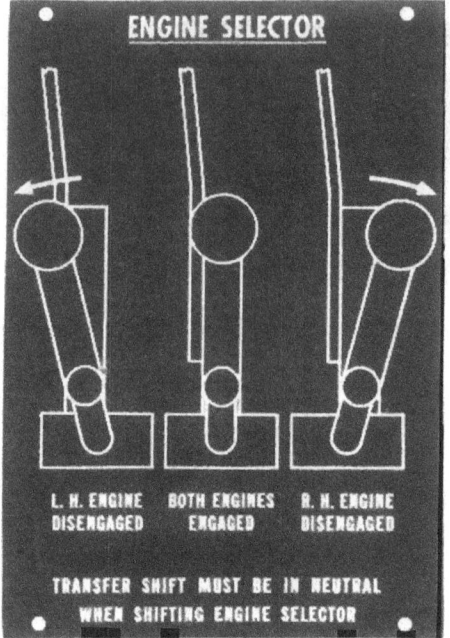

Figure 10 — Engine Selector Lever Instruction Plate

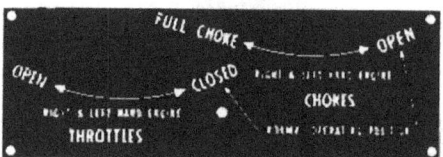

Figure 11 — Choke and Throttle Control Instruction Plate

mixture. When engine starts, push lever forward. If the engine is warm, the use of the choke is unnecessary (fig. 11).

i. **Hand Throttle** (6, fig. 7). These two levers are just in back of the choke levers. The lever next to the hull controls the throttle for the left-hand engine, the outer lever the throttle for the right-hand engine. Pulling back on the lever opens the throttle and can be used when starting the engine and to run engine at a constant speed (fig. 11).

TM 9-741
6

OPERATION AND CONTROLS

Figure 12 — Front Axle Shift Control Instruction Plate

j. **Hand Brake Lever** (7, fig. 7). This lever controls the two-propeller shaft parking brakes at the front of the transfer case. When vehicle is parked, lever should be applied by pulling toward rear as far as possible. Before attempting to move vehicle, lever should be in released position (as far forward as it will go). Lever may be released by pressing release handle (on lever) and pushing forward.

k. **Transfer Case Front Axle Shift Lever** (8, fig. 7). Moving the lever forward disengages the front axle drive; moving lever backward engages front axle drive. When it is necessary to use the front axle drive, the transfer case should be in the low ratio (fig. 12).

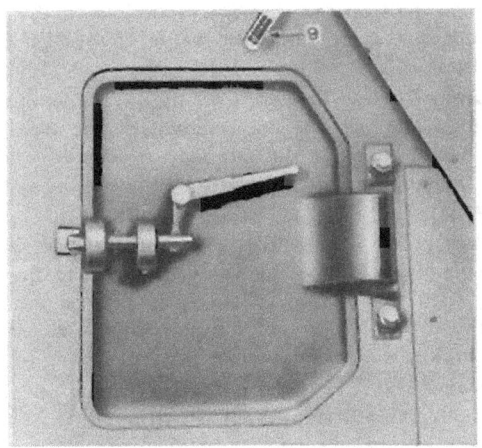

Figure 13 — Jettison Tank Release Lever

15

MEDIUM ARMORED CAR T17E1

Figure 14 — Jettison Tank Release Lever Instruction Plate

l. **Jettison Fuel Tank Release Lever** (9, fig. 13). One lever on each side of hull, when pulled forward, releases the fuel tank mounted on the outside of the hull on the side corresponding to the lever mounting. As the tank drops, an automatic valve closes the gas line from the tank (fig. 14).

m. **Head Lamp Release Button** (10, fig. 7). One button on each side, when pulled out, permits the removal of the head lamps from their sockets. The button on the left side is located underneath the steering gear assembly on the forward bulkhead.

n. **Brake Pedal** (11, fig. 7). Pressing on the brake pedal applies hydraulic brakes at all four wheels and operates the vacuum brake boosters which apply additional hydraulic pressure to the brake shoes. Avoid driving with foot on brake pedal as brakes will be partially applied and cause needless wear of brake linings. Smooth and even application of brakes whenever possible is good driving practice.

o. **Accelerator Pedal** (12, fig. 7). Pressing down on pedal controls the speed of both engines simultaneously through a hydraulic mechanism.

TM 9-741
6

OPERATION AND CONTROLS

p. **Steering Gear Electric Motor Switch** (13, fig. 7). The electric motor that drives the steering gear hydraulic pump is controlled by turning this switch to the "OFF" or "ON" position. This circuit includes a circuit breaker and reset button. The reset button is located on the front side of the switch box. If the motor fails to operate, press the button in to reset the circuit breaker. CAUTION: Do not hold the reset button in to maintain a circuit as damage to the motor or circuit will result.

q. **Siren Foot Control** (14, fig. 7). Pressing down on button sounds siren. This circuit is protected by a circuit breaker. If the siren fails to operate, press the circuit breaker reset button. The reset button is mounted on the right side of the steering gear motor switch box (13, fig. 7). CAUTION: Do not hold this reset button in to maintain a circuit as damage to the circuit will result.

r. **Hand Fire Extinguisher** (15, fig. 7). Remove container from mounting bracket and flood the fire area. Cylinder contains carbon dioxide.

s. **Fire Extinguishing System Control Handles** (16, fig. 7). The two control handles are located above the driver and control the system that is in the engine compartments. To operate the system, pull one control handle. For second fire prior to recharging of first cylinder, pull other control handle.

t. **Periscopes** (17, fig. 7). Two periscopes are provided for use of driver and one for use of assistant driver.

u. **Hand Brake Assist Pedal** (18, fig. 7). This pedal is linked directly to the hand brake lever and provides a means of applying sufficient pressure to the parking brakes to hold the vehicle when parking on extreme grades. Pull on the hand brake lever and depress the pedal with the left foot for additional brake pressure. Use the left foot and hand to apply sufficient pressure to release the lever pawl. NOTE: This brake should not be used as a means of stopping the vehicle.

v. **Choke and Throttle Lever Clamp Bolts** (19 and 20, fig. 7). These clamp bolts provide a means of clamping the levers in any desired position for constant engine speed or to lock the choke levers in the "OFF" position.

w. **Compass** (21, fig. 7). The pioneer compass provides a means of determining the direction of travel.

x. **Caution Plate** (fig. 15 and 12, fig. 17). This plate, located on instrument panel at right of speedometer, shows maximum speeds in the two transfer case gear ratios.

TM 9-741
6

MEDIUM ARMORED CAR T17E1

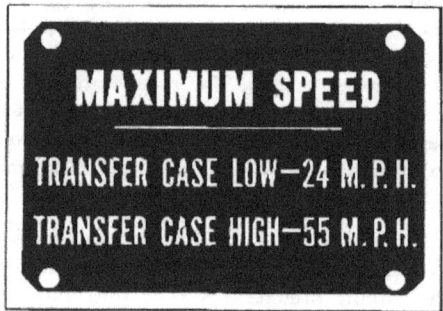

Figure 15 — Speed Caution Plate

y. Identification Plate (fig. 16). This plate covers manufacturer's serial number, ordnance serial number, date of delivery, gross weight, etc. This plate is located forward of the right-hand door at the bottom.

z. Ignition Switches (1, fig. 17). Turn switch to left for "ON" position. Left switch is for left engine and right switch for right engine. Never allow ignition switch to remain turned "ON" when engine is not running, except when making service tests.

aa. Starting Buttons (2, fig. 17). Push starting button to operate starting motor. Left button is for left engine and right button for right engine. Pushing button completes starting motor electrical circuit. When engine starts, release button immediately. CAUTION: Do not push starting button while engine is running. Transmission manual control lever *must* be in neutral position when starting engine, as a safety switch in the starting motor circuit is closed only when this lever is in neutral.

bb. Instrument Panel Lights (3, fig. 17). Pull out on this button to turn on instrument panel lights. Turn switch knob to dim lights.

cc. Lighting Switch (4, fig. 17). This switch controls the head, tail, and stop lights. When the lighting switch is pulled out to the first position, it turns on the blackout head lamps, taillight, and stop light. To turn on the service head lights, taillight, and stop light, depress the safety button on the side of switch and pull the control button out to the second position. When the switch is pulled out to the third position, the service stop light will operate for daylight driving.

dd. Ammeters (5, fig. 17). These two gages register the charging rate of each generator; the gage on right for right engine and gage on left for left engine.

18

TM 9-741
6

OPERATION AND CONTROLS

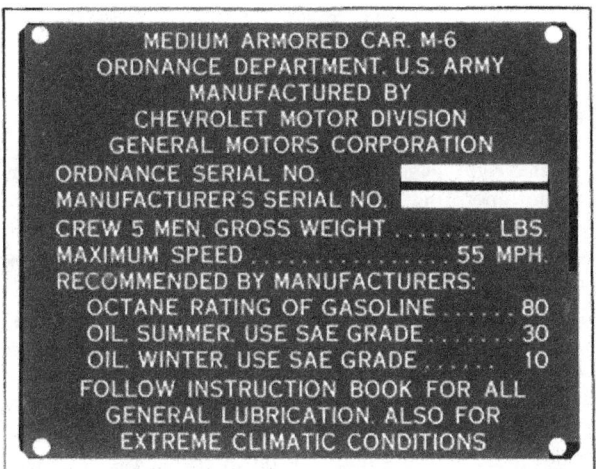

RA PD 32372

Figure 16 — Identification Plate

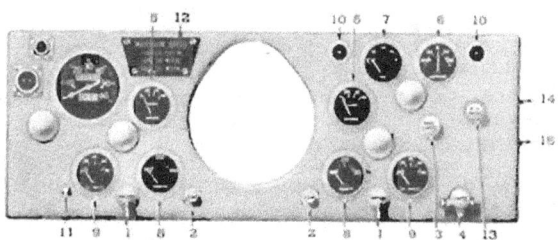

1. Ignition Switches—Left and Right Engines
2. Starting Buttons—Left and Right Engines
3. Instrument Panel Light Switch
4. Lighting Switch
5. Ammeters—Left and Right Engines
6. Ammeter—Total Output, Both Generators
7. Fuel Gage - Main Fuel Tank
8. Engine Temperature Indicators—Left and Right Engines
9. Oil Pressure Gages - Left and Right Engines
10. Electrical Sockets—Left and Right Windshield Wipers
11. Speedometer Trip Mileage Reset
12. Speed Caution Plate
13. Blackout Driving Lamp Lighting Switch Circuit
14. Lighting Switch Circuit Breaker Reset Button
15. Fuel Pump Circuit Breaker Reset Button

RA PD 54264

Figure 17 — Instrument Panel

19

MEDIUM ARMORED CAR T17E1

ee. **Ammeter** (6, fig. 17). This ammeter shows the rate of charge or discharge of the combined electrical system of the vehicle.

ff. **Fuel Gage** (7, fig. 17). This gage registers the amount of fuel in the main tank when either ignition switch is turned on. The dial has graduations for empty, ¼, ½, ¾, and full.

gg. **Engine Temperature Indicator** (8, fig. 17). These instruments indicate the temperature of the water in each cooling system, left gage for left engine, right gage for right engine. Water temperature is dependent upon operating conditions; however, if temperature should reach 240 F, engine should be stopped and trouble corrected. Orange luminous markers are provided at this temperature marking on the instrument dial.

hh. **Oil Pressure Gages** (9, fig. 17). These gages indicate oil pressure in the engine lubrication system, left gage for left engine, right gage for right engine. Pressure reading may vary according to operating conditions; however, if oil pressure falls below normal while engine is running, stop engine immediately and investigate the cause. This gage does not indicate the amount of oil in crankcase.

ii. **Electrical Sockets** (10, fig. 17). These sockets are for plugging in the electrical connections for the windshield wipers on the removable windshields.

jj. **Speedometer Trip Mileage Reset** (11, fig. 17). To reset the trip mileage (lower group of figures on speedometer dial), push in on knob and turn to left until all figures are the same; then turn to right to set to zero.

kk. **Blackout Driving Lamp Switch** (13, fig. 17). When desirable to drive with the blackout head lamp, the lamp must be removed from the storage bracket in the supply compartment and installed in place of the left head lamp and the head lamp placed on its storage bracket in the supply compartment. The main light switch must be in blackout position and the blackout driving lamp switch turned on.

ll. **Lighting Switch Circuit Breaker Reset Button** (14, fig. 17). When the light circuit fails to function, the reset button should be pressed in to reset the circuit breaker. CAUTION: The reset button *should not be held in* to maintain a circuit. If the circuit breaker continues to "throw out," a short circuit or overload is indicated. This would result in a fire or damage to the unit.

mm. **Fuel Pump Circuit Breaker Reset Button** (15, fig. 17). When the fuel pump fails to function, the reset button should be pressed in to reset the circuit breaker. CAUTION: The reset button should not be held *in* to maintain a circuit. This would result in a fire or damage to the unit.

OPERATION AND CONTROLS

Figure 18 — Master Switch Box

nn. **Transmission Low Pressure Signal Lights.** The left light is for the left transmission and the right light for the right transmission oil pressure. Any time the transmission oil pressure drops below a safe pressure, the red lights light. CAUTION: Avoid driving when either of these lights is on as this will damage the transmission.

oo. **Master Electric Switch Box** (fig. 18). This box is located on right side of hull above and to the rear of right entrance door opening. Both tee handles must be pulled out and rotated until tee is horizontal and switch drops into position before vehicle can be operated.

7. PRESTARTING INSPECTION.

a. Before starting either engine the prestarting inspections given in paragraph 16 should be made.

8. STARTING THE ENGINES.

a. Close master switch by pulling each tee handle out and rotating until tee is horizontal and switch drops into position (fig. 18).

b. Pull hand brake lever back to set brakes.

c. Pull right-hand engine hand throttle lever back about ½ inch. This may not be necessary if engine is warm.

d. Pull right-hand engine hand choke lever. NOTE: The distance the choke is to be pulled back and the time required before it can be returned to the forward position depends on the temperature of

TM 9-741
8-9

MEDIUM ARMORED CAR T17E1

the engine. In extremely cold weather, a cold engine may require full choke to start, while in warm weather or with a warm engine, choking should not be required. The choke should be returned to its forward position as soon as the engine is warmed up sufficiently to run without partial choke.

e. Place transmission manual control lever in neutral position.

f. Turn on right-hand ignition switch.

g. Press right-hand starter button. Release starter button as soon as engine starts.

h. Push choke and throttle levers forward until the engine runs smoothly; slightly faster than idling speed. CAUTION: Never race an engine to shorten the warm-up period. Allow engine to run at a fast idle until it is thoroughly warm. Look at the instruments to see that the oil gage registers normal pressure and that the ammeter indicates generator charging.

i. Start left-hand engine in the same manner by using the left controls.

j. After the engines are thoroughly warmed up, they should be tested at idling and under acceleration for oil pressure, generator charging rate, smooth operation, and any unusual noises that might indicate trouble.

9. OPERATING THE VEHICLE.

a. Starting Procedure.

(1) Turn on electric motor steering gear switch.

(2) Move engine selector lever to center position.

(3) Move transfer case lever to "HIGH" or "LOW" speed position (par. 9 c).

(4) Move transmission manual control lever to drive position. The transmissions are automatic and will select the proper gear ratio depending on engine speed and vehicle speed.

(5) Move front axle shift lever forward (par. 9 c).

(6) Release hand brake lever.

(7) Step down on accelerator pedal.

(8) Release accelerator pedal as vehicle speed increases until desired speed is obtained.

b. Shifting Transmissions.

(1) Both transmissions automatically select the proper gear ratio, with the exception of reverse, depending on the speed of the engines

OPERATION AND CONTROLS

and the speed of the vehicle when transmission lever is in drive position. It is not necessary to move the lever for starting, stopping, or accelerating.

(2) When a slow heavy pull is encountered or when descending long steep grades, move the transmission manual control lever to the low position. Transmission will not shift to a higher gear than second but will automatically change from first to second or drop back to first, depending on engine and vehicle speed, permitting better pulling power or braking effort from the engines.

(3) Manual control lever may be moved to low, drive, or neutral position when vehicle is moving. CAUTION: Stop vehicle before moving lever to reverse position. Both engines must be running when vehicle is in reverse or the dead engine must be disengaged with the selector lever as explained in d below.

(4) To accelerate vehicle when driving, and transmission is in high gear, fully depress accelerator pedal to automatically shift from fourth to third gear.

c. **Shifting Transfer Case.**

(1) FRONT AXLE. To engage front axle so that vehicle will drive at all four wheels, move front axle shift lever to rear position. Front axle should be used for heavy duty low gear operation or where additional traction is required. NOTE: Do not use front axle drive when operating on dry, hard surface roads.

(2) "HIGH" AND "LOW" RANGE—TRANSFER CASE. To shift transfer case, stop vehicle and move transmission lever to neutral. Use "HIGH" range except for heavy duty operations and long drives at slow speeds such as encountered in convoys.

d. **Shifting Engine Selector.** To shift engine selector lever, transfer case shift lever must be in neutral position. Operation of the selector lever permits engagement of either right or left engine, or both. Selector lever should be in center position at all times except when one engine may be inoperative or when operating under good road conditions when the power of one engine is sufficient and it is necessary to conserve fuel.

e. **To Stop Vehicle.**

(1) To stop vehicle, remove foot from accelerator pedal and apply brakes by pressing down on brake pedal. If vehicle is to be moved immediately, do not change the other controls but just press down on accelerator pedal.

(2) If vehicle is to be parked, move the transmission manual control lever to neutral position, apply hand brake, and turn off ignition

MEDIUM ARMORED CAR T17E1

switches. Open the two main electrical switches by pulling out on tee handles and rotating until a definite stop is reached.

10. COLD WEATHER PRECAUTIONS.

a. If possible, store the vehicle within a garage or some similar enclosure during extremely cold weather.

b. If freezing temperatures are expected, antifreeze in sufficient quantity to protect the cooling system should be added to the radiators or the complete cooling system must be drained and a tag to this effect placed on the steering wheel. See paragraph 85 for additional information.

c. It may be necessary to change the cooling system thermostat to one of the proper range so that the engine will run at its most efficient operating temperature during cold weather (par. 87).

d. Refer to section V for proper viscosity of lubricant and points of change for cold weather operation. Difficult starting or damage to moving parts may result from the use of the improper viscosity lubricant.

TM 9-741

Section III

ARMAMENT AND AMMUNITION

	Paragraph
Armament and ammunition	11
Armament operation	12
Gun mounts	13
Gyrostabilizer	14

11. ARMAMENT AND AMMUNITION.

a. The armament and ammunition carried with the vehicle and their approximate location is listed below.

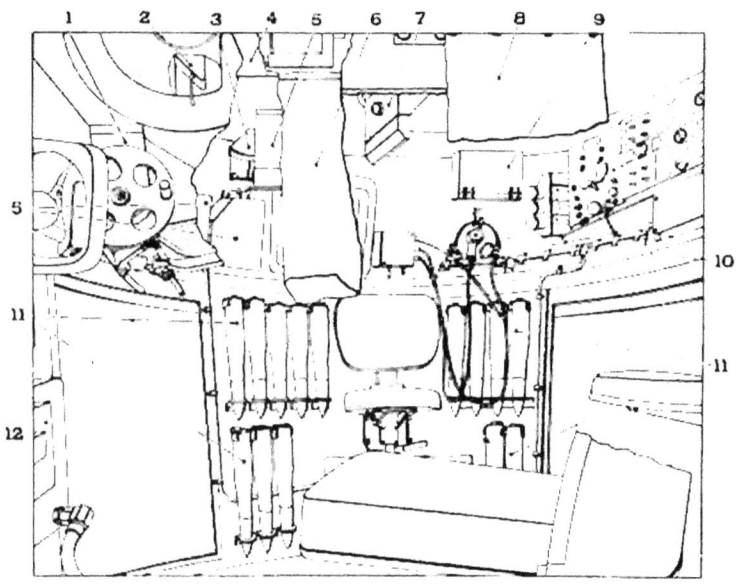

1. 37 M M GUN MOUNT
2. PERISCOPE M2
3. CANTEEN
4. CAL. .30 MACHINE GUN MOUNT
5. CAL. .30 AMMUNITION BOXES
6. CAL. .30 EMPTY CARTRIDGE BAG
7. PERISCOPE M6
8. 37 M M EMPTY CARTRIDGE BAG
9. SPARE PERISCOPE M6
10. PHONE
11. 37 M M AMMUNITION H.E., M63
12. SPARE PERISCOPE M2

RA PD 56272

Figure 19 — Right Side of Turret Basket

TM 9-741

MEDIUM ARMORED CAR T17E1

b. **Armament.**

Material	Quantity	Approximate Location
37-mm gun M6	1	In turret
37-mm gun mount M24	1	In turret
Cal. .30 machine gun (fixed)	1	In turret
Cal. .30 machine gun (flexible)	1	Mounted in hull
Cal. .30 machine gun flexible mount	1	Mounted in hull
Cal. .45 sub-machine gun	1	On right-hand side of hull just ahead of door (accessible to turret crew)

c. **Ammunition.**

Material	Quantity	Approximate Location
37-mm canister M2	10	8—Left turret bulge 2—Left side of hull under door
37-mm A.P., M51 and H.E., M63	91	13—Right side of hull 7—Front bulkhead 33—Left side of hull 35—Turret basket 3—On ammunition rack on tunnel
Boxes, cal. .30 (250 rounds each)	20	4—On floor back of driver 2—On floor back of bow gunner 2—On doors (1 each door) 2—On floor below right door 6—In ammunition rack on tunnel 1—In rack on bow gun mount 2—Right side of turret, forward 1—In gun rack on turret gun
Clips, cal. .45 (30 rounds each)	15	15—In rack on right side of hull just ahead of gun
Boxes, hand grenade (3 grenades each)	4	2—On left hull floor under turret platform 2—Turret basket floor

TM 9-741
11

ARMAMENT AND AMMUNITION

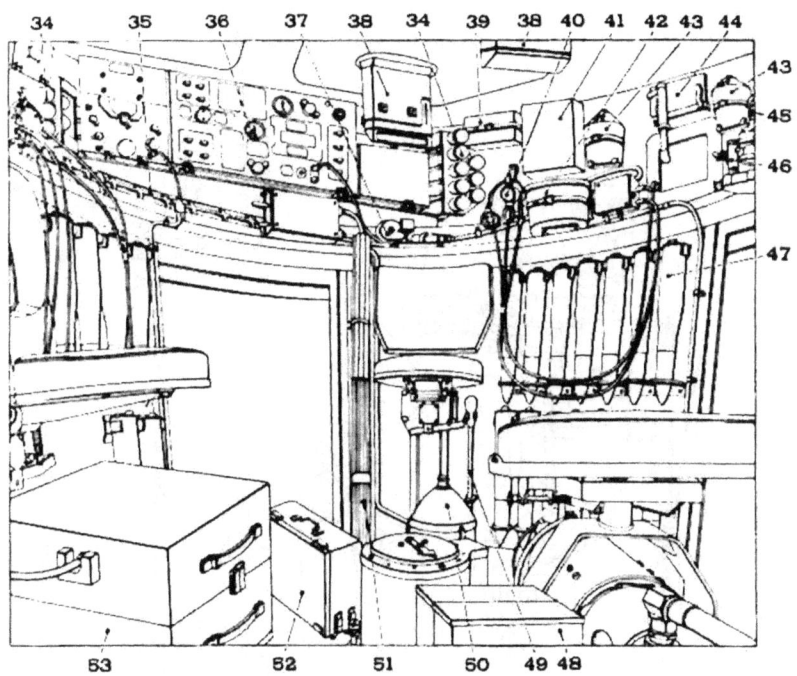

34. 37 M M CANNISTER M2 AMMUNITION
35. SPARE PERISCOPE HEADS M2 AND M6
36. RADIO SCR-506 OR SCR-508
37. SPOTLIGHT HANDLE AND REEL
38. PERISCOPES M6
39. CAL .30 AND 37 M/M SMALL PARTS
40. PHONE
41. SPARE PERISCOPE M6
42. BINOCULARS
43. CANTEENS
44. MAP
45. COMPASS
46. FLASHLIGHT
47. 37 M/M AMMUNITION A.P., M51
48. HAND GRENADE BOXES
49. SPOT LIGHT SHAFT
50. SPOT LIGHT
51. SIGNAL FLAGS
52. 37 M M SPARE PARTS
53. RADIO SCR-510

RA PD 56273

Figure 20 — Left Side of Turret Basket

TM 9-741
11

MEDIUM ARMORED CAR T17E1

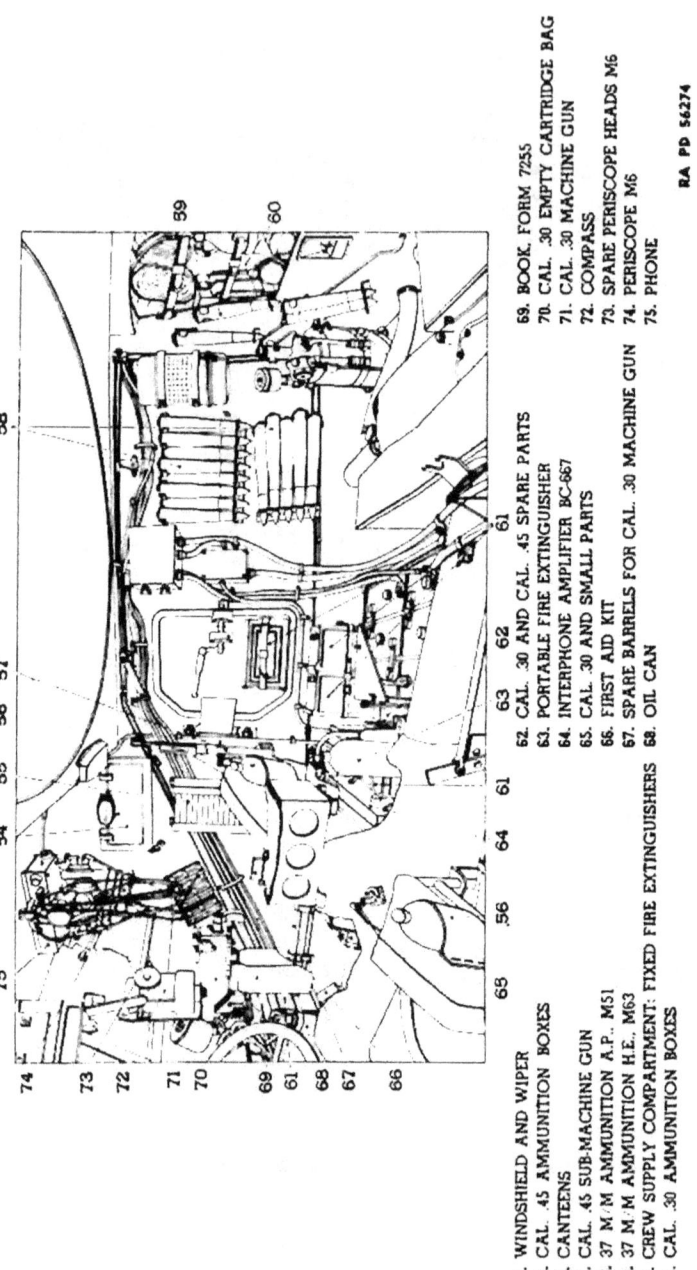

54. WINDSHIELD AND WIPER
55. CAL. .45 AMMUNITION BOXES
56. CANTEENS
57. CAL. .45 SUB-MACHINE GUN
58. 37 M/M AMMUNITION A.P., M51
59. 37 M/M AMMUNITION H.E., M63
60. CREW SUPPLY COMPARTMENT; FIXED FIRE EXTINGUISHERS
61. CAL. .30 AMMUNITION BOXES
62. CAL. .30 AND CAL. .45 SPARE PARTS
63. PORTABLE FIRE EXTINGUISHER
64. INTERPHONE AMPLIFIER BC-667
65. CAL. .30 AND SMALL PARTS
66. FIRST AID KIT
67. SPARE BARRELS FOR CAL. .30 MACHINE GUN
68. OIL CAN
69. BOOK, FORM 7255
70. CAL. .30 EMPTY CARTRIDGE BAG
71. CAL. .30 MACHINE GUN
72. COMPASS
73. SPARE PERISCOPE HEADS M6
74. PERISCOPE M6
75. PHONE

RA PD 56274

Figure 21 — Right Side of Hull Interior

ARMAMENT AND AMMUNITION

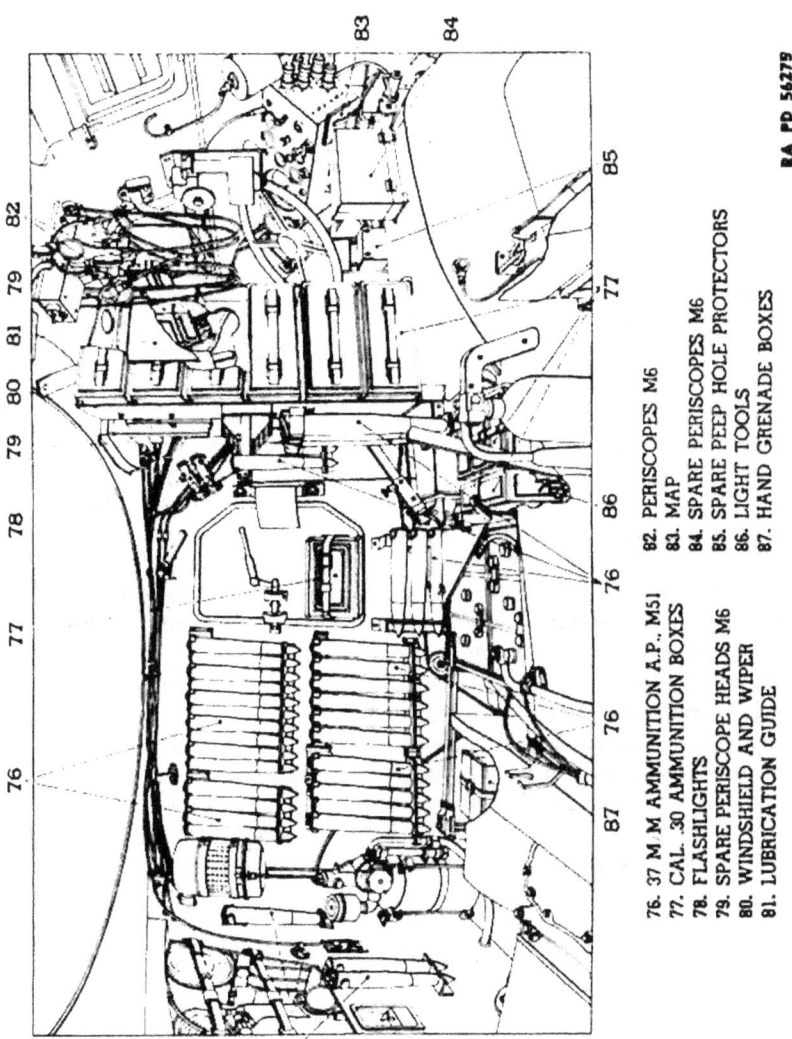

Figure 22 — Left Side of Hull Interior

MEDIUM ARMORED CAR T17E1

12. ARMAMENT OPERATION.

a. Detailed instructions covering operation, care, and maintenance of the guns will be found in various technical manuals on the subject materiel. Refer to section XLI for manual numbers.

13. GUN MOUNTS.

a. **Front Machine Gun Mounting.**

(1) The front machine gun is mounted in the hull in front of the assistant driver.

(2) The mounting assembly (fig. 23) consists of a mounting bracket, ball retainer, and mounting adapter. The mounting bracket is supported in the ball retainer by a ball which is an integral part of the mounting bracket. The ball retainer is secured to the mounting adapter which is part of the vehicle hull.

(3) This mounting permits the rotation of the weapon 17 degrees to the right or left in the horizontal plane and 10 degrees below or 22 degrees 45 minutes above horizontal in the vertical plane.

(4) The upper part of the ball has a milled slot which is engaged by a screw in the upper segment of the ball retainer. This prevents any rotation of the weapon around the axis through its barrel.

(5) When the weapon is installed, the greater part of the weight is back of the ball, therefore, a spring is provided to compensate for this additional weight and permit ease of control.

(6) For protection at the mounting, a shield is attached to a flange on the forward end of the mounting bracket.

(7) To remove the weapon from the mounting, it is only necessary to remove a retaining pin from the right side of the mounting bracket and withdraw the gun from the mounting bracket. This pin is attached to the mounting bracket with a chain to prevent it from being lost.

(8) To hold the pin in place when in use, a retaining device is provided. This consists of a hole drilled diametrically through the pin into which is assembled a spring and two balls. Since the balls extend somewhat beyond the surface of the pin, the spring is compressed as the pin is removed.

b. **37-mm Gun and Cal. .30 Machine Gun.**

(1) The firing of the cal. .30 machine gun and the 37-mm gun is accomplished through solenoids which operate the firing levers.

(2) The 37-mm gunner sits on the left side of the gun in a bucket seat placed over the electric motor and pump of the hydraulic traversing mechanism. The loader sits in a seat which is supported by a bracket on the basket wall; his location is on the right side of the gun.

ARMAMENT AND AMMUNITION

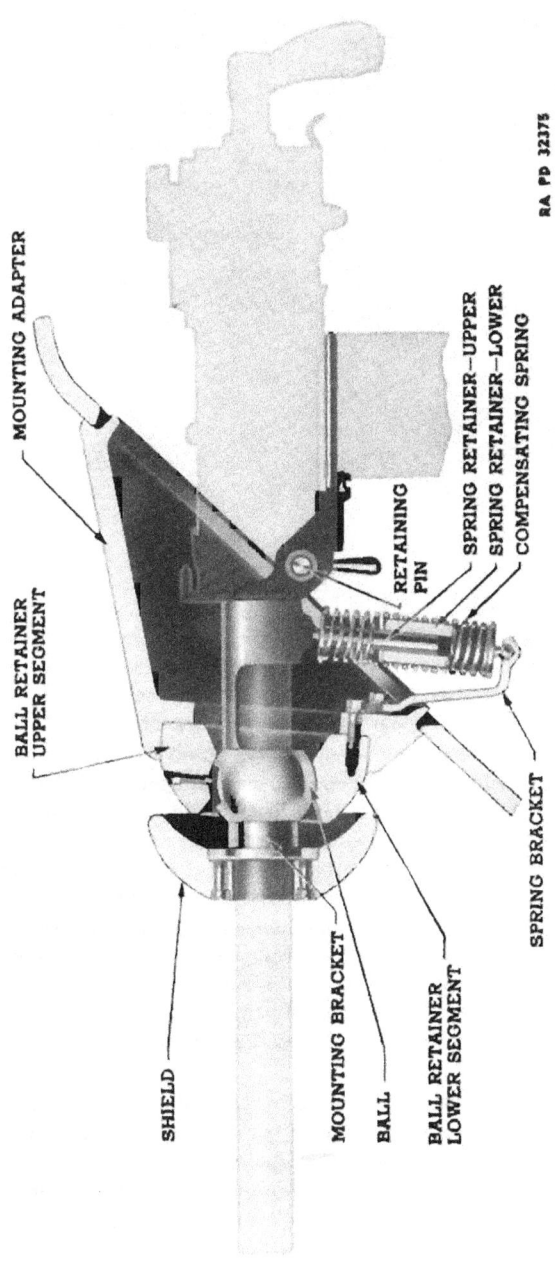

Figure 23 — Front Machine Gun Mount

TM 9-741
MEDIUM ARMORED CAR T17E1

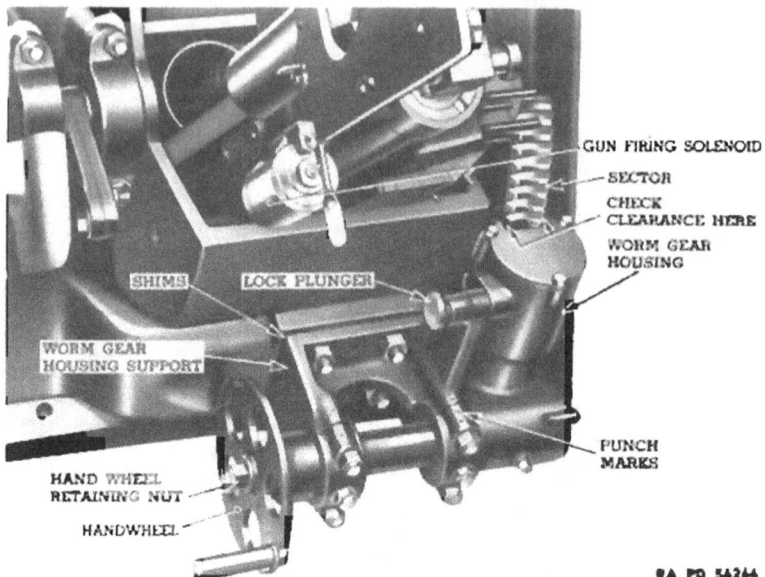

Figure 24 — Gun Elevating Mechanism

Ammunition is obtained from racks suspended from the walls of the basket. The loader's hand should be moving upward as he shoves the projectile "home" in order to clear the automatic breech-block.

(3) The periscopic vision device is located at the eye level of the gunner as he sits in position. A telescope sight is built into the right side of the vision device. Range lines are held in the field of this sight. The entire vision device can be readily removed by pressing in the lever button located in front of the upper left portion of the head rest and then pulling on the ring fastened to the bottom of the housing.

(4) The elevating and depressing handwheel is located on the left side of the 37-mm gun immediately in front of the gunner. Counterclockwise turning depresses the gun a maximum of 7 degrees, while clockwise turning elevates it to 60 degrees from the horizontal. The machine gun moves with the 37-mm gun. Figure 24 shows the gun elevating mechanism.

(5) The 37-mm gun can be removed by removing the bolt which fastens it to the recoil mechanism only after the front plate, gun and mount are removed from the vehicle. This bolt can be reached through a hole in the recoil housing on the right side of the gun. Figure 25 shows the mounting viewed from the turret.

32

TM 9-741
14

ARMAMENT AND AMMUNITION

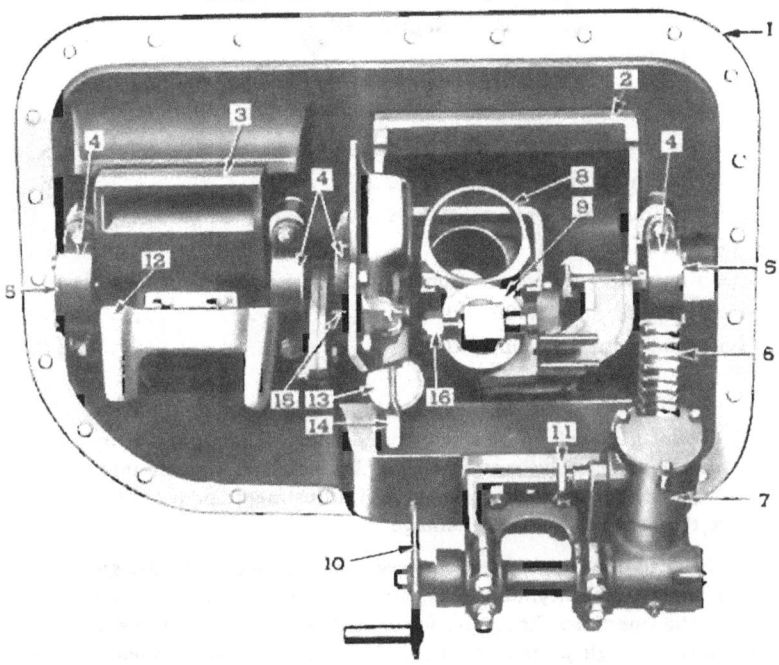

1. FRONT PLATE
2. ROTOR
3. PERISCOPE HOLDER
4. TRUNNION BEARING CAP
5. TRUNNION BEARING THRUST PLATE
6. SECTOR
7. WORM GEAR HOUSING
8. SLEIGH
9. RECOIL MECHANISM
10. HANDWHEEL
11. LOCK PLUNGER
12. HEAD REST
13. GUN FIRING SOLENOID - 37 MM GUN
14. FIRING LEVER
15. SAFETY
16. ATTACHING BOLT 37 MM GUN

RA PD 56267

Figure 25 — Gun Mount Viewed From Inside Turret

14. GYROSTABILIZER.

a. General.

The stabilizer attaches to the combination gun mount for the 37-mm and cal. .30 machine gun and provides a means of maintaining the gun position or aim while the vehicle is in motion. With this device in operation, the gunner may accurately aim and fire the gun while the vehicle is in motion (sec. XXXVIII).

MEDIUM ARMORED CAR T17E1

Section IV

PREVENTIVE MAINTENANCE

	Paragraph
Purpose	15
Prestarting inspection	16
Inspection during operation	17
Inspection at the halt	18
Inspection daily after operation	19
Periodic inspection (1,000 miles)	20
Periodic inspection (6,000 miles)	21

15. PURPOSE.

a. To insure mechanical efficiency and normal service, it is necessary that motor vehicles be systematically inspected at regular intervals in order that the necessary tightening and adjustments or repairs be made before they result in serious damage.

b. Information or suggestions regarding changes in design which might affect durability, efficiency, and economy or the safety and comfort of the operator, should be forwarded to the office of the Chief of Ordnance through proper channels. Such action is encouraged in order that such improvements as necessary can be made in all vehicles.

16. PRESTARTING INSPECTION.

a. Before starting, check the crankcase oil level of both engines with the oil gage rod (fig. 26). If oil level is down to the low mark, add oil.

b. Check solution in both cooling systems. Add water or antifreeze as required. NOTE: Make sure that the radiator caps seat airtight.

c. Check fluid level in oil reservoir for turret traversing mechanism.

d. Check turret traverse in both directions (360 degrees) by hydraulic and manual operation.

e. Check oil level in oil reservoir for M-24 Gun Mount Gyrostabilizer.

f. Check tire inflation. Inflate front tires to 70 pounds, rear tires to 80 pounds.

g. Check fuel supply. Replenish if necessary.

h. Check transmission fluid level.

i. Check transfer case oil level.

PREVENTIVE MAINTENANCE

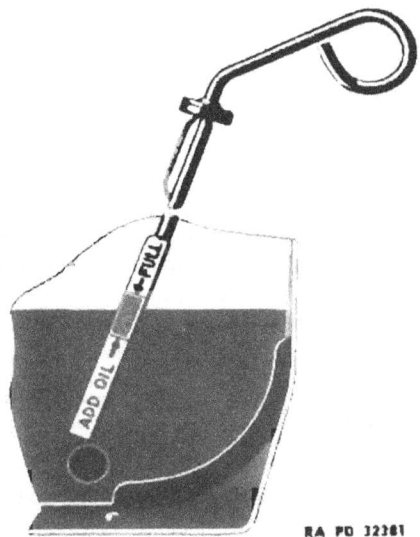

Figure 26 — Engine Oil Level Gage Rod

j. Check power steering operation.

k. Check brake condition by depressing pedal, making sure that it does not go more than half way to floor. If necessary, adjust according to instructions in paragraph 131.

l. Check siren.

m. Check lights. Any troubles should be repaired according to instructions in section **XXXI**.

n. Check operation of all flashlights.

o. Check to see that fire extinguisher is in place.

p. Make sure all fire extinguishers are fully charged and ready for use.

q. Check armament, ammunition, equipment, and tools. Shortages should be reported and replacement made. See section **VII** for lists of materiel and location.

17. INSPECTION DURING OPERATION.

a. The driver should always be alert to detect any abnormal conditions in the operation of the vehicle. Any indications of trouble should be investigated and corrected before proceeding.

MEDIUM ARMORED CAR T17E1

b. Watch the temperature indicators for signs of overheating.

c. Watch the oil gages to make sure the oil pressure is normal.

d. Note readings of the ammeters for output of generators and electrical charging rate.

e. Occasionally check the fuel gage reading for supply of fuel in the main tank. As long as there is fuel in the jettison tanks the main tank should be full.

f. Watch transmission pressure indicator lights. If either light flashes red, stop vehicle and check transmission lubricant level.

18. INSPECTION AT THE HALT.

a. Check for oil or fuel leaks and make the necessary corrections.

b. Check amount of solution in cooling systems. Add water or antifreeze as required. **NOTE**: Make sure that the radiator caps seal airtight.

c. Check engine oil level. Add oil if down to "add oil" mark.

d. Check tires.

e. Clean windshields and headlights.

f. If necessary, open six drain plugs to remove water from hull.

19. INSPECTION DAILY AFTER OPERATION.

a. These inspections should be made at the end of a run in order that the vehicle can be put in condition and ready for use in the least possible time.

b. Check hand and service brake condition. Report, if repairs or adjustments are required.

c. Check the engines for smooth idling and any abnormal noises at different engine speeds. Report any abnormal condition.

d. Check operation of windshield wipers. Report any trouble.

e. Check to see that fire extinguishers are in place, and, if they have been used, they should be replaced or recharged.

f. Check siren. Report any trouble to service personnel.

g. Clean and check lights, head, tail, stop and blackout. If any repairs are required, report to service personnel.

h. Check and tighten wheel and axle flange bolts.

TM 9-741
19

PREVENTIVE MAINTENANCE

RA PD 32382

Figure 27 — Transmission Oil Level Indicator

i. Check and tighten steering gear housing to hull bolts.

j. Check steering gear and steering connections. Report any looseness.

k. Check and inflate all tires. Any damaged tires should be changed or reported.

l. Check batteries, add water, and charge if necessary.

m. Check fluid level in oil reservoir for turret traversing mechanism.

n. Check turret traverse in both directions (360 degrees) by manual and hydraulic operation.

o. Check oil level in oil reservoir for M24 Gun Mount Gyrostabilizer.

p. Check transmission fluid level and fill if necessary (fig. 27).

q. Check transfer case oil level and fill if necessary.

r. Check power steering operation.

s. Visually check jettison tank release mechanism.

t. Fill and check cooling systems for leaks. Any leaks should be repaired or reported.

u. Check fan belts. See paragraph 86 for adjustment instructions.

v. Check engine oil. Add oil if down to "add oil" mark.

w. Check for oil leaks. Any leaks should be corrected or reported.

x. Fill fuel tanks and check for fuel leaks. Any leaks should be corrected or reported.

y. Check armament, ammunition, equipment, and tools. Any shortages should be replaced.

z. Clean and service air cleaners if operating conditions justify this daily.

MEDIUM ARMORED CAR T17E1

20. PERIODIC INSPECTION (1,000 MILES).

a. Make routine after operation inspection, and the following:

b. Check front axle lubricant level (sec. V).

c. Check rear axle lubricant level (sec. V).

d. Check brake lines for fluid leak and the Hydrovac lines for vacuum leak. It is essential that all leaks be corrected at once.

e. Check brake master cylinder reserve tank for sufficient fluid. Add fluid as required.

f. Check brake pedal for excessive travel and adjust brakes if necessary (par. 131).

g. Check steering gear adjustment. It should not require over 5-pound pull on wheel rim to turn wheels if hydraulic system is working. If adjustment is necessary, report to ordnance personnel.

h. Check steering tie rod and connecting rod ends for correct adjustment and being properly locked. If necessary, adjust according to instructions in paragraph 123.

i. Tighten spring to axle U-bolts.

j. Tighten all propeller shaft universal joint U-bolts and nuts.

k. Test all electrical units and tighten connections. See section XIV and sections XXIX to XXXV.

l. Clean air inlet grille in top of hull.

m. Tighten transfer case to hull bolts.

n. Lubricate vehicle according to instructions in section V.

o. **Road Test Vehicle.**

(1) After making the above inspections and adjustments, a final road test should be made at which time the following items should be checked:

(2) Check the engines for proper synchronization.

(3) Test hydraulic throttle operation.

(4) Check steering gear operation.

(5) Check hand and service brake operation.

(6) Check shifting and action of transmission manual control.

(7) Check transfer case shifting to high, low, front wheel engagement and disengagement.

(8) Check the operation of vehicle with each engine individually and compare their efficiency.

TM 9-741

PREVENTIVE MAINTENANCE

21. PERIODIC INSPECTION (6,000 MILES).

a. Make all inspections referred to in the 1,000 mile inspection.

b. Check wheel bearing adjustment.

c. Inspect and adjust brakes (par. 131).

d. Inspect and fill shock absorbers (par. 142).

e. Tune engines completely (par. 62).

f. Drain and flush cooling systems.

g. Remove oil cooler and connections. Clean and flush thoroughly.

h. Check electrical system and wiring and make necessary adjustments.

i. Adjust and service hydra-matic transmission according to instructions in section XIX.

j. Interchange tires to prolong tire life.

k. Inspect entire fire extinguishing system (par. 237).

TM 9-741
22-24

MEDIUM ARMORED CAR T17E1
Section V
LUBRICATION INSTRUCTIONS

	Paragraph
General	22
Lubrication guide	23
Detailed lubrication instructions for using arms	24
Points to be lubricated by ordnance maintenance personnel	25
Reports and records	26
Lubrication information	27

22. GENERAL.

a. The following lubrication instructions are published for the information and guidance of all concerned and supersede all previous instructions.

b. References. Materiel must be lubricated in accordance with the latest instructions contained in technical manuals and/or ordnance field service bulletins. Reference is made to the general instruction section (OFSB 6-1) for additional lubrication information, and to the product guide section (OFSB 6-2) for latest approved lubricants.

23. LUBRICATION GUIDE.

a. Lubrication instructions for all points to be serviced by the using arms are shown in lubrication guide, reproduced on the following pages which specifies the types of lubricants required and the intervals at which they are to be applied. Supplementary instructions appear in the notes.

24. DETAILED LUBRICATION INSTRUCTIONS FOR USING ARMS.

a. Fittings.

(1) Clean before applying lubricant.

(2) Lubricate until new grease is forced from the bearing.

CAUTION: Lubricate chassis points after washing car.

b. Intervals. The intervals indicated at points on lubrication guide are for normal service.

(1) For extreme conditions of speed, heat, water, mud, snow, rough roads, dust, etc., change crankcase oil and lubricate more frequently.

c. Air Cleaners (carburetor and engine ventilator). Proper maintenance of air cleaners is essential to prolonged engine life.

TM 9-741

LUBRICATION INSTRUCTIONS

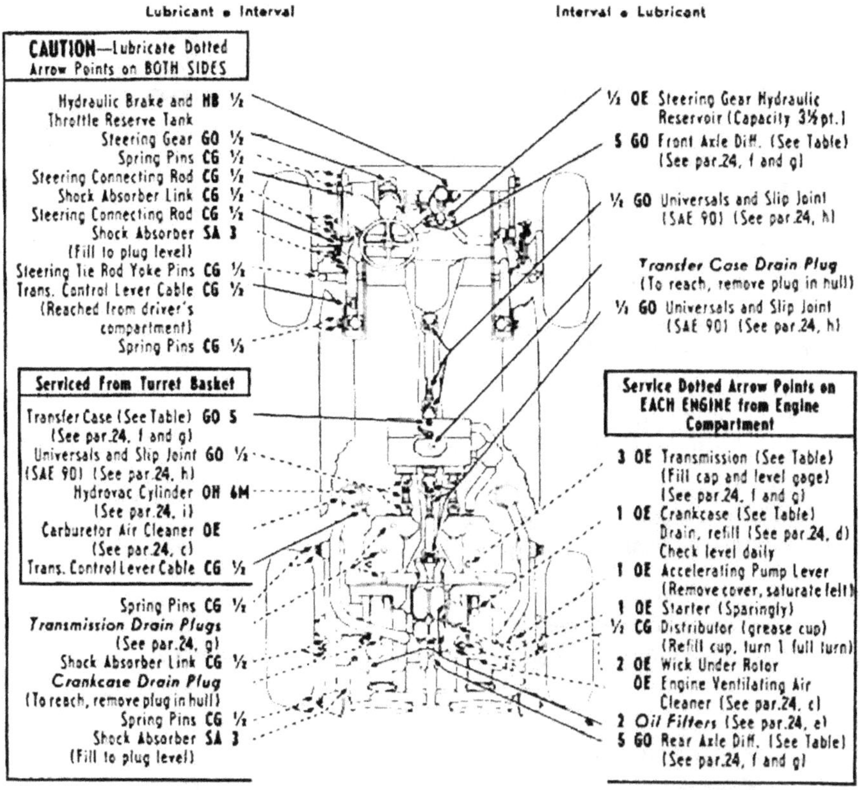

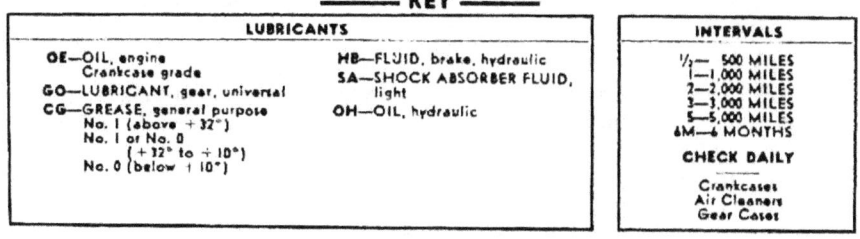

Lubricant • Interval

CAUTION—Lubricate Dotted Arrow Points on BOTH SIDES

Hydraulic Brake and Throttle Reserve Tank — HB ½
Steering Gear — GO ½
Spring Pins — CG ½
Steering Connecting Rod — CG ½
Shock Absorber Link — CG ½
Steering Connecting Rod — CG ½
Shock Absorber — SA 3
(Fill to plug level)
Steering Tie Rod Yoke Pins — CG ½
Trans. Control Lever Cable — CG ½
(Reached from driver's compartment)
Spring Pins — CG ½

Serviced From Turret Basket

Transfer Case (See Table) — GO S
(See par.24, f and g)
Universals and Slip Joint — GO ½
(SAE 90) (See par.24, h)
Hydrovac Cylinder — OH 6M
(See par.24, i)
Carburetor Air Cleaner — OE
(See par.24, c)
Trans. Control Lever Cable — CG ½

Spring Pins — CG ½
Transmission Drain Plugs
(See par.24, g)
Shock Absorber Link — CG ½
Crankcase Drain Plug
(To reach, remove plug in hull)
Spring Pins — CG ½
Shock Absorber — SA 3
(Fill to plug level)

Interval • Lubricant

½ OE Steering Gear Hydraulic Reservoir (Capacity 3½ pt.)
S GO Front Axle Diff. (See Table) (See par.24, f and g)

½ GO Universals and Slip Joint (SAE 90) (See par.24, h)

Transfer Case Drain Plug
(To reach, remove plug in hull)
½ GO Universals and Slip Joint (SAE 90) (See par.24, h)

Service Dotted Arrow Points on EACH ENGINE from Engine Compartment

3 OE Transmission (See Table)
(Fill cap and level gage)
(See par.24, f and g)
1 OE Crankcase (See Table)
Drain, refill (See par.24, d)
Check level daily
1 OE Accelerating Pump Lever
(Remove cover, saturate felt)
1 OE Starter (Sparingly)
½ CG Distributor (grease cup)
(Refill cup, turn 1 full turn)
2 OE Wick Under Rotor
OE Engine Ventilating Air Cleaner (See par.24, c)
2 *Oil Filters* (See par.24, e)
S GO Rear Axle Diff. (See Table) (See par.24, f and g)

COLD WEATHER: For Lubrication and Service below —10°, refer to OFSB 6-11.

KEY

LUBRICANTS	INTERVALS
OE—OIL, engine Crankcase grade	½— 500 MILES
GO—LUBRICANT, gear, universal	1—1,000 MILES
CG—GREASE, general purpose	2—2,000 MILES
No. 1 (above +32°)	3—3,000 MILES
No. 1 or No. 0	5—5,000 MILES
(+32° to +10°)	6M—6 MONTHS
No. 0 (below +10°)	
HB—FLUID, brake, hydraulic	**CHECK DAILY**
SA—SHOCK ABSORBER FLUID, light	Crankcases
OH—OIL, hydraulic	Air Cleaners
	Gear Cases

RA PD 56268

Figure 28 — Lubrication Guide (Chassis)

TM 9-741
24

MEDIUM ARMORED CAR T17E1

(1) Check level and refill oil cup to bead level daily with used crankcase oil or OIL, engine, seasonal grade.

(2) Remove oil cup, clean, and refill every 100 to 1,000 miles, depending on operating conditions. Under extreme dust conditions, service daily.

(3) Every 2,000 miles, also remove air cleaner and wash all parts in SOLVENT, dry-cleaning.

(4) Inspect air outlet hose for leakage and make sure all connections are tight.

(5) Replace air tube if there is evidence of wear or breakage.

d. **Crankcase.** Use grade of engine oil specified in table of capacities and recommendations.

(1) Check level daily; add oil if necessary.

(2) Drain and refill every 1,000 miles.

(a) Drain only when engine is hot.

(b) Refill to "FULL" mark on gage.

(c) Run engine a few minutes and recheck oil level.

(d) CAUTION: Be sure pressure gage indicates oil is circulating.

e. **Oil Filters.** Service two oil filters.

(1) Remove drain plug on bottom of filter every 1,000 miles.

(2) Renew filter element every 2,000 miles or oftener if necessary.

(3) After renewing element, refill crankcase to "FULL" mark on gage.

(4) Run engine a few minutes and recheck oil level.

f. **Transmission, Differentials, and Transfer Case.** Use lubricant specified in table of capacities and recommendations. Check level daily; add lubricant if necessary. Check with car on level ground.

(1) CHECKING TRANSMISSION LEVEL. Each transmission must be checked individually. Use following procedure:

(a) Remove engine compartment covers.

(b) Run engine about five minutes with transmission in driving range and transfer case in neutral. Then stop engine and wait for about one minute. NOTE: If possible, check after vehicle has been driven.

(c) Remove oil indicator plunger, wipe clean, and check level.

(d) Add oil if necessary. Bring level up to "FULL" mark. "FULL" to "LOW" marking on indicator equals one quart.

(2) CHECKING TRANSFER CASE LEVEL.

(a) Remove plug in bottom of hull.

(b) Remove small slotted plug in center of standpipe and drain plug.

TM 9-741

LUBRICATION INSTRUCTIONS

TABLE OF CAPACITIES WITH RECOMMENDATIONS AT TEMPERATURES SHOWN

	Capacity	Above +32°	+32° to +10°	+10° to −10°	Below −10°
		CAUTION: Use only U. S. Army Spec. 2-104a			
Crankcase (each)	8 qt.	OE SAE 30	OE SAE 30 or 10	OE SAE 10	Refer to OFSB 6-11
Transmission (each)	13 qt.				
Differential (front)	27½ qt.	GO SAE 90	GO SAE 90 or 80	GO SAE 80	
Differential (rear)	20¼ qt.				
Transfer Case	4½ qt.				

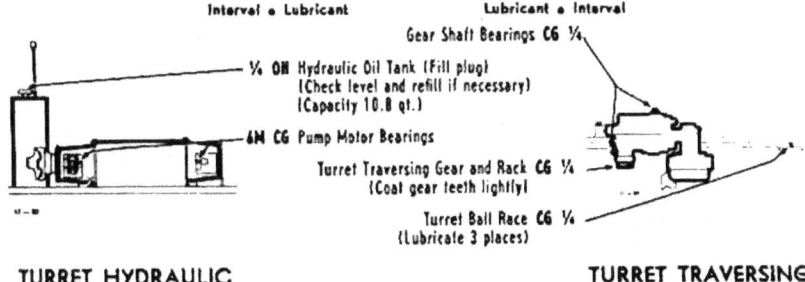

TURRET HYDRAULIC DRIVE
(Located on floor of turret)

TURRET TRAVERSING MECHANISM
(Located on side of turret)

Interval • Lubricant / Lubricant • Interval

- Gear Shaft Bearings CG ¼
- ¼ OH Hydraulic Oil Tank (Fill plug) (Check level and refill if necessary) (Capacity 10.8 qt.)
- 6M CG Pump Motor Bearings
- Turret Traversing Gear and Rack CG ¼ (Coat gear teeth lightly)
- Turret Ball Race CG ¼ (Lubricate 3 places)

COLD WEATHER: For Lubrication and Service below −10°, refer to OFSB 6-11.

——— **KEY** ———

LUBRICANTS		INTERVALS
OE—OIL, engine SAE 30 (above +32°) SAE 10 (below +32°)	OH—OIL, hydraulic	¼—250 MILES 6M—6 MONTHS
CG—GREASE, general purpose No. 1 (above +32°) No. 1 or No. 0 (+32° to +10°) No. 0 (below +10°)		

RA PD 56269

Figure 29 — Lubrication Guide (Turret)

TM 9-741

MEDIUM ARMORED CAR T17E1

(c) Add lubricant through fill plug opening until the lubricant flows, from the slotted plug opening.

g. **Draining Gear Cases.** Drain, flush, and refill at end of first 1,000 miles, except transmission, 3,000 miles; thereafter as indicated at points on guide. When draining, drain immediately after operation.

(1) DRAINING TRANSMISSION AND GEAR REDUCTION ASSEMBLY.

(*a*) Remove the service plates from the bottom of the hull.

(*b*) Remove drain plug at the bottom of the transmision oil pan.

(*c*) Remove drain plug at the bottom of the gear reduction case.

(*d*) Rotate engine by turning engine fan until drain plug in flywheel cover is alined with hole in bottom of flywheel housing.

(*e*) Remove Allen head plug with special "T" Allen wrench in tool kit. All of the above plugs must be removed to completely drain the unit.

(2) DRAINING TRANSFER CASE.

(*a*) Remove plug in bottom of hull.

(*b*) Remove standpipe and drain plug.

h. **Universal Joints and Slip Joints.**

(1) Apply LUBRICANT, gear, universal, SAE 90, to joint until it overflows at relief valve.

(2) Apply LUBRICANT, gear, universal, SAE 90, to slip joint until it is forced from end of slip joint.

i. **Hydrovac Cylinder.**

(1) Remove pipe plug at bottom of cylinder.

(2) Remove atmospheric line connection at the center plate.

(3) Inject one teaspoonful of OIL, hydraulic, in each opening. Replace plug and atmospheric line connection.

j. **Turret Handwheel Lock.** Lubricate lock through oiler with OIL, engine, seasonal grade, every 1,000 miles.

k. **Draining Cooling System.** In order to drain the cooling system completely, follow the procedure below:

(1) Remove the two drain plugs from the bottom of the hull, using special wrench in tool kit.

(2) To drain the cylinder blocks, open the valves located on the right side of each engine near the flywheel housing.

l. **Oil Can Points.** Lubricate peep hole protector slides, door locks and handles, door hinges and latches, shield hinges, control pins, clevises and linkage, cable pulleys, pistol part door operating handle and lock, with OIL, engine, seasonal grade, every 1,000 miles.

TM 9-741
24-27

LUBRICATION INSTRUCTIONS

m. **Points Requiring No Lubrication.** Generators, water pumps, radius rods, front and rear wheel bearings, front axle universal joints, steering gear pump electric motor bearings, turret traversing mechanism gears.

25. POINTS TO BE LUBRICATED BY ORDNANCE MAINTENANCE PERSONNEL.

a. None.

26. REPORTS AND RECORDS.

a. **Reports.** If lubrication instructions are closely followed, proper lubricants used, and satisfactory results are not obtained, a report will be made to the ordnance officer responsible for the maintenance of the materiel.

b. **Records.** A complete record of lubrication servicing will be kept for the materiel.

27. LUBRICATION INFORMATION.

a. **General.**

(1) One of the most vital factors in the performance and durability of any machine is its lubrication. Therefore, it is urged that you exercise care in the application of lubricants and engine oils for these vehicles.

(2) The application at the right time will greatly reduce wear and repairs. It is essential first, that the right lubricants be selected; secondly, that the lubricant be applied at regular intervals; and thirdly, that the lubricant be applied at the right place.

(3) The lubrication in this section should be referred to for instructions on the mileage of application and the grade and quantity of lubricant required for all parts of the vehicle.

b. **SAE Viscosity Numbers.**

(1) The viscosity of a lubricant is simply a measure of its body or fluidity. The oils with the lower SAE numbers are lighter and flow more readily than do the oils with the higher numbers.

(2) The SAE viscosity numbers constitute a classification of lubricants in terms of viscosity or fluidity, but with no reference to any other characteristics or properties.

(3) The refiner or marketer supplying the oil is responsible for the quality of its product. His reputation is your best indication of quality.

(4) The SAE viscosity numbers have been adopted by practically all oil companies, and no difficulty should be experienced in obtaining the proper grade of lubricant to meet seasonal requirements.

TM 9-741
27

MEDIUM ARMORED CAR T17E1

c. Engine Oils. In cold weather operation, starting the engine may prove to be difficult when heavier oils than indicated on the lubrication guide are used. The use of light oils at such times will not only lessen starting difficulties but will result in fuel economy and longer engine life.

d. Changing Lubricating Oil.

(1) The frequency with which crankcase oil must be changed depends upon the type and quality of oil used, the severity of and the type of operation, and the condition of the engine. It is therefore impossible to make a general recommendation concerning mileage intervals between oil changes. The oil should be changed often enough to keep it nonabrasive and noncorrosive and to keep the engine clean. Oil changing is closely related to air cleaner efficiency and to oil filter element cleaning and changing, the frequency of which also depends upon the conditions of operation mentioned above.

(2) If the composition of the oil used is at any time modified, or change made from one brand to another, the engine should first be thoroughly cleaned and should thereafter be examined daily for several days, so that any excessive sludge or carbon which may form will be quickly detected.

(3) When placing a new or overhauled engine in operation, it is recommended that oil be changed at 500-mile intervals for the first 2,000 miles of service. Thereafter, regular oil changing periods are at each 1,000 miles.

e. Crankcase Dilution.

(1) Probably the most serious phase of engine oil deterioration is that of crankcase dilution, which is the thinning of the oil by fuel vapors leaking by the pistons and rings and mixing with the oil.

(2) Leakage of fuel, or fuel vapors, into the oil pan occurs mostly during the warming up period when the fuel is not thoroughly vaporized and burned.

f. Automatic Control.

(1) The engines are equipped with automatic devices which aid greatly in minimizing the danger of crankcase dilution.

(2) Rapid warming up of the engine is aided by the thermostatic water temperature control which automatically prevents circulation of the water in the cooling system until it reaches a predetermined temperature.

(3) The downdraft carburetor is an aid to easy starting, thereby minimizing the use of the choke. Sparing use of the choke reduces danger of raw or unvaporized fuel entering the combustion chamber and leaking into the oil reservoir.

TM 9-741
27

LUBRICATION INSTRUCTIONS

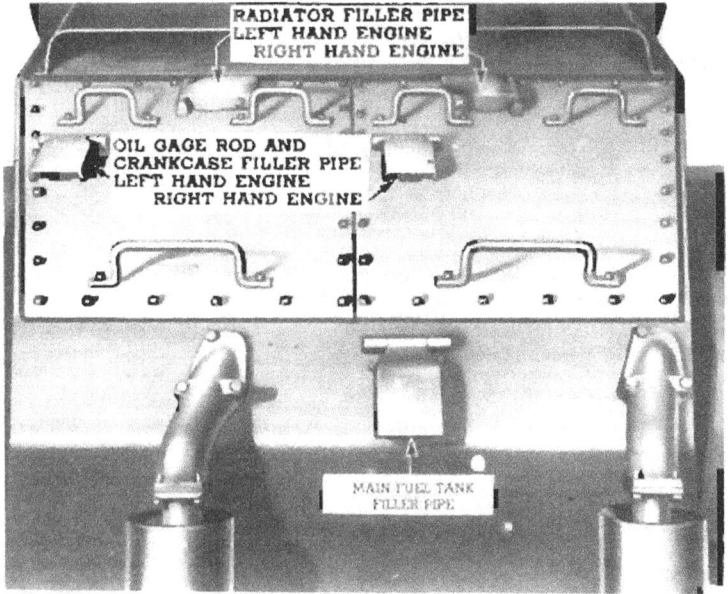

Figure 30 — Location of Filler Pipes

(4) An efficient crankcase ventilating system draws off fuel vapors and aids in the evaporization of the raw fuel and water which may find its way into the oil reservoir.

g. Maintaining Lubricant Levels.

(1) Figure 30 is a rear view of the vehicle showing the location of the filler pipes for the radiators, engines, and fuel tank. Raise the hinge covers to remove the filler caps.

(2) ENGINE.

(a) The oil gage rod (fig. 26), is marked "FULL" and "ADD OIL." These notations have broad arrows pointing to the level lines. The oil level should be maintained between the two lines neither going above the "FULL" line nor under the "ADD OIL" line. Check the oil level frequently and add oil when necessary. Always be sure the crankcase is full before starting on a long drive.

(b) When lowest expected atmospheric temperatures are above 32 F, use OIL, engine, SAE 30. When temperatures are between 32 F and −10 F, use OIL, engine, SAE 10. For temperatures −10 F to −30 F use OIL, engine, SAE 10, and add ½ quart of gasoline to each 4½

47

MEDIUM ARMORED CAR T17E1

quarts of oil. For temperatures –40 F use OIL, engine, SAE 10 and add 1 quart of gasoline to each 5 quarts of oil.

(3) FRONT AND REAR AXLE. Lubricant level should be maintained to the top of the filler plug hole in the housings. Use LUBRICANT, gear, universal, SAE 90, Federal Specification VV-L-761. When low temperatures are encountered, use LUBRICANT, gear, universal, SAE 80. CAUTION: Use a light flushing oil to flush out the housings when draining. Do not use water, steam, kerosene, gasoline, alcohol, etc.

(4) TRANSFER CASE. The transfer case is provided with a stand pipe having a small drain plug which is a part of the unit drain plug. Lubricant level should be maintained to the top of the stand pipe and can be checked by removing the small plug and adding oil at filler hole on top of case until oil runs out the small drain plug hole. Use LUBRICANT, gear, universal, SAE 90, Federal Specification VV-L-761.

(5) HYDRA-MATIC TRANSMISSIONS.

(a) Check level every 500 miles. Each transmission must be checked individually (fig. 27).

(b) Remove engine compartment covers.

(c) Run engines for about five minutes with transmission in driving range and transfer case in neutral. Then stop engine and wait for about one minute. NOTE: If possible, make check immediately after vehicle is driven, otherwise recheck after driving.

(d) Remove oil level indicator and wipe clean.

(e) Replace indicator and check level.

(f) Add oil if necessary to bring up to "FULL" mark. From "FULL" to "LOW" marking on indicator equals one quart.

(g) Each transmission should be drained and refilled every 3,000 miles. Use OIL, engine, seasonal grade, U. S. Army Specification 2-104A. Remove oil pan and check front and rear servo bands as explained in paragraph 98.

(6) AIR CLEANERS. Every 2,000 miles or more often if an unusual amount of dust and dirt is in the air, remove the oil cup assembly. Wash the container with SOLVENT, dry-cleaning, and wipe dry. Fill container to oil level mark, using the same oil as used in the engine and reinstall.

(7) OIL FILTERS. The drain plug on the bottom of the oil filter should be removed periodically to drain off any water or dirt trapped in the filter. The filter element should be replaced every 2,000 miles or when the oil gage rod shows the oil to be dark.

(8) SHOCK ABSORBERS. The shock absorbers should be kept filled with SHOCK ABSORBER FLUID, light, that has a pour test not higher

TM 9-741
27

LUBRICATION INSTRUCTIONS

than 30 degrees below zero. The same fluid is used in both summer and winter.

(9) STEERING GEAR.

(a) The steering gear is filled with LUBRICANT, gear, universal. Seasonal change of this lubricant is unnecessary and the housing should not be drained. Whenever required, additions should be made using LUBRICANT, gear, universal, Federal Specification VV-L-761.

(b) Keep the reservoir, for the hydraulic control system, filled with engine oil.

(10) TURRET TRAVERSE CONTROL (hydraulic unit). Keep reservoir filled with OIL, hydraulic, to the "FULL" mark on the oil gage rod.

h. **Drain Plug Location Major Units.** Refer to figure 31 for location of drain plugs in the bottom of the hull.

(1) COOLING SYSTEM. Remove the two drain plugs from the bottom of the hull, using special wrench in tool kit. To drain the cylinder blocks, open the valves located on the right side of each engine near the flywheel housing.

(2) MAIN FUEL TANK. Remove drain plug in hull and small drain plug in tank.

(3) ENGINE CRANKCASE. Remove drain plug in hull and plug in oil pan. The engine should be run until it reaches normal operating temperature before draining the oil.

(4) FRONT AND REAR AXLE. Remove drain plug from bottom of housings.

(5) TRANSFER CASE. Remove drain plug in hull and plug in bottom of transfer case.

(6) TRANSMISSION AND GEAR REDUCTION ASSEMBLY.

(a) Remove the service plates from bottom of hull.
(b) Remove drain plug at bottom of transmission oil pan.
(c) Remove drain plug at bottom of gear reduction.
(d) Rotate engine by turning engine fan until drain plug in flywheel cover is alined with hole in bottom of flywheel housing.
(e) Remove Allen head plug with special "T" Allen wrench in tool kit. CAUTION: All of the above plugs must be removed for proper draining.

i. **Oil Can Lubrication Points** (OIL, engine, seasonal grade, 1,000 miles).

Transmission manual control cross Shaft to
    cable lever .... Oil clevis pin
Transmission manual control cross Shaft to
    cable lever seal .... Oil sliding plate

TM 9-741

MEDIUM ARMORED CAR T17E1

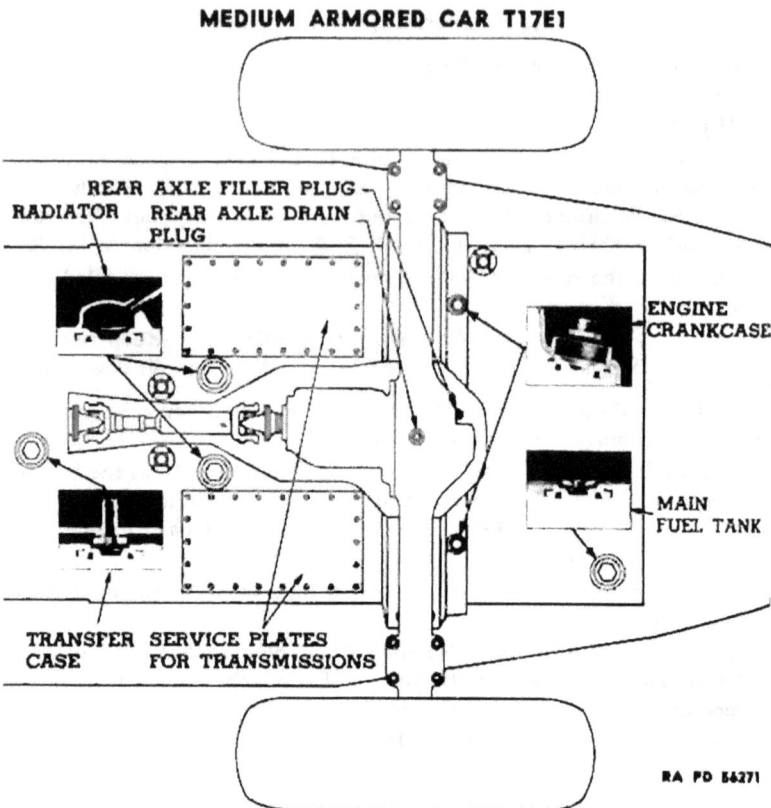

Figure 31 — Bottom of Hull Showing Unit Drain Plug Locations

Jettison tank release hand lever	Oil contacting surfaces (one lever each side)
Jettison tank release lever (inside)	Oil contacting surfaces (one lever each side)
Hull side door lock pin handle	Oil contacting surfaces
Vision door lock assembly	Oil contacting surfaces
Pistol port door operating handle and lock	Oil contacting surfaces
Transfer case shift hand lever and selector	Oil pins
Front axle declutch shift hand lever	Oil pins
Hand brake lever	Oil pins
Hand brake cable pulleys	Oil pins (two pulleys)
Carburetor choke rod and tube hand lever	Oil pins and rods (two)
Throttle rod and tube hand lever	Oil pins and rods (two)
Transmission manual control hand lever	Oil pins

TM 9-741

LUBRICATION INSTRUCTIONS

Accelerator control linkage mounted on
  engines .... Oil clevis pins and rods
Turret handwheel lock .... Fill oil cup

j. Coat Contacting Surfaces (GREASE, chassis, seasonal grade, 250 miles).

Jettison tank release shaft.
Jettison tank release pin and bearing.
Jettison tank release lever (outside) (one each side).
Jettison tank release lever shaft (through hull) (one each side).
Front machine gun mounting swivel.
Turret traversing gear and rack.
Storage battery terminals (4 batteries used).

MEDIUM ARMORED CAR T17E1

Section VI

CARE AND PRESERVATION

	Paragraph
Records	28
Cleaning	29
General	30
Preparing for painting	31
Painting metal surfaces	32
Paint as a camouflage	33
Removing paint	34
Painting lubricating devices	35

28. RECORDS.

a. Use. An accurate record must be kept of each motor vehicle issued by the Ordnance Department. For this purpose the Ordnance Motor Book (O.O. Form No. 7255), generally called "Log Book", is issued with each vehicle and must accompany it at all times. This book furnishes a complete record of the vehicle, from which valuable information concerning operation and maintenance costs, etc., is obtained, and organization commanders must insist that correct entries are made. This book will habitually be kept in a canvas cover to prevent its being injured or soiled.

b. The page bearing a record of assignment must be destroyed prior to entering the combat zone. All other reference which may be posted regarding the identity of the organization must also be deleted.

29. CLEANING.

a. Grit, dirt, and mud are the sources of greatest wear to a vehicle. If deposits of dirt and grit are allowed to accumulate, particles will soon find their way into bearing surfaces, causing unnecessary wear, and if the condition is not remedied, will soon cause serious difficulty. When removing engine parts or any other units, in making repairs and replacements, or if in the course of inspection working joints or bearing surfaces are to be exposed, all dirt and grit that might find its way to the exposed surfaces must first be carefully removed. The tools must be clean, and care must always be taken to eliminate the possibilities of brushing dirt or grit into the opening with the sleeve or other part of the clothing. To cut oil-soaked dirt and grit, hardened grit, or road oil, use **SOLVENT**, dry-cleaning, applied with rags. (not waste) or a brush. The vehicle is so designed that the possibility of interfering

CARE AND PRESERVATION

with its proper operation by careless application of cleaning water is very small. However, care should be taken to keep water from the power unit, as it might interfere with proper ignition and carburetion.

b. Oilholes which have become clogged should be opened with a piece of wire. Wood should never be used for this purpose, as splinters are likely to break off and permanently clog the passages. Particular care should be taken to clean and decontaminate vehicles that have been caught in a gas attack. See section IX for details of this operation.

30. GENERAL.

a. Ordnance materiel is painted before issue to the using arms and one maintenance coat per year will ordinarily be ample for protection. With but few exceptions this materiel will be painted with ENAMEL, synthetic, olive-drab, lusterless. The enamel may be applied over old coats of long oil enamel and oil paint previously issued by the Ordnance Department if the old coat is in satisfactory condition for repainting.

b. Paints and enamels are usually issued ready for use and are applied by brush or spray. They may be brushed on satisfactorily when used unthinned in the original package consistency or when thinned no more than 5 percent by volume with THINNER. The enamel will spray satisfactorily when thinned with 15 percent by volume of THINNER. (Linseed oil must not be used as a thinner since it will impart a luster not desired in this enamel.) If sprayed, it dries hard enough for repainting within ½ hour and dries hard in 16 hours.

c. Certain exceptions to the regulations concerning painting exist. Fire-control instruments, sighting equipment, and other items which require a crystalline finish will not be painted with olive-drab enamel.

d. Complete information on painting is contained in TM 9-850.

31. PREPARING FOR PAINTING.

a. If the base coat on the materiel is in poor condition, it is more desirable to strip the old paint from the surface than to use sanding and touch-up methods. After stripping, it will then be necessary to apply a primer coat.

b. PRIMER, ground, synthetic, should be used on wood as a base coat for synthetic enamel. It may be applied either by brushing or spraying. It will brush satisfactorily as received or after the addition of not more than 5 percent by volume of THINNER. It will be dry enough to touch in 30 minutes, and hard in 5 to 7 hours. For spraying, it may be thinned with not more than 15 percent by volume of THINNER.

TM 9-741
31-33

MEDIUM ARMORED CAR T17E1

Lacquers must not be applied to the PRIMER, ground, synthetic, within less than 48 hours.

c. PRIMER, synthetic, rust-inhibiting, for bare metal, should be used on metal as a base coat. Its use and application is similar to that outlined in paragraph b above.

d. The success of a job of painting depends partly on the selection of a suitable paint, but also largely upon the care used in preparing the surface prior to painting. All parts to be painted should be free from rust, dirt, grease, kerosene, oil, and alkali, and must be dry.

32. PAINTING METAL SURFACES.

a. If metal parts are in need of cleaning, they should be washed in a liquid solution consisting of ½ pound of SODA ASH in 8 quarts of warm water, or an equivalent solution, then rinsed in clear water and wiped thoroughly dry. Wood parts in need of cleaning should be treated in the same manner, but the alkaline solution must not be left on for more than a few minutes and the surfaces should be wiped dry as soon as they are washed clean. When artillery or automotive equipment is in fair condition and only marred in spots, the bad places should be touched with ENAMEL, synthetic, olive-drab, lusterless, and permitted to dry. The whole surface will then be sandpapered with PAPER, flint, No. 1, and a finish coat of ENAMEL, synthetic, olive-drab, lusterless, applied and allowed to dry thoroughly before the materiel is used. If the equipment is in bad condition, all parts should be thoroughly sanded with PAPER, flint, No. 2, or equivalent, given a coat of PRIMER, ground, synthetic, and permitted to dry for at least 16 hours. They will then be sandpapered with PAPER, flint, No. 00, wiped free from dust and dirt, and final coat of ENAMEL, synthetic, olive-drab, lusterless, applied and allowed to dry thoroughly before the materiel is used.

33. PAINT AS A CAMOUFLAGE.

a. Camouflage is now a major consideration in painting ordnance vehicles, with rust prevention secondary. The camouflage plan at present employed utilizes three factors: color, gloss and stenciling.

(1) COLOR. Vehicles are painted with ENAMEL, synthetic, olive-drab, lusterless, which was chosen to blend in reasonably well with the average landscape.

(2) GLOSS. The new lusterless enamel makes a vehicle difficult to see from the air or from relatively great distances over land. A vehicle painted with ordinary glossy paint can be detected more easily and at greater distance.

CARE AND PRESERVATION

(3) STENCILING. White stencil numbers on vehicles have been eliminated because they can be photographed from the air. A blue-drab stencil enamel is now used which cannot be so photographed. It is illegible to the eye at distances exceeding 75 feet.

b. Preserving Camouflage.

(1) Continued friction or rubbing must be avoided, as it will smooth the surface and produce a gloss. The vehicle should not be washed more than once a week. Care should be taken to see that the washing is done entirely with a sponge or a soft rag. The surface should never be rubbed or wiped except while wet, or a gloss will develop.

(2) It is not desirable that vehicles, painted with lusterless enamel, be kept as clean as vehicles were kept when glossy paint was used. A small amount of dust increases the camouflage value. Grease spots should be removed with SOLVENT, dry-cleaning. Whatever portion of the spot cannot be so removed should be allowed to remain.

(3) Continued friction of wax-treated tarpaulins on the sides of a vehicle will also produce a gloss which should be removed with SOLVENT, dry-cleaning.

(4) Tests indicate that repainting with olive-drab paint will be necessary once yearly, with blue-drab paint twice yearly.

34. REMOVING PAINT.

a. After repeated paintings, the paint may become so thick as to crack and scale off in places, presenting an unsightly appearance. If such is the case, remove the old paint by use of a lime-and-lye solution (see TM-850 for details) or REMOVER, paint and varnish. It is important that every trace of lye or other paint remover be completely rinsed off and that the equipment be perfectly dry before repainting is attempted. It is preferable that the use of lye solutions be limited to iron or steel parts. If used on wood, the lye solution must not be allowed to remain on the surface for more than a minute before being thoroughly rinsed off and the surface wiped dry with rags. Crevices or cracks in wood should be filled with putty and the wood sandpapered before refinishing. The surfaces thus prepared should be painted according to directions in paragraph 32.

35. PAINTING LUBRICATING DEVICES.

a. Oil cups, grease fittings, oilholes, and similar lubricating devices, as well as a circle about three-fourths of an inch in diameter at each point of lubrication will be painted with ENAMEL, red, water-resisting, in order that they may be readily located.

TM 9-741
36-37

MEDIUM ARMORED CAR T17E1

Section VII

EQUIPMENT AND TOOLS ON VEHICLE

	Paragraph
Introduction	36
Equipment	37
Vehicle tools	38
Care of equipment and tools	39

36. INTRODUCTION.

a. The express purpose of this section is to itemize the equipment, tools, and supplies (except armament and ammunition) which should be with the vehicle at all times for the care, maintenance, and preservation of the vehicle as well as the safety, protection, and comfort of the crew.

b. The materiel list shows the quantity, description, and approximate location where the materiel will be found. Refer to figures 32 and 33 for equipment and tool stowage.

37. EQUIPMENT.

a. Communication.

Materiel	Quantity	Approximate Location
Case of antenna section	1	Right side of hull outside
Flags, signal	4	In retainer at right of commander's seat
Handle and reel, spot light	1	On turret ledge at right of commander's seat
Hooks, phone	2	1 each side of turret adjacent to interphone boxes
Shaft, spot light	1	On side of turret basket at left of commander's seat
Spot light	1	On floor of turret back of gunner's seat

b. Military.

Materiel	Quantity	Approximate Location
Bag, empty cartridge, bow gun	1	On gun
Bag, empty cartridge, cal. .30 turret gun	1	On gun
Bag, empty cartridge, 37-mm	1	On gun

56

TM 9-741
37

EQUIPMENT AND TOOLS ON VEHICLE

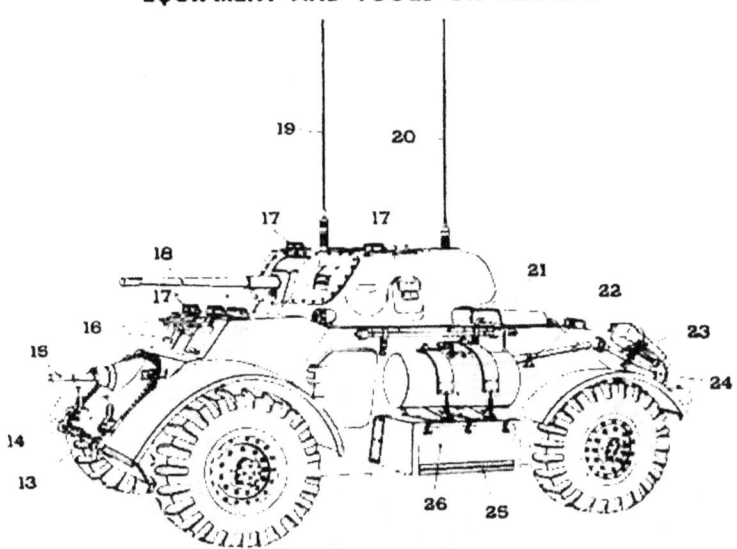

13. TOWING CABLE
14. HEADLIGHT SOCKET PLUGS
15. 30 CAL MACHINE GUN AND MUZZLE COVER
16. FRONT VISION DOORS
17. PERISCOPES M6
18. 37 M M GUN M6 AND MUZZLE COVER
19. ANTENNA MP 48
20. ANTENNA MP 37
21. RAMMER STAFF
22. PICK
23. TRIPOD WITH PINTLE
24. SHOVEL
25. JETTISON TANK
26. LUGGAGE BOX
    SPARE BULBS BOX
    COLLAPSIBLE BUCKET
    RATIONS (TYPE C)
    BLANKET ROLLS
    TIRE PUMP AND HOSE

RA PD 32836

Figure 32 — Equipment Location — Left Side Exterior

Material	Quantity	Approximate Location
Box for spare periscope M6 (1 each)	4	2 — On tunnel ahead of ammunition rack 1 — On right side of turret under escape door 1 — On left side of turret under escape door
Boxes for spare periscope heads (2 heads each)	7	2 — Right ledge of turret near loader 1 — Left ledge of turret near commander

57

TM 9-741

MEDIUM ARMORED CAR T17E1

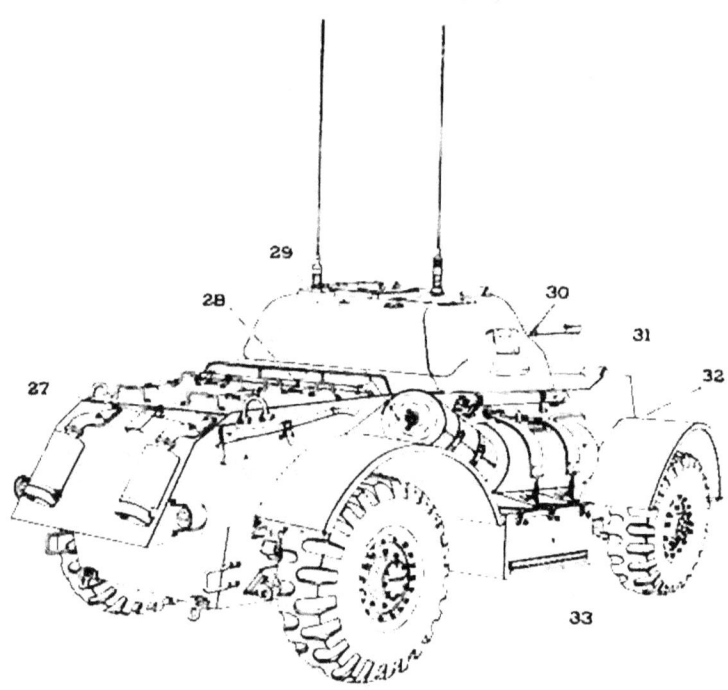

27. AXLE
28. CROWBAR
29. CAMOUFLAGE NET AND TARPAULIN
30. CAL. .30 MACHINE GUN AND MUZZLE COVER
31. ANTENNA SECTIONS
32. JETTISON TANK

33. LUGGAGE BOX:
　　TIRE CHAINS
　　FLEXIBLE NOZZLE
　　(FUEL CAN)
　　HEAVY TOOLS
　　WRENCH AND HANDLE
　　JACK AND HANDLE

RA PD 56276

Figure 33 — Equipment Location — Right Side Exterior

Material	Quantity	Approximate Location
		1 — On left side of hull between transmission control lever and windshield plate
		3 — On right side of hull forward
Can, oil, recoil	1	Stowage compartment between bulkheads
Can, oil, traverse gear	1	Stowage compartment between bulkheads

EQUIPMENT AND TOOLS ON VEHICLE

Material	Quantity	Approximate Location
Covers, muzzle, cal. .30 machine gun	2	On guns
Cover, muzzle, 37-mm	1	On gun
Cover, breech, 37-mm	1	On gun
Cover, receiver, cal. .30 machine gun	1	On gun
Staff, rammer, 37-mm	1	Left side hull, outside, over jettison tank
Tripod with pintle and cover	1	Left side hull, rear

c. **Vehicular.**

Material	Quantity	Approximate Location
Axe with handle	1	Right side hull, outside, rear
Box containing tire pump and hose	1	Luggage box outside
Box, fuses, bulbs, etc.	1	In luggage box on outside
Cable, towing	1	On front, outside
Cans, engine oil (1 quart each)	2	Supply compartment
Case, direct vision slot protectors (6 in case)	1	Left side just above spring housing ledge
Chains, tire	4	Luggage box, outside
Compass (hull)	1	On front of turret just to left of gun mount
Compass (Pioneer)	1	On windshield plate, center
Covers, head light	2	On lamps
Crowbar	1	Across top, just back of air intake grille
Fire extinguisher, fixed (2 10 lb cyl.)	2	In supply compartment between bulkheads
Fire extinguisher, portable (4 lb cyl.)	1	On floor against tunnel back of bow gunner
Lamp, blackout	1	In bracket in supply compartment on front bulkhead
Lamp, head light	2	In brackets in supply compartment on front bulkhead
Net, camouflage	1	In roll on top outside
Nozzle, flexible	2	Luggage box outside
Pick with handle	1	Right side hull, outside, rear
Plugs, lamp socket	2	In lamp stowage brackets
Shovel	1	Left side hull, outside, rear
Tarpaulin	1	In roll on top outside

TM 9-741

MEDIUM ARMORED CAR T17E1

Material	Quantity	Approximate Location
Windshields with wipers	2	1 — Right side of hull over bow gunner's seat
		1 — Left side of ammunition rack on tunnel adjacent driver's seat

d. Service Parts.

Material	Quantity	Approximate Location
Box for spare periscope telescope M2	1	On face of oil reservoir in turret
Sight, bore		In shell rack in right turret bulge
Box, spare parts and accessories cal. .30 and cal. .45	1	Right side of hull below door
Box, cal. .30 small parts	1	Right side of hull—against tunnel under bow gun
Box, cal. .30 and 37-mm small parts	1	On turret ledge near loader
Box, spare parts and accessories, 37-mm	1	On turret platform floor
Barrels, spare, cal. .30 machine gun	2	On spring housing ledge at right of bow gun

e. Miscellaneous.

Material	Quantity	Approximate Location
Bags, musette	5	Supply compartment, loose
Binoculars	1	Left side turret ledge back of commander's seat
Book, ordnance (form 7255)		Supply compartment
Brackets, flash light	4	2 — On left side hull just ahead of door
		2 — In turret
Bucket, collapsible	1	Luggage box, outside
Cans, rations type C	60	30 — In supply compartment strapped together in 5 cans each
		30 — In luggage box, outside, strapped together in 5 cans each
Cans, rations type D	2	Strapped together in supply compartment
Cans, water	2	In supply compartment
Canteens	5	3 — In turret
		2 — In turret pockets on drivers and auxiliary seat back covers

TM 9-741
37-39

EQUIPMENT AND TOOLS ON VEHICLE

Material	Quantity	Approximate Location
Clips for maps	2 pairs	1 pair—On lower edge of instrument panel
		1 pair—In turret near commander
Decontaminator	1	In bracket on rear bulkhead in supply compartment
Harness, retaining	1	Across opening of supply compartment
Helmets (assorted sizes)	5	No stowage
Kit, first-aid	1	On toe board under bow gun
Mittens, asbestos	2 pairs	No stowage
Rolls, blanket	5	Luggage box, outside

38. VEHICLE TOOLS.

a. These tools are part of the equipment furnished by the manufacturer with the vehicle. The vehicle tool list is set up in numerical sequence (fig. 34). The tool location in the vehicle and the method of stowing precedes each group.

(1) The group of tools numbered 1 to 22 inclusive (fig. 34) are stowed in the metal tool box (1, fig. 34). The tool box is carried in one of the side luggage boxes on the outside of vehicle.

(2) The tools numbered 23 through 27 are stowed loose in the same luggage box with the metal tool box, with the exception of the oil can (27, fig. 34) which is carried in a holder inside the hull at the right front corner, behind the wiring conduits.

(3) The tools numbered 28 through 36 (fig. 34) are kept in the tool bag (28, fig. 34) which is stowed inside the hull at the left of the driver.

39. CARE OF EQUIPMENT AND TOOLS.

a. An accurate record of all equipment and tools must be kept in order that their location and condition may be known at all times. Items which have been lost, depleted, or in any way rendered unserviceable must be replaced as soon as possible. All equipment and tools must be cleaned, conditioned (if necessary), and treated against rust or deterioration before returning them to their location.

TM 9-741
39

MEDIUM ARMORED CAR T17E1

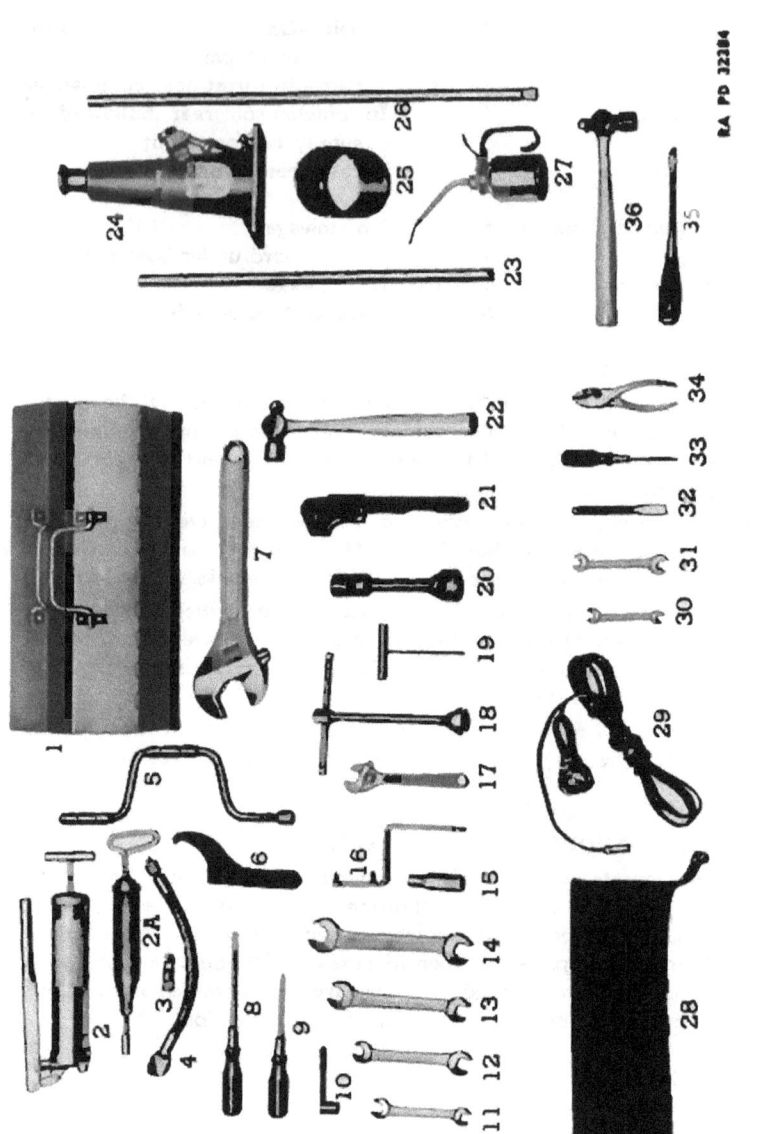

Figure 34 — Vehicle Tools

TM 9-741

EQUIPMENT AND TOOLS ON VEHICLE

Legend for Figure 34

1 — BOX, METAL TOOL, WITH TRAY
2 — GUN, LUBRICATION
2A — GUN, GREASE (UNIVERSAL JOINT)
3 — ADAPTER, LUBRICATION GUN (FOR HYDRAULIC FITTINGS)
4 — HOSE, LUBRICATION GUN
5 — WRENCH, SOCKET SPEED (TO REMOVE ENGINE COMPARTMENT COVERS)
6 — WRENCH, TRANSFER CASE PROPELLER SHAFT SEAL RETAINER NUT
7 — WRENCH, ADJUSTABLE OPEN END (18 IN. CRESCENT TYPE)
8 — SCREWDRIVER (5/16 IN. CLUTCH HEAD SCREW)
9 — SCREWDRIVER (1/4 IN. CLUTCH HEAD SCREW)
10 — WRENCH, GENERATOR TERMINAL NUT
11 — WRENCH, OPEN END (9/16 IN. - 11/16 IN.)
12 — WRENCH, OPEN END (5/8 IN. - 13/16 IN.)
13 — WRENCH, OPEN END (3/4 IN. - 7/8 IN.)
14 — WRENCH, OPEN END (15/16 IN. - 1-1/16 IN.)
15 — WRENCH, SPARK PLUG
16 — GAGE, THROTTLE CONTROL CROSS SHAFT CHECKING
17 — WRENCH, ADJUSTABLE OPEN END 18 IN. - CRESCENT TYPE)
18 — WRENCH, DRAIN PLUG
19 — WRENCH, SPECIAL HEXAGONAL SOCKET HEAD SCREW (FLYWHEEL COVER DRAIN PLUG)
20 — WRENCH, WHEEL NUT
21 — WRENCH, ADJUSTABLE 11 IN.
22 — HAMMER, BALL PEEN, 16 OZ.
23 — HANDLE, AUTO JACK
24 — JACK, AUTO
25 — WRENCH, WHEEL HUB ADJUSTING NUT
26 — HANDLE, WHEEL HUB ADJUSTING NUT WRENCH AND WHEEL NUT WRENCH
27 — CAN, OIL, PUMP TYPE (3/4 PT.)
28 — BAG, TOOL
29 — LAMP, INSPECTION
30 — WRENCH, OPEN END (3/8 IN. - 7/16 IN.)
31 — WRENCH, OPEN END (1/2 IN. - 19/32 IN.)
32 — CHISEL, COLD
33 — SCREWDRIVER (5/32 IN. CLUTCH HEAD SCREW)
34 — PLIERS, COMBINATION 6 IN.
35 — SCREWDRIVER (UTILITY 6 IN. BLADE)
36 — HAMMER, BALL PEEN, 12 OZ.

RA PD 32394A

TM 9-741
40

MEDIUM ARMORED CAR T17E1
Section VIII

OPERATIONS UNDER UNUSUAL CONDITIONS

	Paragraph
Cold weather operation	40
High temperature operation	41

40. COLD WEATHER OPERATION.

　a. The operation and maintenance of automotive vehicles at low temperatures involve factors which do not exist at normal temperatures. Operators and maintenance personnel must spend more time in preventive maintenance. Failure to give this extra service will result in actual damage, unnecessary expenses, and difficulty in starting the engines.

　b. Low temperatures have been divided into 2 ranges: −10 F to −30 F, and below −30 F. Engine and lubricants undergo changes in physical properties below −30 F. In many cases, accessories, equipment for supplying heat to engine, fuel and intake air are required.

　c. The following is a list of things that the operator should be familiar with before operating a vehicle in cold weather.

　(1) FUEL. The formation of ice crystals from small quantities of water in the fuel sometimes causes considerable trouble. The following precautions should be followed to keep water out of the fuel tanks.

　(a) Strain fuel through a suitable strainer. CAUTION: A positive metallic contact must be maintained between the fuel container and fuel tank unless fuel tank and container are independently grounded.

　(b) Keep fuel tanks as full as possible.

　(c) Add ½ pint of **ALCOHOL**, denatured, to a tank of fuel to absorb moisture that might condense.

　(d) Store fuel in clean drums.

　(e) Never pump fuel drums dry. Allow about four inches of fuel to remain.

　(2) CRANKCASE OIL. Engine lubrication at temperatures above −10 F is covered in section V. For temperatures below −10 F, one of the following measures must be taken according to facilities available:

　(a) Keep vehicle in heated inclosure, if possible, when it is not being operated.

　(b) When the engines are stopped, drain the crankcase oil while it is still hot and store it in a warm place until vehicle is to be operated again. If a warm place is not available, heat the oil before putting it in the crankcases. (Do not get the oil too hot. Heat only to a point

64

OPERATIONS UNDER UNUSUAL CONDITIONS

where the bare hand can be inserted without burning.) CAUTION: Tag the vehicle in a conspicuous place on the hull to warn personnel that the crankcases are empty,

(c) Dilute crankcase oil with gasoline. The table below shows the quantities of diluent to be added to the oil prescribed on the lubrication guides for use at −10 F. The quantities of diluents will form mixtures for satisfactory starting at temperatures indicated.

	−10 F to −30 F	Below −30 F
Gasoline	½ qt to each 4½ qt of engine oil	1 qt to each 5 qt of engine oil

1. When the crankcase oil is first diluted, turn the engine over several times to mix oil and diluent thoroughly.

2. Check oil level frequently.

(3) ANTIFREEZE.

(a) ETHYLENE GLYCOL (Prestone) is prescribed for use as antifreeze solution. If ETHYLENE GLYCOL is not available, other materials may be used. The following table gives three permissible materials and the quantity to be added to prevent freezing at indicated temperatures.

Freezing Point	Pints Ethylene Glycol (Prestone) Per Gallon of System Capacity	Pints Glycerine Grade A, U.S.P. Per Gallon of System Capacity	Pints Denatured Alcohol Per Gallon of System Capacity
10 F	2	3	2½
0 F	2½	3½	3
−10 F	3	3½	3½
−20 F	3½	4	4
−30 F	4	5	5
−40 F	4½	—	5½
−50 F	4½	—	6
−60 F	5	—	6½
−70 F	5	—	—

(b) PRECAUTIONS.

1. Do not use alcohol if other materials are available.

2. Flush out the radiators and cylinder heads separately before adding antifreeze material.

3. Do not mix antifreeze solution.

4. Check cooling systems for leaks.

MEDIUM ARMORED CAR T17E1

5. If hot water heater is added to cooling system, add about one gallon to the original volume of the cooling system.

6. Check thermostats.

7. Check adjustment and weakness of fan belts. Replace rubber fan belts with fiber, leather, or synthetic belts below −20 F.

8. Make sure that water pumps are in good operating condition.

(4) BATTERY AND ALL ELECTRICAL PARTS.

(a) Keep batteries fully charged with hydrometer reading between 1.275 and 1.300. A fully discharged battery will freeze and rupture at 5 F.

(b) Clean and repair all electrical wiring accessories (spark plugs, ignition coil, and distributor) to prevent undue resistance. Make sure all connections are tight.

(c) Set spark plug gap 0.005 inch less than that recommended for operation under normal conditions.

(d) Check generators, starter brushes, commutators, and bearings. See that the commutators and brushes are clean.

(e) Be sure that no heavy grease or dirt is on the starter throw-out mechanism.

(5) GEAR LUBRICANTS. Below −15 F, dilute lubricants prescribed for use at −10 F with 10 percent gasoline. If circumstances preclude dilution of lubricants, heat gear cases with a blowtorch. Play the torch lightly under the entire gear case; do not concentrate the heat in one spot.

(6) CHASSIS AND OTHER LUBRICANTS.

(a) Chassis, wheel bearing, and other lubricants prescribed for use at −10 F will furnish satisfactory lubrication as low as −30 F. For sustained temperatures below −30 F, use grease comparable to GREASE, lubricating, special, or GREASE, O.D., No. 00. Use OIL, engine, crankcase grade, in steering gear housing.

(b) Commercial brake and shock absorber fluids remain fluid at temperatures encountered.

(7) SPECIAL OPERATING PRECAUTIONS.

(a) Full choke is necessary to secure the air-fuel ratio required for cold weather starting. Check the butterfly valves to see that they close all the way and otherwise function properly.

(b) Check fuel pump.

(c) Below 10 F, remove oil from the air cleaners. Below −30 F, remove the air cleaners.

(d) Inspect vehicle frequently.

(e) Remove or bypass oil filters at temperatures below −30 F.

OPERATIONS UNDER UNUSUAL CONDITIONS

(*f*) Disconnect oil-lubricated speedometer cables at the drive end for operating vehicles at temperatures of below −30 F.

(*g*) Remove and clean sediment bulb, strainer, etc., in the fuel systems at frequent intervals.

(*h*) Retune engines frequently.

(*i*) Before starting the engines, pull the chokes all the way out and leave them partially pulled out until the engine has warmed up. Turn the engine as rapidly as possible with the starter and release starting pedal as soon as the engine fires. After the engines have been started, idle them until they have warmed up sufficiently to run smoothly. Do not race the engines immediately after starting.

(*j*) To stop the engines, first increase their speed and then turn off the ignition and release the throttle at the same time.

d. **Cold Weather Accessories.** The following list of cold weather accessories is given only as a suggestion and may be employed at the discretion of officers in charge of the material.

(1) Tarpaulins, tents, or collapsible sheds are useful for covering vehicles, particularly the engines.

(2) Fire pots, Primus type, or Van Prag blowtorches, ordinary blowtorches, oil stoves, or kerosene lanterns, can be used for heating vehicles.

(3) Extra batteries and facilities for changing batteries quickly are aids in starting.

(4) Steel drums and suitable metal stands are useful for heating crankcase oil.

(5) Insulation of the fuel line will help prevent ice formation inside the line.

(6) Small quantities of **ALCOHOL**, denatured, about ½ pint to a tank of fuel, will reduce difficulties from water in fuel.

(7) Radiator inlet grille covers can be improvised locally and help to keep the engine running at normal temperatures.

41. HIGH TEMPERATURE OPERATION.

a. To prevent overheating, the following things should be checked:

(1) The cooling systems should be kept clean and full.

(2) The fins in the radiators should be kept free from bugs and foreign material that might affect free circulation of air.

(3) The fan belts should be properly adjusted.

(4) Do not leave the engine compartment open under any conditions, as this reduces the cooling efficiency.

(5) Watch the temperature indicator, and if the needle goes into the red band, investigate the cause.

TM 9-741
42-44

MEDIUM ARMORED CAR T17E1

Section IX

MATERIEL AFFECTED BY GAS

	Paragraph
Protective measures	42
Cleaning	43
Decontamination	44
Special precautions for automotive materiel	45

42. PROTECTIVE MEASURES.

a. When materiel is in constant danger of gas attack, unpainted metal parts will be lightly coated with engine oil. Instruments are included among the items to be protected by oil from chemical clouds or chemical shells, but ammunition is excluded. Care will be taken that the oil does not touch the optical parts of instruments or leather or canvas fittings. Materiel not in use will be protected with covers as far as possible. Ammunition will be kept in sealed containers.

b. Ordinary fabrics offer practically no protection against mustard gas or lewisite. Rubber and oilcloth, for example, will be penetrated within a short time. The longer the period during which they are exposed, the greater the danger of wearing these articles. Rubber boots worn in an area contaminated with mustard gas may offer a grave danger to men who wear them several days after the bombardment. Impermeable clothing will resist penetration more than an hour, but should not be worn longer than this.

43. CLEANING.

a. All unpainted metal parts of materiel that have been exposed to any gas except mustard and lewisite must be cleaned as soon as possible with **SOLVENT**, dry-cleaning, or **ALCOHOL**, denatured, and wiped dry. All parts should then be coated with engine oil.

b. Ammunition which has been exposed to gas must be thoroughly cleaned before it can be fired. To clean ammunition use **AGENT**, decontaminating, noncorrosive, or if this is not available, strong soap and cool water. After cleaning, wipe all ammunition dry with clean rags. *Do not use dry powdered AGENT, decontaminating (chloride of lime), (used for decontaminating certain types of materiel on or near ammunition supplies), as flaming occurs through the use of chloride of lime on liquid mustard.*

44. DECONTAMINATION.

a. For the removal of liquid chemicals (mustard, lewisite, etc.) from materiel, the following steps should be taken:

MATERIEL AFFECTED BY GAS

(1) PROTECTIVE MEASURES. (a) For all of these operations a complete suit of impermeable clothing and a service gas mask will be worn. Immediately after removal of the suit, a thorough bath with soap and water (preferably hot) must be taken. If any skin areas have come in contact with mustard, if even a very small drop of mustard gets into the eye, or if the vapor of mustard has been inhaled, it is imperative that complete first-aid measures be given within 20 to 30 minutes after exposure. First-aid instructions are given in TM 9-850 and FM 21-40.

(b) Garments exposed to mustard will be decontaminated. If the impermeable clothing has been exposed to vapor only, it may be decontaminated by hanging in the open air, preferably in sunlight for several days. It may also be cleaned by steaming for 2 hours. If the impermeable clothing has been contaminated with liquid mustard, steaming for 6 to 8 hours will be required. Various kinds of steaming devices can be improvised from materials available in the field.

(2) PROCEDURE. (a) Commence by freeing materiel of dirt through the use of sticks, rags, etc., which must be burned or buried immediately after this operation.

(b) If the surface of the materiel is coated with grease or heavy oil, this grease or oil should be removed before decontamination is begun. SOLVENT, dry-cleaning, or other available solvents for oil should be used with rags attached to ends of sticks.

(c) Decontaminate the painted surfaces of the materiel with bleaching solution made by mixing one part AGENT, decontaminating (chloride of lime), with one part water. This solution should be swabbed over all surfaces. Wash off thoroughly with water, then dry and oil all surfaces.

(d) All unpainted metal parts and instruments exposed to mustard or lewisite must be decontaminated with AGENT, decontaminating, noncorrosive, mixed one part solid to fifteen parts solvent (ACETYLENE TETRACHLORIDE). If this is not available, use warm water and soap. Bleaching solution must not be used, because of its corrosive action. Instrument lenses may be cleaned only with PAPER, lens, tissue, using a small amount of ALCOHOL, ethyl. Coat all metal surfaces lightly with engine oil.

(e) In the event AGENT, decontaminating (chloride of lime) is not available, materiel may be temporarily cleaned with large volumes of hot water. However, mustard lying in joints or in leather or canvas webbing is not removed by this procedure and will remain a constant source of danger until the materiel can be properly decontaminated. All mustard washed from materiel in this manner lies unchanged on the ground, necessitating that the contaminated area be plainly marked with warning signs before abandonment.

TM 9-741
44-45

MEDIUM ARMORED CAR T17E1

(*f*) The cleaning or decontaminating of materiel contaminated with lewisite will wash arsenic compounds into the soil, poisoning any water supplies in the locality for either men or animals.

(*g*) Leather or canvas webbing that has been contaminated should be scrubbed thoroughly with bleaching solution. In this event the treatment is insufficient, it may be necessary to burn or bury such material.

(*h*) Detailed information on decontamination is contained in FM 21-40, TM 9-850, and TC 38, 1941, Decontamination.

45. SPECIAL PRECAUTIONS FOR AUTOMOTIVE MATERIEL.

a. When vehicles have been subjected to gas attack with the engine running, the air cleaner should be serviced by removing the oil, flushing with SOLVENT, dry-cleaning, and refilling with the proper grade of oil.

b. Instrument panels should be cleaned in the same manner as outlined for instruments.

c. Contaminated seat cushions will be discarded.

d. Washing the compartments thoroughly with bleaching solution is the most that can be done in the field. Operators should constantly be on the alert, when running under conditions of high temperatures, for slow vaporization of the mustard or lewisite.

e. Exterior surfaces of vehicles will be decontaminated with bleaching solution. Repainting may be necessary after this operation.

PART TWO — ORGANIZATION INSTRUCTIONS

Section X

ORGANIZATION MAINTENANCE

	Paragraph
Scope	46
Definition of terms	47
Allocation of repair	48

46. SCOPE.

a. The scope of maintenance of repairs by the crew and other units of the using arms is determined by the ease with which the project can be accomplished, the amount of time available, available parts, the nature of the terrain, weather conditions, temperatures, concealment, shelter, proximity to hostile fire, the equipment available and the skill of the personnel. All of these are variable and no exact system of procedure can be prescribed.

47. DEFINITION OF TERMS.

a. The definitions given below are included in order that the operation name may be correctly interpreted by those doing the work.

(1) SERVICE. Consists of cleaning, lubricating, tightening bolts and nuts, and making external adjustments of subassemblies or assemblies and controls.

(2) REPAIR. Consists of making repairs to, or replacement of a part, subassembly or assembly that can be accomplished without completely disassembling the subassembly or assembly and does not require heavy welding or riveting, machining, fitting and alining.

(3) REPLACE. Consists of removing the part, subassembly or assembly from the vehicle and replacing it with a new or reconditioned part, subassembly or assembly, whichever the case may be.

(4) REBUILD. Consists of completely reconditioning and placing in serviceable condition any unserviceable part, subassembly or assembly of the motor vehicle including welding, riveting, machining, fitting, alining, assembling and testing.

48. ALLOCATION OF REPAIR.

a. The following are the maintenance duties for which tools and parts have been provided the using arm personnel. Other replacements and repairs are the responsibility of the other maintenance personnel, but may be performed by the using arm personnel, when circumstances permit, within the discretion of the pertinent ordnance officer.

TM 9-741
48

MEDIUM ARMORED CAR T17E1

ENGINES

Manifold gasket (par. 57)..............................Replace
Valve clearance (par. 58)..............................Adjust
Cylinder head or gasket (par. 59)......................Replace
Oil filter (par. 60)...................................Replace
Engine ....Tune up
Engine (par. 63)......................................Replace

ENGINE IGNITION SYSTEM

Spark plug (par. 66).........................Service or replace
Distributor (par. 67).................................Replace
Coil (par. 68)..Replace
Filter (par. 69)......................................Replace
Suppressor (par. 70).........................Service or replace
Breaker point (par. 71)......................Service or replace
Condenser (par. 72)...................................Replace

FUEL SYSTEM

Carburetor (par. 77).........................Service or replace
Air cleaner (par. 78)........................Service or replace
Fuel pump (par. 79)...................................Replace
Fuel filter (par. 80)........................Service or replace

COOLING SYSTEM

Fan belts (par. 86)..........................Service or replace
Thermostat (par. 87)..................................Replace
Water pump (par. 88)..................................Replace
Radiator core (par. 89)...............................Replace
Oil cooler (par. 90)..................................Replace

EXHAUST SYSTEM

Muffler (par. 93).....................................Replace
Exhaust pipe (par. 94)................................Replace

TRANSMISSION

Servo bands (par. 98).................................Adjust
Hydraulic throttle control (par. 99)..................Adjust
Manual shift control (par. 100).......................Adjust

PROPELLER SHAFTS AND UNIVERSAL JOINTS

Propeller shaft (gear case to transfer case) (pars. 108-110)..Replace
Propeller shaft (transfer case to axle) (pars. 111-113).......Replace
Universal joints (pars. 109-112)......................Replace

TRANSFER CASE

Transfer case (pars. 116-117).........................Replace

72

ORGANIZATION MAINTENANCE

FRONT AXLE
Drive flange (par. 120) .... Replace
Axle shaft (par. 121) .... Replace
Tie rod (par. 122) .... Replace
Toe-in (par. 123) .... Adjust

REAR AXLE
Axle shaft (par. 126) .... Replace

BRAKE SYSTEM
Brakes (par. 131) .... Service
Parking brake (par. 132) .... Adjust
Brakes (par. 133) .... Bleed
Brake shoe (par. 134) .... Replace
Wheel cylinders (par. 135) .... Replace
Main cylinder (par. 136) .... Replace
Hydrovac (par. 137) .... Replace
Brake lines (par. 138) .... Replace

SPRINGS, RADIUS RODS, AND SHOCK ABSORBERS
Springs (par. 140) .... Replace
Radius rods (par. 141) .... Replace
Shock absorbers (par. 142) .... Replace

STEERING GEAR
Steering gear (par. 146) .... Service
Steering gear motor (par. 147) .... Replace
Steering connecting rod (par. 148) .... Replace

WHEELS, TIRES, WHEEL BEARINGS, TIRE PUMP
Wheel and tire (par. 152) .... Replace
Wheel bearings (par. 153) .... Adjust
Tire pump (par. 154) .... Replace

BATTERIES AND STARTING SYSTEM
Batteries (pars. 156-159) .... Service or replace
Starting motors (pars. 157-160) .... Service or replace
Solenoid switches (par. 161) .... Replace

GENERATORS AND CONTROLS
Generator (par. 165) .... Replace
Regulator unit (par. 166) .... Replace

LIGHTING SYSTEM
Head lamp (par. 169) .... Replace
Head lamp sealed beam assembly (par. 170) .... Replace
Head lamp aim (par. 171) .... Aim

MEDIUM ARMORED CAR T17E1

LIGHTING SYSTEM—Cont'd

Marker lamp bulb (par. 172)	Replace
Auxiliary blackout drive lamp sealed beam assembly (par. 173)	Replace
Tail and stop lamp (pars. 174-175)	Replace
Instrument panel bulb (par. 176)	Replace
Dome lamp bulb (par. 177)	Replace

INSTRUMENTS AND GAGES

Instrument panel (pars. 179-181)	Replace
Instrument panel units (par. 180)	Replace
Fuel gage (par. 182)	Replace
Engine heat indicator (par. 183)	Replace
Engine oil pressure gage (par. 184)	Replace

ELECTRICAL ACCESSORIES

Siren (par. 186)	Replace
Siren switch (par. 187)	Replace
Electric windshield wipers (par. 188)	Replace
Wiper motor (par. 189)	Replace

RADIO SUPPRESSION

Radio suppression (par. 191)	Service

TURRET AND TRAVERSING SYSTEM

Turret electric motor (par. 199)	Replace
Control valve (par. 200)	Replace
Traversing gear (par. 201)	Replace
Hydraulic motor (par. 202)	Replace
Hydraulic pump (par. 203)	Replace

FIRE EXTINGUISHING SYSTEM

Cylinders (par. 238)	Replace
Controls (par. 238)	Reset

GYROSTABILIZER

Control gear box (par. 221-222)	Adjust or replace
Oil pump and/or motor (par. 224)	Replace
Control box (par. 225)	Replace
Piston and cylinder (par. 226)	Replace
Recoil switch (par. 227)	Replace
Gear box (par. 228)	Replace
Wiring and conduit (par. 229)	Replace
Clutch (par. 230)	Replace
Flexible shaft (par. 231)	Replace
Flexible shaft gear box (par. 232)	Replace
Control link (par. 233)	Replace
Oil lines (par. 234)	Replace

TM 9-741
49-50

Section XI

TOOLS AND EQUIPMENT

Paragraph

Organization 49
Special tools 50

49. ORGANIZATION.

a. Tank Crew Tools and Equipment. The tools and equipment ordinarily required for operations performed by the using arms are included as regular equipment with each vehicle, and are listed in section VII.

b. Company Tools and Equipment. The tools and equipment ordinarily required for inspection and maintenance in the field include many items not used by the crew and are not provided with each armored car. They are a regular part of company equipment and material to be used on all Medium Armored Cars T17E, in the company.

c. Regimental Tools and Equipment. A still more extensive group of regular and special tools, and special equipment is provided for use of the regimental maintenance unit. They cover all requirements of first and second echelon maintenance.

d. Care of Tools and Equipment. An accurate record of all tools and equipment must be kept in order that their location and condition may be known at all times. Items becoming lost or unserviceable should be immediately replaced. All tools and equipment should be cleaned and in proper condition for further use before being returned to their location. Care must be used in fastening the tools carried on the outside of the vehicle, and frequent inspection and oiling is necessary to prevent corrosion.

50. SPECIAL TOOLS.

a. The following is a list of the special tools and tool numbers.

KM-J2280	CLAMP, wheel cylinder
KM-KM03A	CUTTER, tube
MTM-M3-27	GAGE, 1,000-pound pressure
KM-J2252	HOSE, brake bleeder
KM-KM0142	PLIERS, brake spring
KM-J1618M6	PULLER, steering wheel
KM-J2289	REAMER, spring pin bushing
KM-KM0204	TESTER, fuel gage
KM-J1885	TOOL, flaring
KM-J2274	TOOL, spring pin and bushing replacer

TM 9-741

MEDIUM ARMORED CAR T17E1
Section XII

ORGANIZATION SPARE PARTS AND ACCESSORIES

	Paragraph
Organization spare parts	51
Accessories	52

51. ORGANIZATION SPARE PARTS.

a. A set of organization spare parts is supplied to the using arms for field replacement of those parts most likely to become broken, worn, or otherwise unserviceable. The set is kept complete by requisitioning new parts for those used. Organization spare parts are listed in pertinent SNL'S.

b. Care of organization spare parts is covered in the section of this manual entitled "Care and Preservation."

52. ACCESSORIES.

a. Accessories include tools and equipment required for such disassembling and assembling as the using arms are authorized to perform, and for the cleaning and preservation of the gun mount, sighting and fire-control equipment, ammunition, etc. They also include chests, covers, tool rolls and other items necessary to protect the material when it is not in use, or when traveling. Accessories should not be used for purposes other than as prescribed, and when not in use, should be properly stored.

TM 9-741
53-54

Section XIII

ENGINES

	Paragraph
General description	53
Design and construction	54
Data	55
Trouble shooting	56
Manifold gasket replacement	57
Valve clearance adjustment	58
Cylinder head or gasket replacement	59
Oil filter service	60
Crankcase ventilating system	61
Engine tune-up	62
Engine replacement	63

53. GENERAL DESCRIPTION (figs. 35, 36, 37, and 38).

a. The two heavy duty engines used in the Medium Armored Car T17E1 are identical with the exception of certain external equipment or fittings. They are four-cycle, six-cylinder-in-line, valve-in-head engines, mounted side by side in the rear compartment of the hull. Due to this mounting, the flywheel and transmission end of the engine is toward the front of the vehicle.

b. The cylinders are numbered with No. 1 at the fan end of the engine.

c. The direction of rotation of the engine is clockwise as you face the fan end.

d. The right and left side of an individual engine is determined by facing the engine from the flywheel end.

54. DESIGN AND CONSTRUCTION.

a. These engines are the three story type—the cylinder head, the cylinder block, and the oil pan.

b. The cylinder head, as it is installed on the cylinder block, includes the valve guides, valves, valve springs, rocker arm assemblies with shafts, thermostat housing, water outlet, manifolds, carburetor, temperature indicator fitting, valve cover with air cleaner, and miscellaneous small parts.

c. The cylinder block assembly is the major section as it is fitted with the crankshaft, camshaft, timing gear plate, timing gears, timing gear cover, harmonic balancer, piston assemblies, connecting rods,

MEDIUM ARMORED CAR T17E1

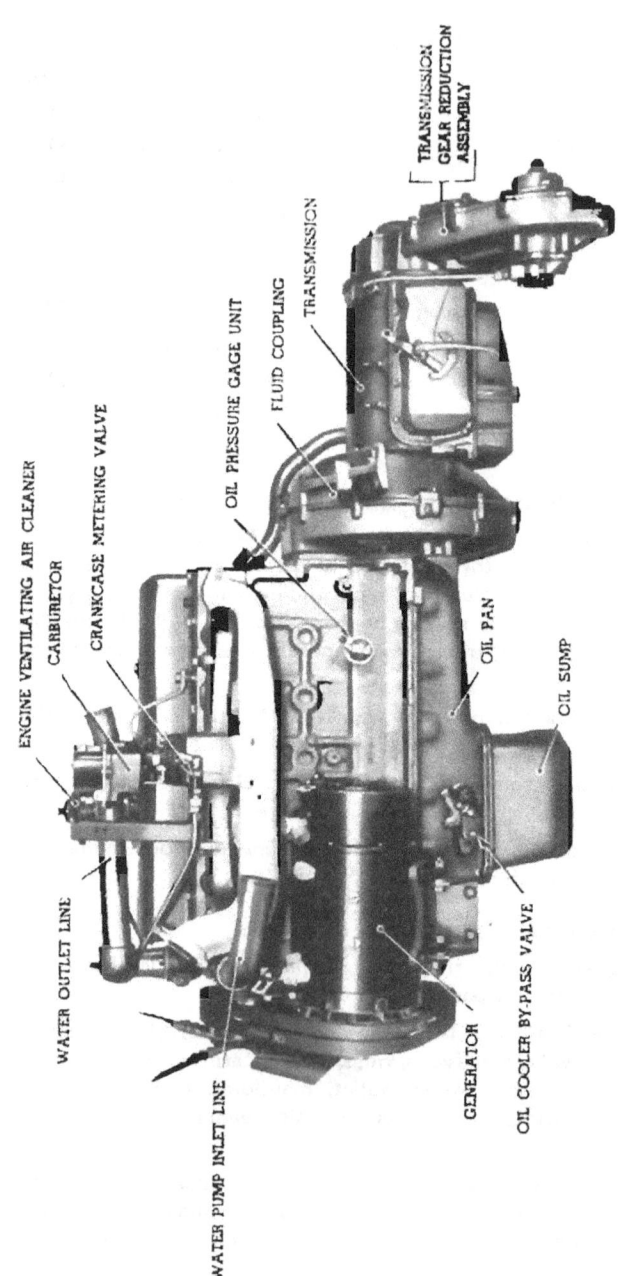

Figure 35 — Left Side of Left Engine

TM 9-741
54

ENGINES

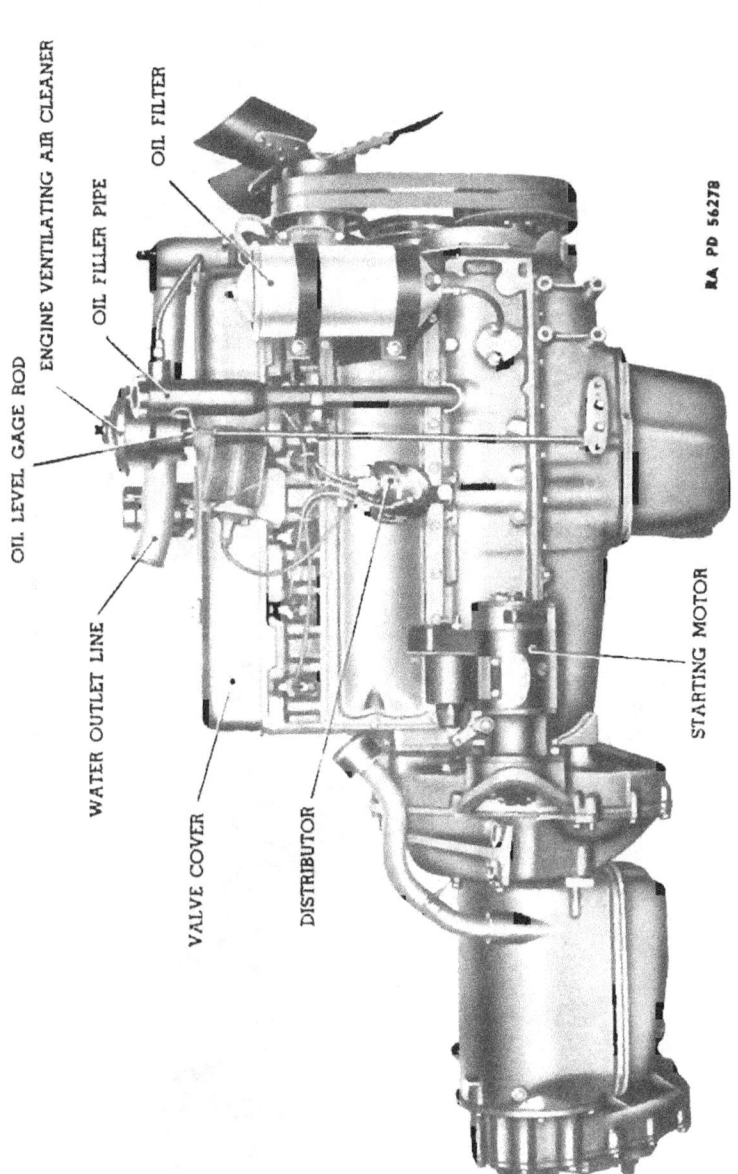

Figure 36 — Right Side of Left Engine

TM 9-741
54

MEDIUM ARMORED CAR T17E1

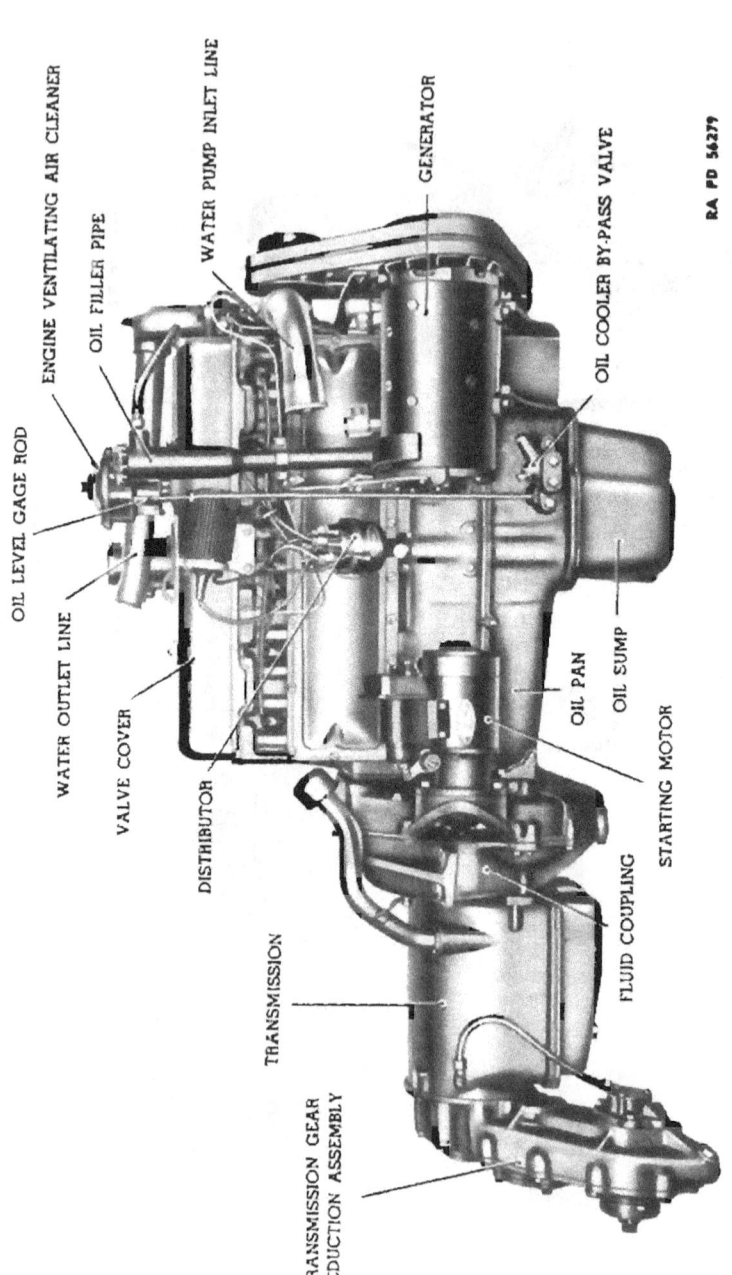

Figure 37 — Right Side of Right Engine

ENGINES

TM 9-741
54

RA PD 32462

Figure 38 — Fan End of Left Engine

TM 9-741
54-56

MEDIUM ARMORED CAR T17E1

water pump, flywheel, flywheel housing, oil pump, electrical equipment, and miscellaneous other parts.

d. The oil pan, when fitted with the oil sump, provides a closure for the bottom of the engine as well as clearance for the oil pump, connecting rods, and space for the oil supply.

e. The fan, water pump, and generator are driven by two V-type fan belts from the crankshaft harmonic balancer.

f. The oil level of the engines and the water supply of the cooling systems can be checked, and, if necessary, replenished through the four hinged covers on the engine compartment covers (fig. 30).

g. Other service operations will require the removal of one or both of the engine compartment covers (figs. 30 and 39) and in some cases the removal of the fan shrouds to permit working on the fan end of the engines from under the rear of the hull.

55. DATA.

a. The following data refers to one engine only.

```
Type ................................................ Valve-in-head
Number of cylinders ........................................... 6
Bore ................................................... 3 25/32-in.
Stroke ...................................................... 4-in.
Piston displacement (cu in.) ............................. 269.52
Compression ratio ..................................... 6.75 to 1
Horsepower, SAE ........................................... 34.35
Horsepower, rated (at 3,000 rpm) ............................ 97
Firing order (from fan end of engine) .............. 1-5-3-6-2-4
Engine weight (with accessories) .......................... 830 lb
Number of main bearings ....................................... 4
Oil capacity ................................................ 8 qt
```

56. TROUBLE SHOOTING.

a. The engines must be watched closely for any indications of trouble. Possible troubles are broken down here into major symptoms, their probable causes and probable remedies.

b. Lack of Power.

Probable Cause	Probable Remedy
Incorrect valve clearance.	Adjust valve clearance according to instructions under valve adjustment procedure (par. 58).
Leaky valves.	Report to ordnance personnel.
Valve stems or lifters sticking.	Replace cylinder head.
Valve springs weak or broken.	Replace cylinder head.

TM 9-741
56

ENGINES

Figure 39 — Engine Compartment

83

TM 9-741
56

MEDIUM ARMORED CAR T17E1

Probable Cause	Probable Remedy
Valve timing incorrect.	Report to ordnance personnel.
Leaking cylinder head gasket.	Tighten or replace.
Piston rings broken.	Report to ordnance personnel.
Poor fits between pistons, rings, and cylinders.	Report to ordnance personnel.
Exhaust system partly restricted.	Replace or clean (par. 93).
Ignition not properly timed.	Set ignition according to instructions under engine tune-up (par. 67).
Octane selector not adjusted for grade of fuel being used.	Set octane selector (par. 62).
Spark plugs faulty.	Replace or clean and test spark plugs (par. 66).
Distributor points not set correctly.	Set distributor points and time engine (par. 62).
Dirt or water in carburetor.	Clean carburetor (par. 77).
Fuel lines partly plugged.	Clean fuel lines.
Air leaks in fuel line.	Tighten and check fuel lines.
Fuel pump not functioning properly.	Replace fuel pump (par. 79).
Air cleaner dirty.	Service air cleaner (par. 78).
Carburetor choke partly closed.	Adjust or replace choke mechanism (par. 77).
Cooling system troubles.	See paragraph 83 for procedure.
Improper grade and viscosity of oil being used.	Change to correct oil.
Fuel mixture too lean.	See paragraph 77 for corrections.
Restricted air cleaner.	Service air cleaner (par. 78).
Transfer case in four-wheel drive on hard surface roads.	Shift transfer case to two-wheel drive.
Dragging brakes.	Adjust brakes (par. 131).

c. **Excessive Oil Consumption.**

Probable Cause	Probable Remedy
Oil pan drain plug loose.	Tighten drain plug.
Oil pan retainer bolts or oil pump bolts loose.	Report to ordnance personnel.
Oil pan gaskets damaged.	Report to ordnance personnel.
Timing gear cover loose or gasket damaged.	Tighten timing gear cover or report to ordnance personnel.
Oil return from timing gear case to block restricted causing leak at crankshaft fan pulley hub.	Report to ordnance personnel.
Push rod or rocker arm cover gaskets damaged or covers loose.	Tighten push rod and rocker arm covers, or replace gaskets.

ENGINES

Probable Cause	Probable Remedy
Rear main bearing leaking oil into clutch housing.	Report to ordnance personnel.
Broken piston rings.	Report to ordnance personnel.
Rings not correctly seated to cylinder walls.	Report to ordnance personnel.
Piston rings worn excessively or stuck in ring grooves.	Report to ordnance personnel.
Piston ring oil return holes clogged with carbon.	Report to ordnance personnel.
Excessive clearance between piston and cylinder walls due to wear or their being improperly fitted.	Report to ordnance personnel.
Cylinder walls scored, tapered or out-of-round.	Report to ordnance personnel.

d. **Popping, Spitting, and Spark Knock.**

Loose wiring connections.	Tighten all wire connections.
Faulty wiring.	Replace faulty wiring.
Faulty spark plugs.	Clean or replace spark plugs (par. 66).
Lean combustion mixture.	Clean and adjust carburetor (par. 77).
Dirt in carburetor.	Clean carburetor (par. 77).
Restricted fuel supply to carburetor.	Clean fuel lines and check for restrictions.
Leaking carburetor or intake manifold gaskets.	Tighten carburetor to manifold and manifold to head bolts or replace gaskets.
Carburetor metering rod hole cover not in place.	Replace metering rod hole cover (par. 77).
Valves adjusted too closely.	Adjust valve clearance (par. 58).
Valves sticking.	Lubricate and free up, or report to ordnance personnel.
Weak valve springs.	Replace cylinder head.
Valves timed early.	Report to ordnance personnel.
Excessive carbon deposits in combustion chamber.	Remove head and clean carbon (par. 59).
Cylinder head water passages partly clogged causing hot spot in combustion chamber.	Remove cylinder head and clean water passages (par. 59).

TM 9-741
56

MEDIUM ARMORED CAR T17E1

Probable Cause	Probable Remedy
Partly restricted exhaust ports in cylinder head.	Remove cylinder head and clean exhaust ports (par. 59).
Cylinder head gasket blown between cylinders.	Replace cylinder head gasket (par. 59).
Spark plugs glazed.	Clean or replace spark plugs (par. 66).
Wrong heat range plug being used.	Change to correct type spark plugs (par. 66).
Exhaust manifold or muffler restricted, causing back pressure.	Clean or replace manifold and muffler (pars. 57 and 93).

e. Rough Engine Idling.

Improper idling adjustment.	Adjust according to instructions in fuel system section (par. 77).
Carburetor needle valve not seating.	Replace (par. 77).
Carburetor to manifold gasket leaks.	Tighten carburetor to manifold bolts or replace gasket (par. 77).
Manifold to head gasket leaks.	Tighten manifold to head bolts or replace gasket (par. 57).
Air leaks in the Hydrovac lines.	Check Hydrovac lines and correct leaks (pars. 137 and 138).
Improper valve clearance.	Check and adjust valves (par. 58).
Valves not seating properly.	Replace cylinder head.
Cracks in exhaust ports.	Replace cylinder head (par. 59).
Head gasket leaks.	Replace cylinder head gasket (par. 59).

f. Engine Noises.

(1) It is often very difficult to determine the exact cause of certain engine noises. If unusual noises develop which the driver is unable to trace definitely and correct, the vehicle should not be driven until an experienced officer or mechanic has checked the vehicle and given instructions regarding its use or repairs.

(2) When any engine noise develops, quickly check for low or no oil pressure and abnormally high temperature indicator reading. Check for sufficient oil in the engines and sufficient solution in the cooling systems.

TM 9-741
57

ENGINES

Figure 40 — Manifold Gasket Replacement

57. MANIFOLD GASKET REPLACEMENT.

a. Right Engine.

(1) Remove the twenty-five retaining bolts and remove the right engine compartment cover (fig. 30).

(2) Remove the three exhaust pipe-to-manifold flange stud nuts (fig. 39). The three muffler-to-hull bracket bolts and the muffler strap swing the muffler and pipe out of the way.

(3) Disconnect crankcase ventilator suction pipe at the oil filler and manifold ends.

(4) Disconnect temperature indicator wire connection from thermostat housing fitting.

(5) Disconnect accelerator rod from throttle arm and Hydrovac vacuum line from manifold.

(6) Remove plug from back end of manifold to provide clearance at generator regulator.

(7) Disconnect air cleaner pipe and fuel feed line from carburetor.

(8) Remove the water outlet pipe-to-brace clamp nuts and remove the clamp.

(9) Remove the eight manifold stud nuts, washers, and clamps.

(10) Support the manifold by tying it to an engine compartment cover bolt hole in the right side of hull (fig. 40) and slide the manifold off the studs.

(11) Remove the old gaskets and clean the gasket flanges on the cylinder head and manifold.

TM 9-741

MEDIUM ARMORED CAR T17E1

(12) Install the new gaskets over the studs; slide the manifold on the studs and against the gaskets. NOTE: Make sure the gaskets are in place and that the manifold slides up against them.

(13) Install the washers, clamps, and the eight retaining nuts, and tighten securely.

(14) Install the water outlet to brace clamp and tighten securely.

(15) Connect air cleaner pipe and fuel feed line to carburetor.

(16) Attach accelerator rod to throttle arm and install washer and cotter pin.

(17) Install pipe plug in back end of intake manifold.

(18) Connect temperature indicator wire connection to thermostat housing fitting.

(19) Connect and tighten both ends of the crankcase ventilator suction pipe.

(20) Install a new exhaust manifold-to-pipe flange gasket. Place the muffler and exhaust pipe in position; install the three pipe to manifold stud nuts; the three muffler support-to-hull bolts and the muffler strap.

(21) Replace the engine compartment cover.

b. Left Engine.

(1) Remove the twenty-five retaining bolts and remove the left engine compartment cover.

(2) Remove the three exhaust pipe-to-manifold flange stud nuts; the three muffler-to-hull bracket bolts and the muffler strap. Pull exhaust pipe and muffler away from manifold.

(3) Disconnect crankcase ventilator suction pipe at manifold end and disconnect gas line at carburetor.

(4) Disconnect accelerator rod from throttle arm and air cleaner pipe from carburetor.

(5) Remove the water outlet pipe-to-brace clamp nuts and remove the clamp.

(6) Disconnect engine oil cooler oil lines from cooler.

(7) Remove the eight manifold stud nuts, washers, and clamps.

(8) Slide the manifold off the studs and carefully let it rest on the generator.

(9) Remove the old gaskets and clean the gasket seats on the head and manifold.

(10) Install the new gaskets over the studs (fig. 40); slide the manifold on the studs and up against the gasket. NOTE: Make sure the gaskets are in place and that the manifold slides up against them.

TM 9-741
57-58

ENGINES

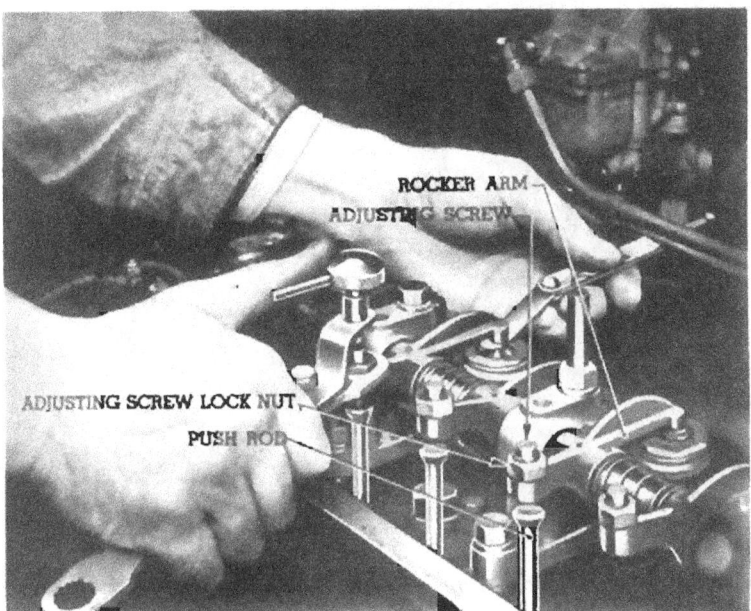

Figure 41 — Valve Clearance Adjustment   RA PD 56280

(11) Install the washers, clamps, and eight manifold nuts. Tighten securely.

(12) Connect engine oil cooler oil lines to cooler.

(13) Install the water outlet pipe to brace clamp and tighten the nuts securely.

(14) Connect accelerator rod to throttle arm and install washer and cotter pin.

(15) Connect air cleaner pipe to carburetor.

(16) Connect crankcase ventilator suction pipe to manifold fitting and connect fuel line to carburetor.

(17) Install a new gasket between the exhaust pipe flange and manifold; install the three stud nuts, muffler-to-hull bolts and muffler strap. Tighten securely.

(18) Replace the engine compartment cover and tighten the twenty-five bolts.

58. VALVE CLEARANCE ADJUSTMENT (fig. 41).

a. Before adjusting valve clearance on these valve-in-head engines, they must be thoroughly warmed up to normalize the expansion of

TM 9-741
58

MEDIUM ARMORED CAR T17E1

RA PD 32448

Figure 42 — Tightening Cylinder Head Bolts

all parts and stabilize the oil temperature. This is very important because during the warm up period the valve clearance varies considerably. It is advisable to adjust valves after the vehicle has been on a normal run, otherwise the engine (or engines) should be run for about thirty minutes before adjusting valve clearance. Run the engine at idling speed while adjusting clearance.

b. Right Engine.

(1) Remove the twenty-five bolts which retain the right engine compartment cover and remove the cover.

(2) Remove the valve cover air cleaner by turning counterclockwise (fig. 39). CAUTION: Keep the cleaner upright or the oil will leak out.

(3) Remove the two valve cover retaining nuts and remove the cover.

(4) Tighten rocker arm shaft bolts and nuts to 25-30 pound-feet and the cylinder head bolts to 60-70 pound-feet, using a torque wrench (fig. 42).

(5) Adjust the valves by loosening the adjusting screw lock nut and turning the adjusting screw clockwise or counterclockwise to decrease or increase the clearance between the rocker arm and valve, to

ENGINES

0.012 inch on the intake and 0.016 inch on the exhaust valves (fig. 41). If the special wrench shown is not available, a box wrench and screwdriver can be used.

(6) Tighten the lock nut while holding the adjusting screw to prevent a change in adjustment. Recheck to see that the clearance is still 0.012 inch on the intake and 0.016 inch on the exhaust after lock nut is tight.

(7) Follow the same procedure on the remaining eleven valves.

(8) Make sure the valve cover gasket is in good condition, if not, replace it. Clean the valve cover inside and out. Install cover and check to see that it seats on the gasket; replace two nuts and tighten securely.

(9) Install the valve cover air cleaner and check the oil level in cleaner. If necessary, add oil.

(10) Install the engine compartment cover and tighten the twenty-five retaining bolts.

c. **Left Engine.** The valve adjustment procedure on the left engine is identical to that of the right engine except that the air cleaner tube must be removed from the carburetor.

59. CYLINDER HEAD OR GASKET REPLACEMENT (fig. 43).

a. **Cylinder Head Description.**

(1) The cylinder head of a valve-in-head engine plays an important part in its operation. It contains the inlet ports, exhaust ports, valve seats, valve assemblies, rocker arm assemblies, spark plugs, combustion chambers, and necessary water passages to maintain proper temperatures of these important parts.

(2) Due to this construction, it is possible for the using arms to remove a cylinder head and exchange it for one that has had the valves conditioned.

b. **Right Engine Cylinder Head Removal.**

(1) Remove the twenty-five bolts which attach the right engine compartment cover and remove the cover.

(2) Drain the cooling system by removing the drain plug from the bottom of hull (fig. 31).

(3) Remove the three nuts which attach the manifold to exhaust pipe, the three bolts which attach the muffler support to the hull and in some cases, it may be necessary to loosen the muffler strap. Swing the exhaust pipe to the right to clear the manifold.

(4) Disconnect hand throttle and choke cables from carburetor.

(5) Remove the two bolts which attach water outlet - support bracket to intake manifold (fig. 39).

MEDIUM ARMORED CAR T17E1

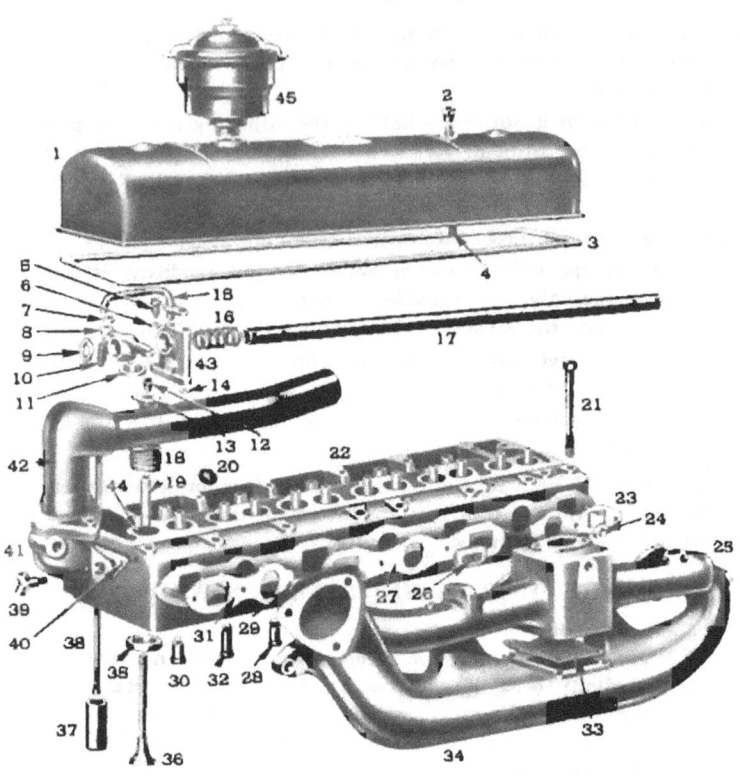

1. VALVE COVER
2. COVER NUT
3. COVER GASKET
4. COVER STUD
5. LOCATING SCREW
6. SCREW NUT
7. ADJUSTING SCREW LOCK NUT
8. ROCKER ARM
9. RETAINING WASHER
10. ADJUSTING SCREW
11. OIL SHIELD – INTAKE ONLY
12. VALVE KEY
13. SPRING RETAINER
14. PIPE GASKET
15. OVERFLOW PIPE
16. ROCKER SHAFT SPRING
17. ROCKER ARM SHAFT
18. VALVE SPRING
19. VALVE STEM GUIDE
20. PLUG
21. CYLINDER HEAD BOLT
22. CYLINDER HEAD
23. MANIFOLD GASKET
24. MANIFOLD WASHER
25. INTAKE MANIFOLD
26. MANIFOLD CLAMP
27. MANIFOLD GASKET
28. WATER NOZZLE
29. MANIFOLD PILOT
30. WATER NOZZLE
31. MANIFOLD GASKET
32. WATER NOZZLE
33. MANIFOLD GASKET
34. EXHAUST MANIFOLD
35. VALVE SEAT INSERT
36. VALVE
37. VALVE TAPPET
38. VALVE PUSH ROD
39. OIL INLET
40. HOUSING GASKET
41. THERMOSTAT HOUSING
42. WATER OUTLET
43. SHAFT BRACKET
44. SPRING SEAT
45. ENGINE VENTILATING AIR CLEANER

RA PD 56281

Figure 43 — Cylinder Head Exploded View

ENGINES

(6) Disconnect radiator upper hose from radiator, remove two water outlet-to-thermostat housing bolts and lay the water outlet and carburetor control cables out on the hull.

(7) Disconnect crankcase ventilator suction pipe at both ends and remove pipe.

(8) Disconnect the foot accelerator rod from the throttle arm by removing the cotter pin and washer.

(9) Disconnect Hydrovac vacuum line from manifold, air cleaner delivery pipe from carburetor, and fuel feed line from carburetor.

(10) Disconnect accelerator pull-back spring, remove the two retaining nuts, and remove carburetor.

(11) Remove temperature indicator wire connector from thermostat housing fitting.

(12) Remove valve cover air cleaner by turning counterclockwise. CAUTION: Keep the cleaner upright as it is an oil-bath type cleaner.

(13) Disconnect oil filter and rocker arm oil lines from fitting at front of cylinder head.

(14) Disconnect oil gage rod clip from oil filler tube bracket, remove the two oil filler tube bracket bolts, and remove the filler.

(15) Disconnect the ignition wires from the coil and remove the coil.

(16) Remove the two coil bracket bolts and remove the bracket (fig. 39).

(17) Disconnect the spark plug wires.

(18) Remove the eighteen screws which attach the push rod cover and remove cover and gasket.

(19) Remove the two valve cover retaining nuts and remove the valve cover and gasket (fig. 43).

(20) Remove valve rocker arms and shaft as a complete assembly by removing the bolts from the six shaft brackets. Remove the six intake valve shields.

(21) Lift the twelve push rods out.

(22) Remove the fifteen cylinder head bolts and remove the cylinder head assembly and gasket.

(23) Remove the eight nuts, washers, and clamps which retain the manifold to head and remove the manifold (figs. 40 and 43).

(24) Remove the six spark plugs, the thermostat housing, and the rocker arm oil inlet fitting (fig. 43).

c. **Carbon Removing.** Each time a cylinder head is removed, the carbon should be removed from the piston heads and the combustion chambers in the cylinder head. Figure 44 shows the method of removing

TM 9-741
59

MEDIUM ARMORED CAR T17E1

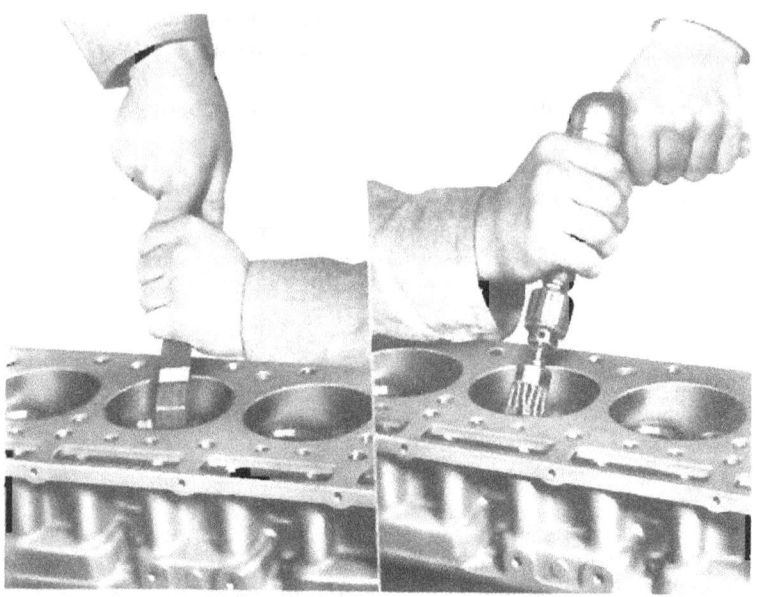

Figure 44 — Cleaning Carbon From Combustion Chamber

carbon from the cylinder head. Clean the water passages in the cylinder head and inspect the head for cracks.

d. Rocker Arm and Shaft Assemblies.

(1) In order for the lobes of the camshaft to operate the valves, the following parts are used: valve lifters (which contact the lobes of the camshaft), push rods, rocker arms and rocker arm shaft (fig. 45). In order to maintain correct adjustment of this mechanism, an adjusting screw and lock nut is used at the push rod end of each rocker arm. It is important that correct adjustment be maintained to keep the mechanism normally quiet and provide correct opening and closing of the valves.

(2) Sludge and gum formation in the hollow shaft may prevent normal lubrication of the rocker arms and valves. Each time the rocker arm and shaft assembly is removed, it should be disassembled and thoroughly cleaned.

(3) To disassemble the rocker arm shaft, remove the locking screw, the two cotter pins, the retaining washers, rocker arms, brackets, and springs.

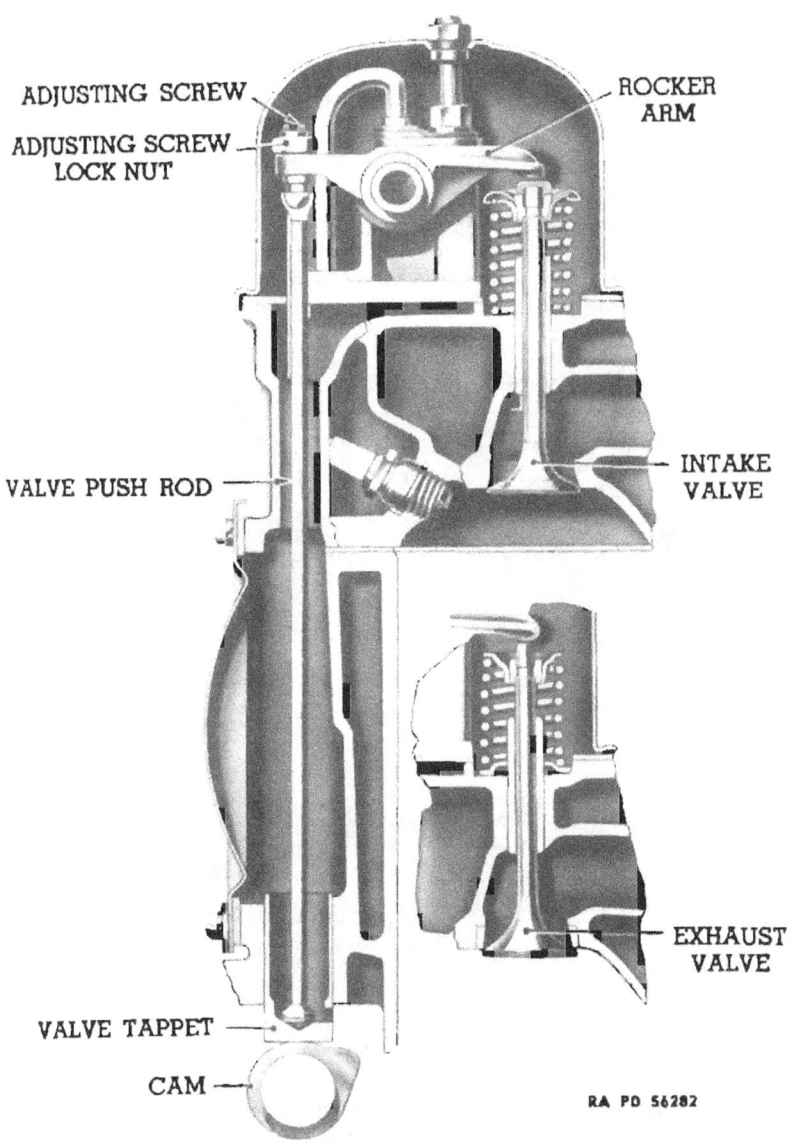

Figure 45 — Valve Operating Mechanism

MEDIUM ARMORED CAR T17E1

(4) Remove the end plugs and clean the sludge from the shaft. Clean the oilholes and grooves in the rocker arms and shaft.

(5) Check the shaft and rocker arms for excessive wear. Replace if necessary.

(6) Assemble the rocker arms, supports, springs, washers, and lock keys to the shaft in the order shown in figure 46.

e. **Right Engine Cylinder Head Installation.**

(1) Install the rocker arm oil inlet fitting and the thermostat housing, using a new gasket.

(2) Clean, adjust, and install the six spark plugs (par. 62).

(3) Install new manifold gaskets, the manifold assembly, washers, clamps, and the eight nuts, and tighten securely.

RA PD 32467

Figure 46 — Cylinder Head — Valve Cover Removed

(4) Clean the top of the block and the cylinder head, install a new gasket; place the head assembly on the block, and install the fifteen cylinder head bolts.

RA PD 32469

Figure 47 — Cylinder Head Bolts — Sequence Tightening

(5) Tighten the cylinder head bolts with a tension wrench (fig. 42). They should be tightened to a torque load of 60 to 70 pounds-feet, following the sequence shown in figure 47.

(6) Install the twelve valve push rods.

TM 9-741
59

ENGINES

(7) Place the six intake valve shields on the top of intake valve stems, install the rocker arm shaft assembly, and tighten the shaft bracket bolts to 25 to 30 pound-feet tension.

(8) Replace the push rod cover, using a new gasket. Install the cover screws, place the oil filler tube in position so that the two cover screws will go through the bracket; make sure the gasket is in position. Tighten the eighteen screws securely and attach the oil gage rod tube bracket to the oil filler bracket.

(9) Connect the spark plug wires.

(10) Install the coil bracket and tighten the two bracket bolts securely.

(11) Attach the coil to the bracket and connect the ignition wires to the coil.

(12) Connect oil filter and rocker arm oil lines to fitting at front of cylinder head.

(13) Connect temperature indicator wire connection to thermostat housing fitting.

(14) Install carburetor, hook the accelerator spring clip to the carburetor flange bolt, install the two nuts, and tighten securely (fig. 39).

(15) Attach fuel feed line to carburetor, Hydrovac vacuum line to manifold fitting, and air cleaner delivery pipe to carburetor.

(16) Attach foot accelerator rod to throttle arm and install washer and cotter pin.

(17) Install crankcase ventilator suction pipe and attach it to oil filler tube and manifold fitting.

(18) Place the thermostat in the thermostat housing; install a new water outlet gasket, the water outlet, and the two retaining bolts. Tighten the bolts securely.

(19) Install the radiator upper hose and tighten the clamps securely.

(20) Install and tighten the two bolts which attach the water outlet brace to the intake manifold.

(21) Connect the hand throttle and choke cables to carburetor. Install or tighten any clips or brackets which might have loosened up.

(22) Attach the exhaust pipe to manifold flange, using a new gasket (fig. 39). Install and tighten the muffler support bolts and tighten the muffler strap.

(23) Install the drain plug, fill the cooling system, and check thoroughly for leaks.

(24) Check to see that the rocker arm adjusting screws are not adjusted to hold any valves open, start the engine, warm it up

MEDIUM ARMORED CAR T17E1

thoroughly; tighten the cylinder head, manifold, carburetor flange, exhaust flange, and push rod cover screws. Adjust the valve according to the procedure given in paragraph 58.

f. Left Engine Cylinder Head Removal.

(1) Remove the twenty-five bolts which attach the left engine compartment cover and remove the cover.

(2) Drain the cooling system by removing the drain plug from the bottom of hull (fig. 31).

(3) Disconnect air cleaner delivery pipe from carburetor.

(4) Remove the three stud nuts which attach the manifold to exhaust pipe, the three bolts which attach the muffler support to the hull, and in some cases it may be necessary to loosen the muffler strap. Swing the exhaust pipe to the right to clear the manifold flange.

(5) Disconnect hand throttle and choke cables from carburetor.

(6) Remove the two bolts which attach water outlet bracket to intake manifold.

(7) Disconnect radiator upper hose from radiator and remove two bolts which attach water outlet to thermostat housing (fig. 39).

(8) Remove choke and throttle cable clips from water outlet and remove outlet (figs. 39 and 43).

(9) Disconnect crankcase ventilator suction pipe at both ends, loosen clips, and remove pipe.

(10) Disconnect fuel feed line at carburetor end, loosen clips, and lay pipe back out of the way.

(11) Disconnect the foot accelerator rod from the throttle arm by removing the cotter pin and washer.

(12) Disconnect the Hydrovac vacuum line from manifold.

(13) Disconnect accelerator pull-back spring, remove the two retaining nuts, and remove carburetor.

(14) Disconnect temperature indicator wire connection fitting from thermostat housing fitting.

(15) Remove valve cover air cleaner by turning counterclockwise. CAUTION: Keep the cleaner upright as it is an oil-bath type cleaner.

(16) Disconnect oil filter and rocker arm oil lines from fitting at front of cylinder head.

(17) Remove two bolts which attach generator shield to manifold.

(18) Disconnect oil filter return line, remove filter bracket screws, and remove filter and bracket.

(19) Disconnect oil gage rod tube clip from oil filler tube bracket, remove the bracket screw, and remove filler tube and bracket.

TM 9-741
59

ENGINES

(20) Disconnect wires from coil, remove the two coil-to-bracket bolts, and remove the coil (fig. 39).

(21) Remove the two coil bracket bolts and remove the bracket.

(22) Disconnect the spark plug wires.

(23) Remove the remaining push rod cover screws and remove the cover and gasket.

(24) Remove the two valve cover retaining nuts and remove the cover and gasket (fig. 43).

(25) Remove valve rocker arms and shaft as an assembly by removing the bolts from the six shaft brackets. Remove the six inlet valve shields.

(26) Lift the twelve push rods out.

(27) Remove the fifteen cylinder head bolts; remove the cylinder head assembly and gasket.

(28) Remove the eight nuts, washers, and clamps which retain the manifold to head and remove the manifold (fig. 40).

(29) Remove the six spark plugs, the thermostat housing, and the rocker arm oil inlet fitting.

g. **Carbon Removing — Rocker Arm and Shaft Assemblies.** Refer to paragraphs c and d above.

h. **Left Engine Cylinder Head Installation.**

(1) Install the rocker arm oil inlet fitting and the thermostat housing, using a new gasket.

(2) Clean, adjust, and install the six spark plugs (par. 62).

(3) Install new manifold gaskets, the manifold assembly, washers, clamps, and the eight nuts. Tighten securely.

(4) Clean the top of block and cylinder head; install a new gasket; place the head assembly on the block and install the fifteen cylinder head bolts.

(5) Tighten the cylinder head bolts with a tension wrench (fig. 42). They should be tightened to a torque load of 60 to 70 pound-feet, following the sequence shown in figure 47.

(6) Install the twelve valve push rods.

(7) Place the six intake valve shields on the top of intake valve stems, install the rocker arm shaft assembly, and tighten the shaft bracket bolts to 25 to 30 pound-feet tension.

(8) Replace the push rod cover, using a new gasket. Install the retaining screws, placing the oil filler tube and oil filter in position so that the screws will attach the brackets to the block. Make sure the gasket is in position and tighten the screws securely.

MEDIUM ARMORED CAR T17E1

(9) Attach the oil gage rod tube brace to the oil filler bracket.

(10) Connect the spark plug wires.

(11) Install the coil bracket and tighten the two bolts.

(12) Install the two bolts which attach the coil to coil bracket.

(13) Connect oil filter and rocker arm oil lines to fitting at from of cylinder head. Connect oil filter return line.

(14) Attach the generator shield to manifold with two bolts.

(15) Connect temperature indicator wire connector to thermosta housing fitting.

(16) Install carburetor; hook the accelerator spring clip to car buretor flange bolt; install the two nuts and tighten securely (fig. 39)

(17) Attach Hydrovac vacuum line to manifold.

(18) Connect the foot accelerator rod to carburetor throttle arm anc install washer and cotter pin.

(19) Connect fuel feed line to carburetor and tighten securely.

(20) Connect crankcase ventilator suction pipe at both ends anc install clips.

(21) Install a new water outlet gasket, the thermostat and wate outlet. Install the two nuts and tighten securely.

(22) Attach the water outlet bracket to the intake manifold witl two bolts. Connect the choke and throttle cable clip to the water outlet

(23) Install and tighten radiator upper hose.

(24) Attach hand throttle and choke cables to carburetor.

(25) Attach the exhaust pipe to manifold flange, using a new gasket Install and tighten the muffler support bolts and tighten the muffler support strap.

(26) Install the drain plug; fill the cooling system and check for leaks.

(27) Check to see that the rocker arm adjusting screws are not adjusted tightly enough to hold any valves open. Start the engine anc warm it thoroughly, tighten the cylinder head, manifold, carburetor flange, exhaust flange, water outlet, push rod cover and radiator hos bolts.

(28) Adjust the valves according to the procedure given in para graph 58.

60. OIL FILTER SERVICE.

a. Description. Oil filters for each engine are mounted in the engin compartment (fig. 39). Oil is taken from the oil gallery through line to the oil filter, through the filter, and returned to the oil pan. Th filter element, shown in figure 48, filters dirt and impurities from the oi

TM 9-741
60-61

ENGINES

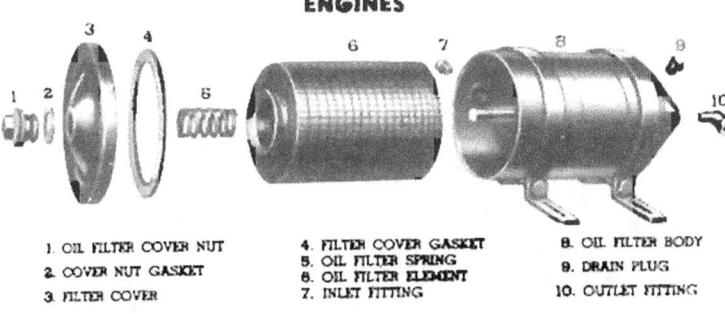

1. OIL FILTER COVER NUT
2. COVER NUT GASKET
3. FILTER COVER
4. FILTER COVER GASKET
5. OIL FILTER SPRING
6. OIL FILTER ELEMENT
7. INLET FITTING
8. OIL FILTER BODY
9. DRAIN PLUG
10. OUTLET FITTING

RA PD 32470

Figure 48 — Oil Filter Parts

b. Maintenance.

(1) Oil filters, in good condition will adequately remove dust and dirt which enters the oil system, yet the filter elements must be replaced whenever they become clogged, regardless of mileage the vehicle has traveled since the last change. Vehicles operating in dusty areas require replacements more often than those which do not encounter such conditions.

(2) Oil filter element changing periods are directly related to type and quality of oil used, severity and type of engine operation, and oil change periods.

(3) Remove plug and drain settlement from bottom of each oil filter every 1,000 miles.

(4) Renew filter element every 2,000 miles, or oftener, if necessary.

c. Filter Element Replacement.

(1) Remove cover retaining nut, cover, cover gasket, and spring (fig. 48).

(2) Remove the filter element and clean the filter body thoroughly.

(3) Install a new filter element, the spring, gasket, cover, and retaining nut. NOTE: If the gasket is damaged a new one should be installed.

(4) After changing the element, refill crankcase to "FULL" mark on gage, run engine a few minutes, and recheck oil level.

61. CRANKCASE VENTILATING SYSTEM.

a. A vacuum-type crankcase ventilator is provided on each engine assembly. The purpose of this unit is to create a partial vacuum in the crankcase which will draw harmful vapors from the crankcase, preventing contamination of the engine oil.

b. This is accomplished by means of a vacuum line extending from

101

MEDIUM ARMORED CAR T17E1

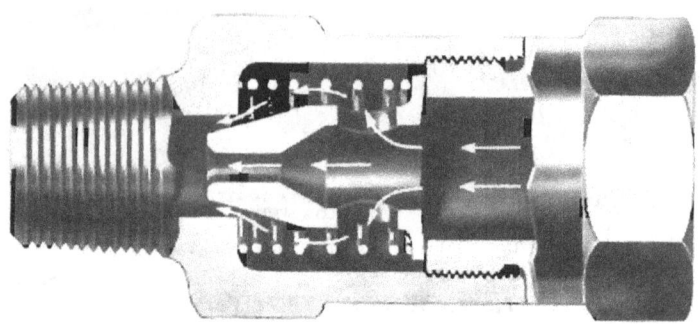

Figure 49 — Crankcase Metering Valve Assembly

the engine intake manifold to the crankcase filler pipe, creating a partial vacuum in the crankcase.

c. A baffle is provided in the oil filler pipe to prevent drawing of liquid oil from the crankcase. The filler cap is sealed.

d. An oil-bath air cleaner is provided at the point of air entrance to the engine. This is located at the top of the valve cover (fig. 43), and cleans the air entering the valve cover. The air passes through the valve cover, into the push rod chamber, and then enters the crankcase through nine cored holes between the valve lifters.

e. In order to have approximately the same vacuum on the crankcase at all speeds and all throttle positions, a metering valve assembly is installed in the vacuum line at the intake manifold (fig. 49).

f. This valve allows full opening at low vacuum or high speeds but closes, restricting the opening at periods of high vacuum at low speeds.

62. ENGINE TUNE-UP.

a. General.

(1) One of the most important operations in the maintenance of the engines is proper engine tune-up. This operation, more than any other, determines whether or not the engines deliver the maximum in performance and economy. Each engine should be tuned according to the following procedure.

(2) Before making any checks on an engine, it should be run for several minutes to warm it up and lubricate the valve mechanism. The compression of the engine should be checked first when tuning an

ENGINES

engine, because an engine with uneven compression cannot be tuned successfully.

b. Compression Check.

(1) Remove the engine compartment cover.

(2) Remove all spark plugs.

(3) Open the throttle and see that the ignition is turned off. NOTE: In some cases it may be necessary to remove the coil or oil filler tube, to test compression with certain types of compression gages.

(4) Insert the compression gage in a spark plug hole and hold it tightly. Crank the engine with the starting motor until the gage reaches its highest reading which requires only a few turns of the engine. Repeat the same test on all cylinders and make a note of the compression on each cylinder.

(5) The compression on all cylinders should be 110 pounds or better; all cylinders should read alike within 5 to 10 pounds for satisfactory engine performance.

(6) Should you have a low compression reading on two adjacent cylinders, it indicates a possible inter-cylinder leak, usually caused by a leak at a cylinder head gasket.

(7) If the compression readings are low, or vary widely, the cause of the trouble may be determined by injecting a liberal supply of oil on top of the pistons of the low reading cylinders.

(8) Crank the engine over several times and then take a second compression test. If there is practically no difference in the readings when compared with the first test, it indicates sticky or poorly seating valves. However, if the compression reading on the low reading cylinders is about uniform with the other cylinders, it indicates compression loss past the pistons and rings.

(9) Correct the cause of low or uneven compression before proceeding with an engine tune-up.

c. Spark Plugs.

(1) Clean the spark plugs thoroughly, using an abrasive-type cleaner. If the porcelains are badly glazed or blistered, the spark plugs should be replaced. All spark plugs must be of the same make and heat range.

(2) Adjust the spark plug gaps to 0.025 inch. Check with round feeler gage (fig. 50). CAUTION: Do not bend the center electrode.

(3) Care must be used when installing the fourteen millimeter spark plugs, or the setting of the gap may be changed. If a tension wrench is used when installing the plugs, the proper tension is 25 to 30 pound-

TM 9-741

MEDIUM ARMORED CAR T17E1

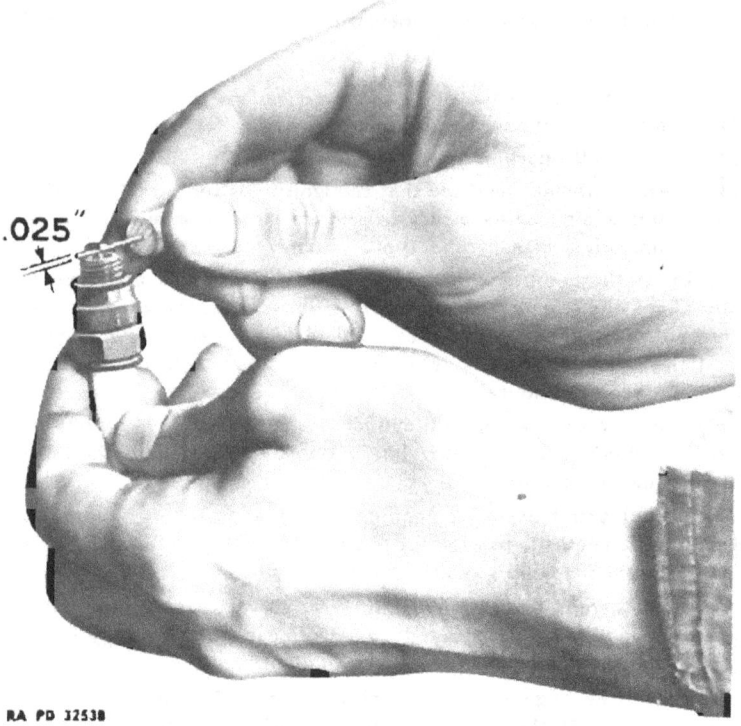

RA PD 32538

Figure 50 — Checking Spark Plug Gap

feet. If a tension wrench is not available, the following procedure should be used.

(4) Install new gaskets on the plugs, screw them in until they "bottom" and then tighten with a wrench ¼ to ½ turn.

d. **Battery and Cables.**

(1) Clean the battery end of each battery cable thoroughly and tighten all connections.

(2) Check voltage through entire primary circuit for possible voltage drop.

(3) Inspect ignition system high and low tension cables. Terminals on each end must be clean and tight. If insulation shows evidence of deterioration, cables should be replaced.

e. **Distributor.**

(1) Remove the distributor cap and check the cap and distributor rotor for cracks or burned contacts.

TM 9-741
62

ENGINES

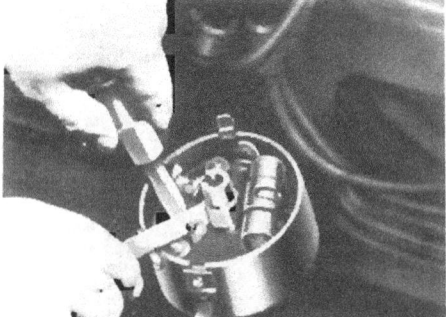

RA PD 32471
Figure 51 — Setting Distributor Points

(2) Check the automatic advance mechanism by turning the distributor cam in a clockwise direction as far as possible, then release the cam and see if the springs return it to its retarded position. If the cam does not return readily, the distributor must be replaced.

(3) Remove the dust cap and examine the distributor points. Dirty points should be cleaned with a flat point file. Pitted or worn points should be replaced. Check the points for alinement, and aline them if necessary. NOTE: Never use CLOTH, emery, on the distributor points.

(4) Crank the engine by rotating fan until the cam follower rests on the peak of the cam. Adjust the point gap to 0.018 inch, using a feeler gage (fig. 51). This operation must be performed very accurately because it affects point dwell. Crank the engine until the cam follower is located between the cams. Hook the end of a point scale over the movable point and pull steadily on the spring scale until the points just start to open. At this point the reading on the scale should be between 17 and 21 ounces.

(5) Set the manual timing adjustment at zero on the scale (fig. 53). Reassemble distributor cap and spark plug wires. Make sure that the terminals of the primary wires from the ignition coil to the distributor are clean and tight.

f. Air Cleaner.

(1) Remove the oil cups, empty the oil, and scrape the dirt out (fig. 52).

105

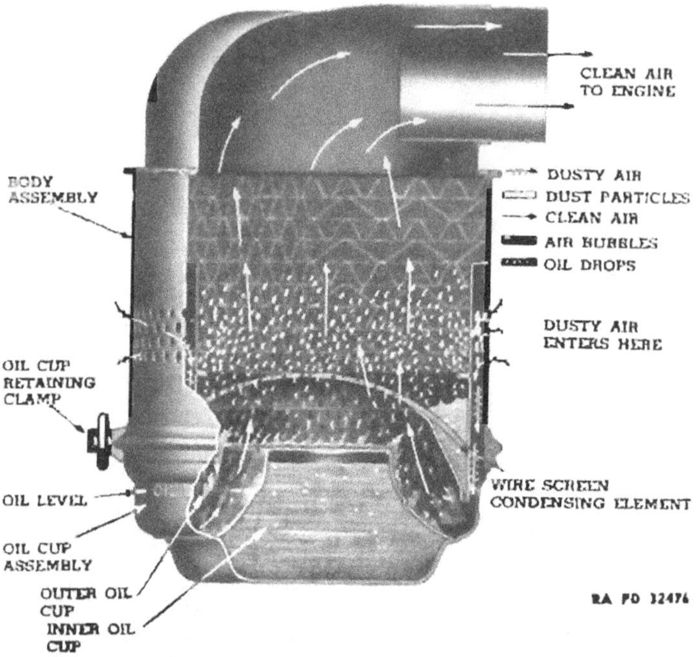

MEDIUM ARMORED CAR T17E1

Figure 52 — Carburetor Air Cleaner

(2) Remove the entire air cleaner. Thoroughly flush the wire screen condensing element in solvent, dry-cleaning. Clean the air cleaner body and cups in SOLVENT, dry-cleaning, and make sure the air inlet passages are clean.

(3) Fill the oil cups to the correct level and assemble the air cleaner.

(4) Install the cleaner and check all connections between the air cleaner and the carburetor to see that there are no air leaks.

g. Carburetor.

(1) Test carburetor flange and intake manifold gaskets for leaks.

(2) Adjust carburetor according to instructions in paragraph 77.

(3) If any other service is required, the carburetor should be removed and sent to responsible ordnance service for repairs.

h. Ignition Timing.

(1) Set manual timing adjustment at zero (midway between advance and retard) (fig. 53).

(2) Attach a neon timing light according to the instructions furnished with the light.

ENGINES

(3) Run engine at slow idling speed.

(4) Hold light close to flywheel timing opening, loosen distributor clamp bolt, and rotate distributor clockwise or counterclockwise until the timing ball in the flywheel is in line with the pointer on the housing when the neon light flashes (fig. 54). Tighten distributor clamp bolt (fig. 53) and recheck timing to make sure it did not change while tightening the clamp.

i. **Valve Clearance Adjustment.** Adjust the valve clearance according to the instructions in paragraph 58.

j. **Idling Adjustment.** Adjust the carburetor idle and throttle stop screw (fig. 60) in combination with each other to secure the best idling performance. The engine should idle at about 600 revolutions per minute.

k. **Cooling System.** Tighten all hose connections and examine the cooling system for any indication of air or water leaks. Make sure that the radiator cap seals air tight.

l. **Crankcase Ventilator.**

(1) Make sure that the oil filler cap seals air tight.

(2) Check the ventilator vacuum line connections to see that they are tight.

(3) Service the ventilator air cleaner (on the valve cover) by removing it, clean thoroughly, and refill to proper level.

m. **Road Test.** After the completion of the above operations on both engines, the vehicle should be road tested for performance. Test it with each engine individually and then with both engines. During this test, the manual spark control should be advanced or retarded for the grade of fuel being used. NOTE: For peak performance and maximum gasoline economy, the manual control should be set to produce a slight "ping" upon accelerating at wide open throttle.

63. ENGINE REPLACEMENT (figs. 55 and 56).

a. **General.**

(1) Due to the construction of this vehicle, it is necessary to remove the engines in order to perform any major service on the engines, fluid couplings, or transmissions.

(2) The removal and installation procedure varies slightly between the left and right engines. Procedure items which pertain to both engines will only carry the item number, while items which pertain to the left or right engine only, will carry the item number followed by an "L" or "R" to indicate that they pertain to the left or right engine only.

Figure 53 — Manual Timing Adjustment

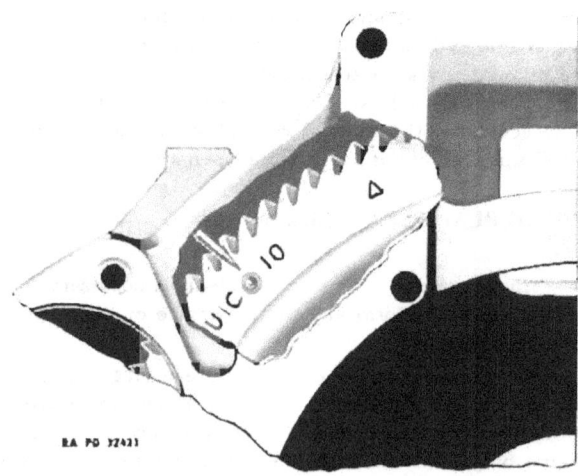

Figure 54 — Flywheel Timing Marks

ENGINES

TM 9-741
63

Figure 55 — Engine Replacement

TM 9-741

MEDIUM ARMORED CAR T17E1

Figure 56 — Engine Replacement

h. Engine Removal.

(1) Remove the twenty-five retaining bolts from each engine compartment cover and remove the covers (figs. 30 and 39).

(2) Remove radiator filler cap from radiator on engine being removed.

(3) Drain cooling system by removing drain plug from bottom of hull (fig. 31).

(4) Drain engine oil by removing plugs from bottom of hull and removing oil pan drain plug (fig. 31).

(5) Remove transmission service plate for engine being removed (fig. 31). Remove radiator hose from fitting in bottom of hull and remove brass fitting from hull drain fitting.

(6) Drain transmission, transmission reduction case, and flywheel

ENGINES

cover. NOTE: It is necessary to rotate flywheel until drain plug is down to drain flywheel cover.

(7) Remove the four fan blade retaining bolts and remove the fan blades (fig. 38).

(8) Remove the fan shroud retaining bolts and remove the four pieces of the shroud from engine being removed.

(9) Remove the three muffler bracket-to-hull bolts, the muffler strap bolt nut, the three exhaust pipe-to-manifold stud nuts (fig. 39), and slide the muffler and pipe assembly back out of the way.

(10) Loosen the two generator braces and force the generator toward the engine (fig. 38).

(11) Loosen the clamp and remove air cleaner inlet tube from top of carburetor (fig. 39).

(12) Remove the two bolts which attach the water outlet support to intake manifold; remove the two water outlet flange-to-thermostat housing bolts and gas line clip. Loosen throttle and choke clamp nuts at carburetor levers. Loosen hose clamp at radiator inlet (fig. 39).

(13R) Slide water outlet assembly off radiator inlet and lay outlet assembly back on hull.

(14L) Loosen throttle and choke clamp nuts at carburetor levers on right engine; loosen throttle and choke rod clamp bolt nuts at right engine water outlet support; remove throttle and choke clamp at water header of right engine and remove clamp.

(15L) Disconnect left engine primary ignition lead from ignition filter box and pull primary wire, choke, and throttle cables through division panel.

(16L) Slide water outlet assembly off radiator inlet and lay outlet assembly out on hull.

(17) Remove water outlet gasket and thermostat (fig. 43).

(18) Remove crankcase ventilator air cleaner from rocker arm cover (fig. 39). NOTE: Do not spill oil from cleaner.

(19) Disconnect ignition primary lead from coil and release clip.

(20) Disconnect lower end of right engine oil filter inlet line from engine line; disconnect filter outlet line from bottom of filter. Remove the four attaching bolts and remove the filter. NOTE: The two top bolts also attach the engine compartment junction box to fire wall.

(21) Disconnect all engine and transmission oil cooler oil inlet and outlet lines from the cooler (fig. 68). Wire the oil lines to the engine to hold them out of the way while removing the engine.

(22) Disconnect Hydrovac vacuum hose from fitting on the intake manifold.

TM 9-741

MEDIUM ARMORED CAR T17E1

(23) Disconnect the leads from the four terminals on the top of filter and regulator box (fig. 123). Remove the four filter and regulator box-to-hull bracket bolts and remove the box. NOTE: Mark the regulator box and wires so that they can be installed correctly, or damage to the unit will result.

(24) Remove the two bolts which attach forward fire extinguisher nozzle to bracket. Unscrew nozzle and remove bracket bolts.

(25) Disconnect the line from center fire extinguisher nozzle and remove the nozzle bracket-to-fire wall bolts and nozzles (fig. 39).

(26) Disconnect throttle rods at cross shaft end (8, fig. 74).

(27) Disconnect transmission throttle control rods from cross shaft (9, fig. 74).

(28) Disconnect throttle cross shaft brackets from engines and swing the cross shaft out of the way of the radiator (3 and 13, fig. 74).

(29) Remove the five universal joint seal retaining ring bolts. Remove the ring and work the seal through the bulkhead.

(30) Disconnect transmission manual control rod from transmission manual control lever and tie up to hull (fig. 82).

(31) Remove generator terminal plugs and disconnect generator terminal leads.

(32) Disconnect temperature indicator lead from thermostat housing fitting and remove clip; disconnect oil pressure gage lead from gage unit on engine.

(33) Disconnect right jettison and main fuel tank vent lines at each side of fire wall and at main fuel tank filler neck; remove main tank to fire wall fitting vent line.

(34) Remove the three bolts which attach the engine compartment cover center support to hull (fig. 39), the bolts which attach it to fire wall, and remove the support.

(35) Remove the bolts which attach the rear fire wall panel to center panel and to hull. Remove panel.

(36) Remove the bolts which attach center fire wall panel to front panel and swing the center panel up on the hull out of the way.

(37) Remove the front fire wall panel-to-hull bolts and remove panel.

(38) Loosen radiator-to-cooler hose clamps at radiator end.

(39) Remove the two bolts which attach each water inlet connector to water pump (fig. 39).

(40) Remove the three oil cooler bracket bolts and remove the oil cooler by pulling it toward the rear until the hoses are off the radiator (fig. 68).

(41) Remove starter relay terminal cover plate and disconnect

TM 9-741

ENGINES

starter relay connections. Remove terminal plug and disconnect starter motor cables.

(42) Remove the eight bolts which retain the air inlet grille and remove the grille with a hoist.

(43) Slide a piece of 2-x 4-inch beam down along the fire wall side of the engine and onto the top of flywheel housing to support the radiator. Remove the six bolts which retain the radiator and work the radiator up out of the engine compartment. Support the radiator on the 2-x 4-inch beam to avoid damaging the core (fig. 67).

(44R) Disconnect fuel line from carburetor.

(45L) Disconnect fuel line from fuel filter.

(46) Remove the bolts from the four engine mountings.

(47) Remove the starter cable from center rear push rod cover screw.

(48) Disconnect the gear reduction case to transfer case propeller shaft universal joint at the gear reduction case end by removing the universal clamps. Slide the universal joint forward and tape the trunnions to the yoke to hold them in place. Turn the universal joint yoke to a horizontal position.

(49) Remove transmission throttle control rod and throttle valve lever from transmission throttle control shaft (10, fig. 74).

(50) Hook hoist to engine lifting hook, which should be attached to No. 3 and No. 5 cylinder head bolts. Raise engine slowly, watching to see that all lines and connections are clear. When the fan end of the engine clears the lower division panel, it should be turned toward the center of vehicle to get clearance for the gear reduction case (figs. 55 and 56). Raise the engine and pull it toward the rear of the hull as soon as the harmonic balancer clears the hull. Watch to see that the rocker arm cover is not damaged on the radiator support upper panel. NOTE: As the engine clears its mountings, make sure that it does not move toward the front of the vehicle or the flywheel housing will bind between the hull engine mountings. CAUTION: When removing the right engine, watch the transmission manual control lever and sector to see that it does not bind at the side of hull or engine mounting, as this would damage the control lever and shaft.

(51) Place the engine in a suitable stand to avoid damaging the oil pan or other engine equipment.

(52) When the engine is being exchanged for an overhauled engine or being sent to a higher echelon shop for repairs, all fittings, connections, and equipment, which would not be furnished with an exchange engine, should be removed.

MEDIUM ARMORED CAR T17E1

c. **Engine Installation.**

(1) Due to the serious consequences that might result from leaks in the oil, fuel, or cooling system, all connections in these systems are sealed with permatex. In reassembling, care should be taken to use a small amount of permatex on all joints and fittings which indicate the use of permatex in production.

(2) Shift transmission to neutral (fig. 82), set universal joint yoke horizontal, and shift transmission to reverse (the top of lever toward the engine).

(3) Loosen generator braces and swing the generator toward engine (fig. 38).

(4) Remove the four bolts which retain the fan and remove the fan (fig. 38).

(5) Attach a hoist to the engine lifting hook (which should be attached to No. 3 and No. 5 cylinder head bolt from the fan end of engine). Swing the engine assembly over the engine compartment and start the gear reduction case down into the hull.

(6) Swing the fan end of the engine toward the center of vehicle to provide clearance for the gear reduction case (figs. 55 and 56).

(7) Keep the gear reduction case end as low as possible so that it will go down into position. Watch to see that the engine accessories or engine compartment equipment are not damaged. **CAUTION:** Watch to see that the transmission manual control lever or sector are not damaged when installing the right engine; also watch the rocker arm cover at the radiator support top panel and the harmonic balancer at rear of hull.

(8) Enter the universal joint yoke through the opening in hull.

(9) Line up the engine mountings; install and tighten the mounting bolts at all four mountings.

(10) Work the universal joint seal through the hull opening. Install universal joint seal retaining ring and bolt securely.

(11) Connect the propeller shaft universal joint to flange at gear reduction case and tighten U clamp nuts securely.

(12) Connect the transmission manual control rod to transmission manual control lever (fig. 82).

(13) Install the throttle valve lever on transmission and attach throttle control rod (10, fig. 74).

(14) Attach starter cable clip to center rear push rod cover screw and attach cable to starter terminal.

(15) Install radiator drain hose on radiator and tighten clamp. Place a piece of 2-x 4-inch beam down along the fire wall side of

TM 9-741
63

ENGINES

engine and over the flywheel housing. Slide the radiator down on the 2-x 4-inch beam to avoid damaging the radiator (fig. 67). Hold the cables, lines, and throttle cross shaft out of the way as the radiator is lowered into position. Install the six radiator retaining bolts.

(16) Install brass fitting at hull radiator drain fitting. Install the radiator drain hose and tighten clamp securely.

(17) Attach throttle cross shaft brackets to the engines, hook up the cross shaft to carburetor throttle rods, and to transmission rods (fig. 74).

(18R) Connect fuel line to carburetor.

(19L) Connect fuel line to fuel filter.

(20) Attach forward fire extinguisher bracket, screw extinguisher nozzle on line and bolt to bracket.

(21) Bolt the terminal and regulator box to hull and attach the leads to the four terminals (fig. 123). CAUTION: Be sure to install the leads according to the marking made when removing the unit or serious damage will result.

(22) Permatex the oil cooler to radiator outlet hoses and attach them securely to cooler. Put a light coat of permatex on the radiator outlet and in the cooler hose. Slide the cooler into position (fig. 68). Raise the engine end of cooler and work the two lower hoses over the radiator outlet.

(23) Install the three cooler bracket bolts and tighten the cooler-to-radiator hose clamps securely.

(24) Attach inlet connectors to water pumps, using new gaskets (fig. 38).

(25) Attach the transmission and engine oil lines to cooler, using permatex on the fittings.

(26) Install the fire wall front panel and bolt it loosely to radiator support center panel.

(27) Swing the fire wall center panel down into position and bolt it loosely to front panel.

(28) Install the fire wall rear panel and bolt it loosely to the center panel.

(29) Install engine compartment cover center support and bolt it to hull and fire wall (fig. 39). Tighten all fire wall panel bolts securely.

(30) Install right engine oil filter and engine compartment terminal box to fire wall panel (fig. 39).

(31) Connect the oil filter lines.

MEDIUM ARMORED CAR T17E1

(32) Connect right jettison and main fuel tank vent lines to fitting at front of fire wall.

(33) Install main fuel gas tank vent line under wires on left side of fire wall center panel and attach it to main tank filler neck and fitting at division panel.

(34) Attach all wiring clips to fire wall.

(35) Connect temperature gage terminal to fitting on thermostat housing and oil gage terminal to oil gage unit on left side of engine block.

(36) Connect the two wires to starting motor relay and install cover plate.

(37) Remove generator terminal plugs; install lead wires; attach conduit terminal fittings, and install plugs.

(38) Connect fire wall fire extinguisher nozzles to lines and attach the nozzle brackets to fire wall (fig. 39).

(39) Attach Hydrovac line to fitting on intake manifold.

(40) Attach ignition primary lead to coil and install clip which attaches wire to gas line.

(41) Install crankcase ventilator air cleaner on rocker arm cover and fill with oil (fig. 39).

(42) Place thermostat in thermostat housing; install a new gasket; feed the throttle and choke rods into the fittings on carburetor arms and place the water outlet in position.

(43) Bolt the water outlet to thermostat housing. **NOTE:** The fuel line clip is held in place by the housing bolt on the carburetor side of engine.

(44) Attach the water outlet hose to radiator and tighten clamp securely (fig. 39).

(45) Install the two bolts which attach water outlet support to intake manifold (fig. 35).

(46L) Place right engine choke and throttle cables and left engine ignition wire through grommet in fire wall. Attach ignition wire to terminal on bottom of filter box.

(47L) Install right engine choke and throttle cables through clamps on water outlet brace and enter the ends into the fittings on carburetor arms. Tighten cable clamps on water outlet support.

(48) Set choke and throttle levers in driver's compartment in the forward position and tighten lever clamps. Place the carburetor throttle in closed position and tighten the nut which attaches the throttle cable to carburetor throttle lever.

TM 9-741

ENGINES

(49) Place the carburetor choke in the open position and tighten the choke cable to choke lever by tightening clamp nut.

(50) Install carburetor air cleaner inlet tube on carburetor (fig. 39).

(51) Install fan shroud and tighten securely.

(52) Install fan blade assembly and adjust fan belt tension as instructed in paragraph 86.

(53) Place a new gasket on the manifold flange; place the muffler and exhaust pipe in position; install the three muffler-to-hull bracket bolts, the three exhaust pipe-to-manifold stud nuts (fig. 39), and the muffler strap bolt nut.

(54) Place the air inlet grille on the hull and bolt it in position with the eight bolts.

(55) Tighten the drain plugs securely and fill the transmission according to instructions given in paragraph 24.

(56) Install the radiator drain plug and fill cooling system with a suitable cooling solution for the temperature expected.

(57) Install radiator filler caps and tighten securely.

(58) Install oil pan drain plug and put new oil in the engine (par. 24).

(59) Start the engine or engines and warm them up thoroughly. Make sure oil pressure gage indicates normal pressure. **NOTE:** While the engines are warming up, a thorough check should be made for any indications of oil, water, or fuel leaks.

(60) Check the operation of temperature indicators, oil gages, fuel gage, ammeter, and other instruments that might be affected by engine removal.

(61) Adjust the valve clearance according to instructions in paragraph 58.

(62) Tune the engine according to the instructions in paragraph 62.

(63) Adjust the hydraulic throttle control mechanical linkage according to instructions in paragraph 99.

(64) Install the transmission service plate to bottom of hull and hull engine oil drain plug.

(65) Install engine compartment covers.

(66) Road test vehicle.

TM 9-741
64-65

MEDIUM ARMORED CAR T17E1

Section XIV

ENGINE IGNITION SYSTEM

	Paragraph
Description	64
Trouble shooting	65
Spark plug replacement, cleaning, and adjustment	66
Distributor replacement	67
Coil replacement	68
Filter replacement	69
Suppressor replacement	70
Breaker point replacement and adjustment	71
Condenser replacement	72

64. DESCRIPTION.

a. Each engine is equipped with its own ignition system. The purpose of the ignition system is to ignite the fuel and air mixture in the cylinder at exactly the proper time. The ignition system consists of a battery to supply the electrical energy, the coil which induces a high tension current to jump the gap in the spark plug, a mechanical breaker in the distributor which opens and closes the primary circuit at the proper time, a distributor to distribute the high tension current to the various cylinders (fig. 53), the spark plugs which provide the gap in the cylinders. An ignition switch is used to open or close the battery circuit when it is desired to stop the engine and the necessary wiring to connect the various units, complete the system. All the wires in the ignition system, except the high tension wires, are shielded to reduce interference with the operation of the radio. The shielding of the wires is explained in paragraph 190.

65. TROUBLE SHOOTING.

a. Ignition System Fails to Operate.

Probable Cause	Probable Remedy
Short circuit in the ignition filter.	Replace filter (par. 69).
Open circuit in the ignition filter.	Replace filter (par. 69).
Shielding shorted to lead wires.	Check and replace necessary wiring.

b. Radio Interference Due to Ignition System.

Filters noisy.	Check and replace faulty filter (par. 69).

TM 9-741
65

ENGINE IGNITION SYSTEM

Probable Cause	Probable Remedy
c. Hard Starting.	
Distributor points burned or corroded.	Replace points (par. 71).
Points improperly adjusted.	Adjust points (par. 71).
Spark plugs improperly gapped.	Adjust gap (par. 66).
Spark plug wires loose or corroded in distributor cap.	Tighten terminals and clean cap terminals.
Loose connections in primary circuit.	Tighten all connections.
Corroded battery terminals.	Clean all battery terminals (par. 156).
Series resistance in condenser circuit.	Clean and tighten all connections in condenser circuit.
Low capacity condenser.	Replace condenser (par. 72).
d. Engine Misses at All Speeds.	
Spark plug broken or fouled.	Clean or replace plug (par. 66).
High tension wire broken or loose connection.	Replace wire or tighten connection.
Distributor cap or rotor cracked or burned.	Replace parts affected (par. 71).
e. Lack of Power — Overheating.	
Ignition timing late.	Check and adjust timing (par. 62).
f. Engine Rough When Idling.	
Ignition timing early.	Check and adjust timing (par. 62).
Carburetor out of adjustment.	Adjust carburetor (par. 77).
Valves not seating properly.	Refer to ordnance personnel.
g. Engine Backfires Continuously.	
Spark plug wires reversed.	Check firing order and correct (par. 62).
h. High Speed Intermittent Miss.	
Spark plug gap incorrect.	Check and adjust gap to 0.025 inch (par. 66).
Distributor breaker arm spring tension weak.	Replace points (par. 71).
Distributor point gap incorrect.	Adjust to 0.018 inch (par. 71).
Spark plugs oxidized, fouled, or broken.	Clean and adjust to 0.025 inch, or replace plugs (par. 66).
Ignition coil weak.	Replace coil (par. 68).
Condenser weak or with high series resistance.	Check and clean terminals or replace condenser (par. 72).

MEDIUM ARMORED CAR T17E1

Probable Cause   **Probable Remedy**

i. Excessive Spark Knock on Acceleration.

Ignition timing early.   Check and adjust (par. 62).
Distributor governor springs weak.   Refer to ordnance personnel.

66. SPARK PLUG REPLACEMENT, CLEANING, AND ADJUSTMENT.

a. Procedure.

(1) REMOVE SPARK PLUGS. Remove the twenty-five bolts that attach the compartment cover and remove the cover. Disconnect the spark plug wires and remove the six spark plugs.

(2) CLEAN AND ADJUST SPARK PLUGS (fig. 50). Clean the spark plugs thoroughly and set gap at 0.025 inch, using a round feeler gage. Never bend the center electrode when adjusting the gap.

(3) INSTALL SPARK PLUGS. Install the spark plugs and screw them in until they bottom against the shoulder, then with a wrench tighten them between ¼ and ½ a turn. Use extreme care when tightening the plugs to prevent breaking them with the wrench. Place the compartment cover in position and tighten the twenty-five bolts securely.

67. DISTRIBUTOR REPLACEMENT.

a. Removal Procedure.

(1) REMOVE COIL (fig. 36). Remove the twenty-five bolts that attach the compartment cover and remove the cover. Remove the two bolts that attach the coil to the bracket, remove the high tension wire from the center of the coil and the primary wire from the coil and remove coil.

(2) REMOVE DISTRIBUTOR. Remove the distributor cap, and, using a sharp tool, scratch a line on the rim of the distributor body opposite the center of rotor arm. Also scratch a line on the outside of the body at a right angle to the engine. Disconnect the primary wire from the side of the distributor. Loosen the clamp screw which is located between the distributor and cylinder block. Lift the distributor assembly up and out of the cylinder block.

b. Installation Procedure.

(1) INSTALL DISTRIBUTOR (fig. 53).

(a) If the engine has not been turned since removing the distributor, thread the distributor part way into the cylinder block and place distributor body and rotor in the same position they were in before the distributor was removed. Then, while holding distributor body, turn rotor one tooth or about ¼ inch in clockwise direction and push distributor into place, make sure that the rotor and distributor body are

TM 9-741
67-69

ENGINE IGNITION SYSTEM

lined up as they were before the distributor was removed. Tighten clamp screw. Connect the primary lead to the side of the distributor and install the distributor cap.

(b) If the engine has been turned since removing the distributor, it will be necessary to retime the engine. Remove the No. 1 spark plug and remove the valve cover. Hand-crank the engine until the second valve (No. 1 intake) closes. Then, turn the engine about ½ turn until the steel ball in the flywheel is in line with the pointer on the flywheel housing above the starting motor. Place distributor cap on distributor and scratch a line on the edge of the distributor housing opposite the terminal that leads to the No. 1 spark plug. Remove the distributor cap, and, while holding the distributor body, turn the rotor one tooth or about ¼ inch in a clockwise direction past the mark on the distributor housing and push the distributor into place. Tighten clamp screw. Install primary wire to side of distributor and install distributor cap, install No. 1 spark plug and connect the wire to the spark plug.

(2) INSTALL COIL (fig. 36). Connect high tension wire to center of coil. Connect primary wire to terminal on coil. Place coil in position on mounting bracket and install the lower attaching bolt. Tighten bolts securely. Start engine and with one lead of a timing light connected to No. 1 spark plug wire and the other lead grounded, have engine running at idling speed and check in the flywheel housing opening to see that the ball is at the pointer as the light flashes (fig. 54). If the ball does not line up with the pointer, loosen the clamp screw and rotate the distributor to the right or left until the ball is in line with the pointer, then tighten clamp screw (fig. 53). Install the compartment cover and tighten the twenty-five bolts securely.

68. COIL REPLACEMENT.

a. Removal Procedure (fig. 36). Remove the twenty-five bolts that attach the compartment cover and remove the cover. Remove the two bolts that attach the coil to the mounting bracket. Disconnect the high tension wire from the center of the coil and the primary wire from the terminal. Remove the coil.

b. Installation Procedure (fig. 36). Connect the primary wire to the terminal and the high tension wire to the center of the coil. Place coil in position and install the mounting bolts. Tighten bolts securely. Place compartment cover in position and install the twenty-five mounting bolts. Tighten bolts securely.

69. FILTER REPLACEMENT.

a. Removal Procedure (fig. 39). Remove the twenty-five bolts that attach the right compartment cover and remove the cover. Remove

MEDIUM ARMORED CAR T17E1

the four screws that attach the cover to the filter box on the right side of the partition in the engine compartment and remove the cover. Remove the terminal screw at each end of the filter. Remove the four screws that attach the filter to the box and lift out the filter.

b. Installation Procedure (fig. 39). Place the filter in position in the box and install the mounting screws and shakeproof washers. Connect the wires to the terminals, using the screws and shakeproof washers. Install the cover on the box. Install compartment cover.

70. SUPPRESSOR REPLACEMENT.

a. Removal Procedure. Remove the terminal from the spark plug and unscrew the spark plug suppressor from the wire. Remove the terminal from the center of the distributor cap and unscrew the distributor suppressor from the wire. NOTE: Pull the terminal straight out to prevent breaking the distributor cap.

b. Installation Procedure. With the wire flush with the end of the insulation, start the screw of the suppressor in the center of the strands of wire. Screw the suppressor on the wire until it is tight. Install the terminal of the spark plug suppressor on the spark plug. Install the terminal of the distributor suppressor in the center terminal of the distributor cap.

71. BREAKER POINT REPLACEMENT AND ADJUSTMENT (fig. 51 and 57).

a. Removal Procedure. Remove the twenty-five bolts that attach the compartment cover and remove the cover. Remove the two bolts that mount the coil to the mounting bracket and lay coil to one side. Remove the distributor cap rotor and dust cap. Loosen the nut that retains the movable point spring to the terminal and lift out movable point. Remove the lock screw from the stationary point and lift out the point.

b. Installation Procedure. Place stationary point in position and start lock screw. Place movable point in position on terminal and tighten nut.

c. Adjustment Procedure.

(1) ADJUST POINTS (figs. 51 and 57). Turn engine until one of the lobes on the cam centers on the fiber block on the movable point. Turn adjusting screw so that the gap between the two points is 0.018 inch and tighten lock screw.

(2) INSTALL ROTOR, DISTRIBUTOR CAP AND COIL. Place dust cap and rotor on distributor and install the distributor cap. Install the coil on bracket.

ENGINE IGNITION SYSTEM

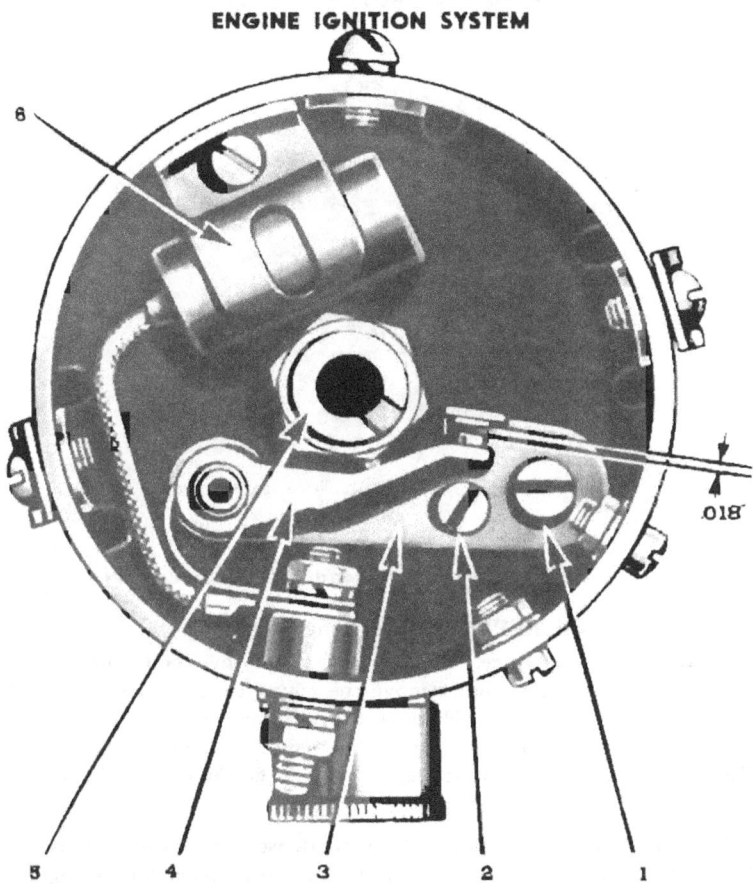

1. ADJUSTMENT LOCK SCREW
2. ECCENTRIC ADJUSTING SCREW
3. BREAKER PLATE
4. BREAKER ARM AND SPRING
5. DISTRIBUTOR CAM
6. CONDENSER

RA PD 32424

Figure 57 — Distributor Point Adjustment

(3) CHECK TIMING (fig. 53). Set manual setting on the clamp at the bottom of the distributor body at zero. Attach timing light to No. 1 spark plug and to ground. Run engine at idling speed and hold neon light to opening in flywheel housing above the starting motor. If steel ball in flywheel is not opposite the pointer, loosen clamp screw and turn distributor to right or left until the ball lines up with the pointer (fig. 54). Tighten clamp screw (fig. 53). Install compartment cover and tighten the twenty-five bolts.

MEDIUM ARMORED CAR T17E1

Figure 58 — Distributor Condenser Replacement

72. CONDENSER REPLACEMENT (fig. 58).

a. Removal Procedure. Remove the twenty-five bolts that attach the compartment cover and remove the cover. Remove the distributor cap, rotor, and dust cover. Remove the terminal nut (B, fig. 58) that attaches the pigtail to the terminal and remove the pigtail connection. Remove the condenser attaching screw (A, fig. 58) and lift out the condenser.

b. Installation Procedure. Place condenser in position in distributor and install attaching screw. Place pigtail on terminal and tighten nut securely. Install dust cap, rotor, and distributor cap. Place compartment cover in place and install the twenty-five bolts.

TM 9-741
73-74

Section XV

FUEL SYSTEM

	Paragraph
Description	73
Trouble shooting	74
Fuel tanks	75
Venting system	76
Carburetor	77
Air cleaner	78
Fuel pump	79
Fuel filter	80

73. DESCRIPTION.

a. The fuel system consists of one main and two jettison fuel tanks, tank venting system, one electric fuel pump, two carburetors, two fuel filters, two air cleaners, and the necessary fuel lines and connections.

74. TROUBLE SHOOTING.

a. Excessive Fuel Consumption.

Probable Cause	Probable Remedy
Idle and stop screw out of adjustment.	Adjust according to procedure outlined in paragraph 77.
Improper float level.	Replace carburetor (par. 77).
Metering rod not synchronized with throttle valve.	Replace carburetor (par. 77).
Dirty air cleaner.	Clean air cleaner (par. 78).
Fuel leaks.	Check all lines and connections for leaks and repair.
Sticking controls.	Choke not returning to "OFF" position.
Excessive idling.	Stop engine when vehicle is not to be moved for long periods of time.
Improper engine temperature.	Refer to cooling system section (par. 83).
Dragging brakes.	Refer to cooling system.
Tires under inflated.	Inflate front tires to 70 pounds and rear tires to 80 pounds.
Engine improperly tuned.	Tune engine (par. 62).

b. Fast Idling.

Improper adjustment.	Adjust idle and throttle stop screws (par. 77).

MEDIUM ARMORED CAR T17E1

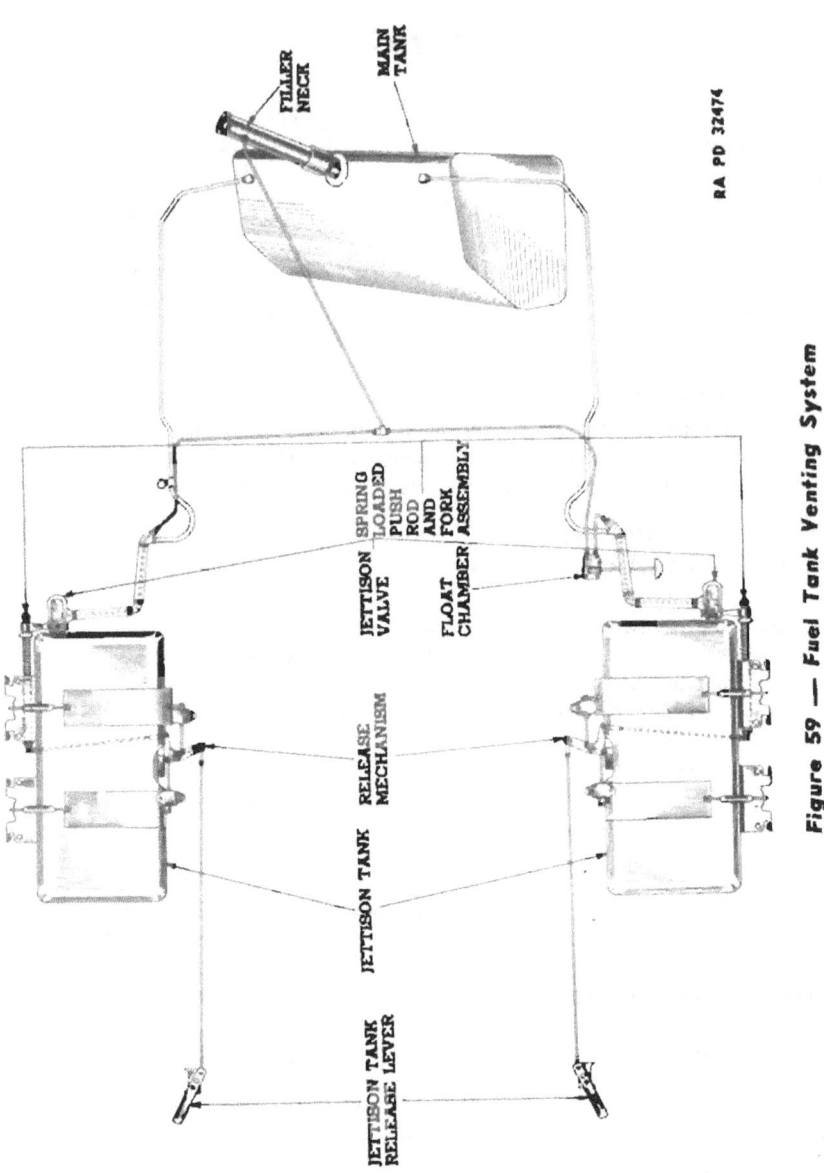

Figure 59 — Fuel Tank Venting System

FUEL SYSTEM

Probable Cause	Probable Remedy
Carburetor controls sticking.	Free up and lubricate linkage.

c. Engine Stops.

Will not idle at low speed.	Jets plugged with dirt. Replace carburetor (par. 77).

d. Engine Misses on Acceleration.

Improper spark plug adjustment.	Adjust according to instructions given in paragraph 66.
Improper tappet adjustment.	Adjust valves (par. 58).
Carburetor jets plugged with dirt.	Replace carburetor (par. 77).

e. Stalling of Engines, Erratic Operation, Popping Back Through Carburetors, or Requiring Excessive Choking to Keep Engines Operating Probably Due to Lack of Fuel at Carburetors.

Insufficent fuel in tank.	Fill tank.
Fuel pump inoperative.	See g below.
Fuel line from pump to carburetor clogged or kinked.	Blow out or replace.
Vapor lock.	Use correct grade of fuel.
Fuel vent valve stuck or vent lines clogged or kinked.	Replace valve and blow out or replace lines.

f. Poor Fuel Economy, Excessive Smoking at Exhaust Due to Rich Mixture, or Actual Flooding of Carburetors.

Needle valve not seating.	Replace carburetor (par. 77).
Sticking relief valve in fuel pump.	Replace fuel pump (par. 79).

g. Fuel Pump Inoperative.

Loose or faulty connection at ignition switch.	Tighten or replace wire.
Loose or poor connection at circuit breaker.	Tighten or replace wire and/or circuit breaker (par. 180).
Short circuit in wiring.	Replace.
Loose connection at pump.	Tighten.
Defective fuel pump.	Replace (par. 79).

75. FUEL TANKS (fig. 59).

a. Capacity. Three fuel tanks are used, a 62-gallon main tank and two 25-gallon (each) jettison tanks.

b. Location. The main tank is located inside the hull under the engine assemblies. The filler opening is located between the two mufflers and covered by a special cover. The two jettison tanks are located on the outer sides of the hull above the level of the main tank and feed by gravity to the main tank. As long as fuel remains in either jettison tank the main tank will be full. When the jettison tanks are released from the vehicle a shut-off valve closes both the air vent and fuel passages preventing loss of fuel from the tanks.

   c. **Filling Jettison Tanks.** The jettison tanks are filled through an opening in the top of each tank. The opening is covered by a screw-tight cap which has a wire handle to facilitate removal.

   d. **Jettison Tank Installation.**

   (1) Place the jettison tank on the support with filler cap up. Hold it on the support while positioning the valve as outlined in (2) below.

   (2) Position the two ends of the valve by pushing the yoke towards the tank, compressing the release spring. Rotate the lever on the end of the spring rod until the stop pin can be removed from the support. The release valve can then be returned to its normal position.

   (3) Loosen the turn buckle at the bottom of each of the two straps until the link in the top of the strap can be worked over the release pin. Retighten the turn buckle to hold the tank securely in place.

76. VENTING SYSTEM (fig. 59).

   a. **Description.** Loss of fuel when traveling over uneven ground is prevented by use of a venting system incorporating a float, float chamber and float-operated needle valve vent.

   b. **Location.** This assembly is located inside the hull on the left side above the level of the three tanks. Vent lines from this float chamber lead to each of the jettison tank shut-off valves. From the venting passage in the valves, vent lines lead to the filler openings inside of each tank. A main tank vent line from the main tank filler neck is connected in the vent line between the two jettison tanks.

   c. **Operation.** Under normal vehicle operation, no fuel will reach the vent float chamber so the float will be down and the vent needle valve open, venting the system. When operating on rough roads or steep grades, fuel may surge in the tanks or lines filling the float chamber and raising the float, closing the needle valve which seals the entire fuel system, preventing loss of fuel.

   d. **Replacement.** The fuel line from the jettison tank to the hull is supported and protected by a casting bolted to the side of the hull. The casting is open at the bottom to make the fuel line accessible for replacement or service.

   e. **Air Cleaner.** An air cleaner for the float chamber is located on the outside of the hull. This cleaner is a cast container filled with cow

TM 9-741
76-77

FUEL SYSTEM

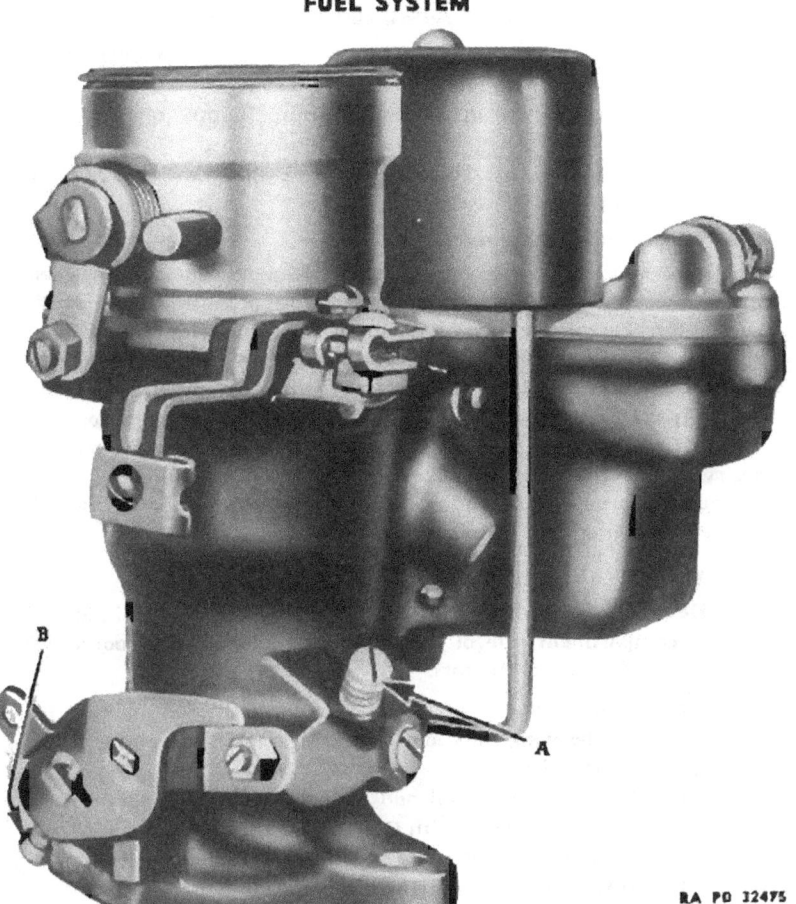

Figure 60 — Carburetor Idle Adjustment

tail hair which acts as a cleaner for the incoming air. A short pipe leads from the float chamber to the cleaner.

77. CARBURETOR

a. **Description.** The carburetors used are of the Carter downdraft type, model W1.

b. **Adjustments** (fig. 60). Both the adjustments for idling mixture and idling must be made together to obtain proper operation. To adjust idling mixture:

(1) Open idle adjustment screw A, 1½ turns.

(2) Start engine and let it run at idling speed.

TM 9-741

MEDIUM ARMORED CAR T17E1

(3) Adjust the idle and throttle stop screw B, in combination with each other, to secure the best idling performance.

(4) Idling speed should be approximately 600 revolutions per minute.

(5) After making this adjustment, check the transmission control rod adjustment (par. 99).

c. Carburetor Replacement.

(1) REMOVE CARBURETOR (fig. 39). Remove the engine compartment cover. Disconnect the fuel line and air cleaner tube from the carburetor. Disconnect accelerator rod from throttle arm, pull-back spring and choke control from carburetor. Remove the two carburetor-to-manifold nuts and lock washers and lift off the carburetor.

(2) INSTALL CARBURETOR. Place new gasket and carburetor on manifold and install the two lock washers and nuts. Connect the pull-back spring and accelerator rod. Connect the throttle and choke cables. Connect air cleaner tube and fuel line. Adjust carburetor as instructed in previous paragraph. Install compartment cover.

78. AIR CLEANER (fig. 52).

a. Description. Oil-bath type air cleaners are located inside the driver's compartment side of the bulkhead so that clean, cool air will be available for the carburetors.

b. Oil Replacement.

(1) Loosen the wing nuts and turn cup bracket to unhook it from the cup retaining bolts.

(2) Empty out the dirty oil and wash in SOLVENT, dry-cleaning.

(3) Refill to the oil level with OIL, engine, SAE 30. Check to make sure both the outer and inner cups are filled to the same level.

(4) Install the oil cups and tighten securely.

c. Air Cleaner Servicing.

(1) REMOVE AIR CLEANER. Disconnect the flexible hose from the cleaner. Remove the bolts that attach the cleaner to the hull and remove the cleaner.

(2) CLEAN AIR CLEANER. Loosen oil cup retaining clamp and remove the oil cup. Flush the body assembly in SOLVENT, dry-cleaning until all dirt is removed from the screen. Pour the oil out of the oil cups and remove all the dirt from the oil cups. Refill the oil cups with OIL, engine, SAE 30, to the oil level and install the body on the oil cups and tighten the retaining clamp.

(3) INSTALL AIR CLEANER. Place air cleaner in position on the hull and install the mounting bolts. Connect the flexible hose and tighten the clamp securely.

TM 9-741
79

FUEL SYSTEM

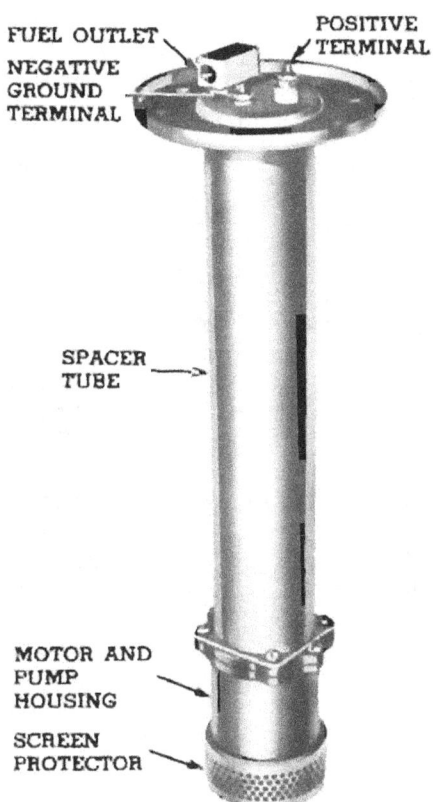

Figure 61 — Electric Fuel Pump

79. FUEL PUMP (fig. 61).

a. Description. An electric fuel pump is mounted in the main fuel tank and pumps fuel from the tank through the fuel filters to the carburetors. It extends inside the tank and is mounted through a hole in the top of the tank.

b. Fuel Pump Replacement (fig. 62).

(1) REMOVE FUEL PUMP. Remove the left engine compartment cover. Remove the right half of the fan shroud in the left engine compartment. Disconnect the jettison valves from both jettison tanks and drain enough fuel from the main tank to bring the fuel level below the top of the main tank. Remove the two cap screws that attach the cover

TM 9-741
79-80

MEDIUM ARMORED CAR T17E1

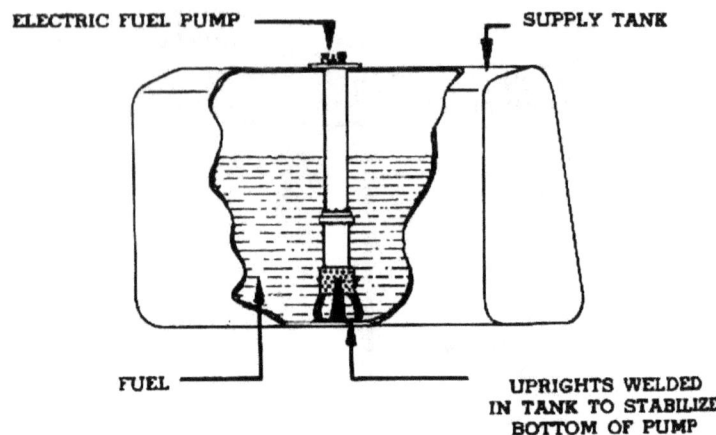

Figure 62 — Fuel Pump Installation

over fuel pump. Remove the two fuel connections and the two wires from the pump. Remove the six cap screws that attach the fuel pump to the tank and lift out the fuel pump.

(2) INSTALL FUEL PUMP. Coat a new gasket with permatex and place in position on tank. Place fuel pump in tank and install the six cap screws. Tighten the cap screws securely. Connect the two wires and the two fuel lines. Place cover in position and install the two cap screws. Install the shroud and compartment cover.

80. FUEL FILTER (fig. 63).

a. Description. The two fuel filters are of the multiple-disk type. They are placed in the fuel lines between the main fuel tank and each carburetor, one being located on each side wall of the hull near the carburetor which it feeds.

b. Draining. To drain water or dirt from the filter bowl, remove the compartment cover, remove the drain plug in the bottom of each filter, drain off the water or dirt, and then install the drain plug. Install the compartment cover.

c. Cleaning Filter (fig. 63).

(1) DISASSEMBLE FILTER. Remove compartment cover. Remove the bolt in the top of the filter and pull off the bowl with filter. Remove the gasket and filter from bowl. Clean all parts thoroughly in **SOLVENT**, dry-cleaning.

TM 9-741
80

FUEL SYSTEM

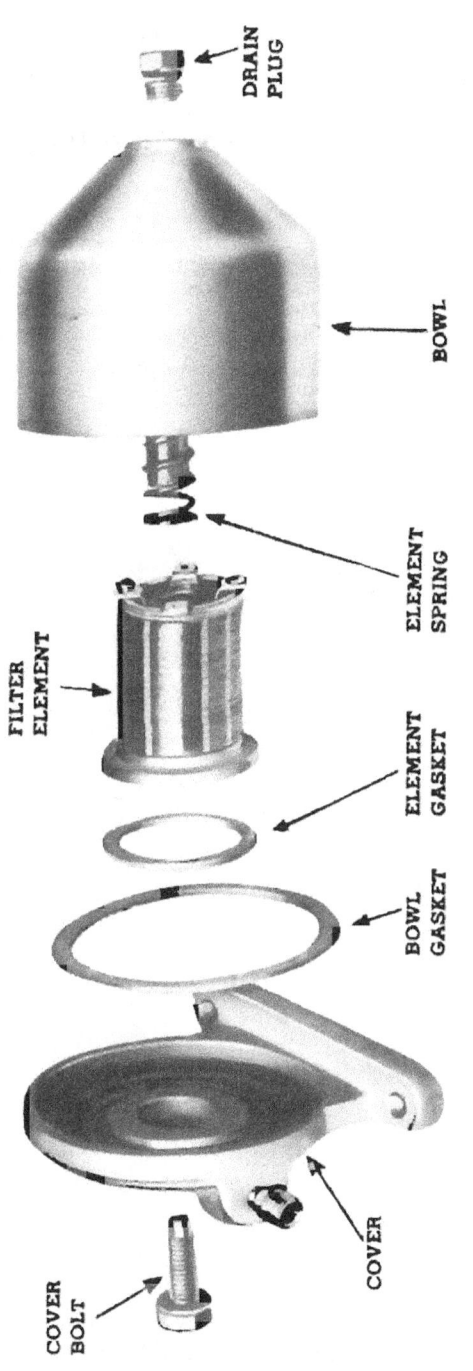

Figure 63 — Fuel Filter Parts

MEDIUM ARMORED CAR T17E1

(2) ASSEMBLE FILTER (fig. 63). Place element in bowl. Place a new element gasket and bowl gasket in position. Place bowl assembly in position in cover and install the cover bolt. Tighten bolt securely. Install compartment cover.

d. Filter Replacement (fig. 63).

(1) REMOVE FILTER. Remove compartment cover. Disconnect the two fuel lines from the filter. Remove the bolts that attach the filter to the hull and remove the filter.

(2) INSTALL FILTER. Place filter in position on hull and install the mounting bolts. Connect the two fuel lines. Install the compartment cover. Tighten fuel lines and mounting bolts securely.

TM 9-741
81-82

Section XVI

COOLING SYSTEM

	Paragraph
Description	81
Inspection	82
Trouble shooting	83
Pressure filler cap	84
Draining and refilling	85
Fan belts	86
Thermostat	87
Water pump	88
Radiator	89
Oil cooler	90

81. DESCRIPTION (figs. 64 and 65).

a. General. Each engine has its own complete cooling system consisting of a radiator core, fan shroud, fan, thermostat, centrifugal water pump, two fan belts, hoses, and connections.

b. Cooling Liquid. Water or a mixture of water and antifreeze is normally used as the cooling fluid.

c. Capacity. The fluid capacity of each system is twenty-five quarts.

82. INSPECTION.

a. Fluid Leaks. Hose connections and drain points should be checked periodically for leaks. Permatex or a similar compound must be used at all hose and screw or pipe thread connections. It is important that leakproof joints be obtained due to the difficulty of proper cleaning of the floor of this vehicle.

b. Clogged Passages. The hull air inlet grille should be checked at regular intervals to insure adequate opening for air intake. The front surface of the radiator cores should be inspected frequently to prevent accumulation of dirt which will retard cooling.

c. Flushing Out System. The cooling system should be flushed out at least twice a year to remove rust and scale.

d. Head Bolt Tightening. Cylinder head bolts should be checked to make sure they are tight and no water leaks into the cylinders.

e. Fan Belt Adjustment. Fan belts should be checked for proper tension (par. 86).

MEDIUM ARMORED CAR T17E1

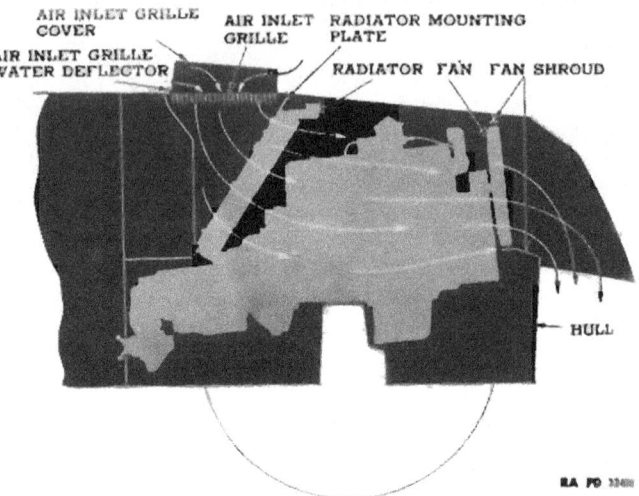

Figure 64 — Air Circulation — Engine Compartment

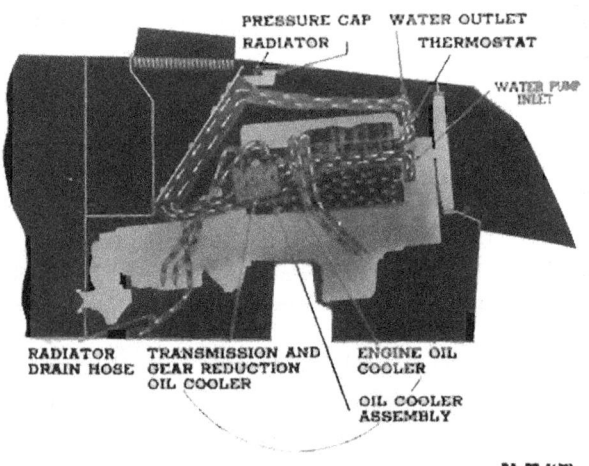

Figure 65 — Cooling System Circulation

TM 9-741
83-84

COOLING SYSTEM

83. TROUBLE SHOOTING.

a. Overheating.

Probable Cause	Probable Remedy
Lack of water.	Refill system.
Fan belts loose.	Adjust or replace (par. 86).
Fan belts worn or oil soaked.	Replace (par. 86).
Thermostat sticking closed.	Replace (par. 87).
Water pump shaft seized.	Replace water pump (par. 88).
Cooling system clogged.	Flush and clean system (par. 89).
Incorrect ignition timing.	Retime engine (par. 62).
Brakes dragging.	Adjust brakes (par. 131).
Air inlet grille clogged.	Clean grille.
Radiator core air passages clogged.	Blow out air passages (par. 89).
Radiator cap not sealing properly.	Tighten cap or replace gasket (par. 84).
Air leak in system.	Check for air leaks.

b. Overcooling.

Thermostat sticking open.	Replace (par. 87).

c. Loss of Cooling Liquid.

Loose hose connections.	Tighten all connections.
Damaged hose.	Replace hose.
Leaking water pump.	Replace pump (par. 88).
Radiator core leaks.	Replace radiator.
Pressure cap not seating properly.	Install new gasket (par. 84).
Leaks at cylinder head gasket.	Tighten head bolts or replace gasket (par. 59).

84. PRESSURE FILLER CAP (fig. 66).

a. Purpose. The radiator maintains pressure on the cooling liquid after the engines warm up to operating temperature. The pressure cap reduces evaporation of the cooling liquid and prevents loss through surging into the upper tank of the radiator after the engines have been shut off following a hard drive in hot weather.

b. Operation. As the overflow pipe is located above the valve which is built into the cap, no liquid or air can escape until the cooling system pressure rises to 8¼ to 9¾ pounds and forces the valve off its seat. It is important that the fiber gasket on which the pressure cap valve seats be in good condition so that there will be no leaks at this point.

c. Cap Removal.

(1) Turn the cap to the left until the tabs on the cap come in contact with the safety stop on the radiator filler neck.

137

TM 9-741
84-85

MEDIUM ARMORED CAR T17E1

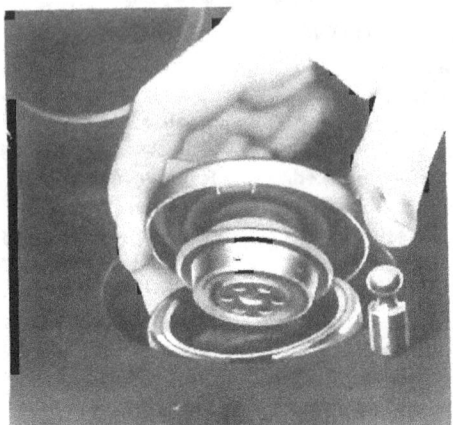

Figure 66 — Radiator Pressure Cap

(2) Pause at this point for a few seconds to allow any pressure or steam to pass off through the overflow pipe.

(3) After the pressure has been relieved, press the cap down and turn it to the left as far as it will go, then lift it off the filler neck.

85. DRAINING AND REFILLING.

a. Draining.

(1) REMOVE FILLER CAP. Remove the filler caps to provide proper venting.

(2) DRAIN RADIATORS (fig. 31). Remove the two radiator drain plugs from the bottom of the hull, using the special wrench provided with the tool kit.

(3) DRAIN CYLINDER BLOCKS (fig. 35). Open the valves located on the left side of each engine block near the flywheel housing.

b. Refilling Cooling System.

(1) CLOSE VALVES (figs. 31 and 35). Close the valves on the cylinder blocks and install the drain plugs in the bottom of the hull.

(2) FILL RADIATOR. Fill the radiators with clean water or, during cold weather, with antifreeze solution. As the two cooling systems are not connected between the two engines, each system must be filled separately.

(3) PRECAUTION.

(a) Do not overfill the radiator when antifreeze solutions are being used, as the solution expands and a quantity will be lost through the overflow.

138

TM 9-741
85-87

COOLING SYSTEM

(*b*) Do not add water when the engines are hot, as there is a possibility of the cold water cracking hot cylinder heads.

86. FAN BELTS.

a. Adjustment.

(1) CHECK PULLEY ALINEMENT. Remove the twenty-five bolts that attach the compartment cover and remove the cover. Check to see that the generator pulley lines up with the crankshaft and water pump pulleys. If they are not in line, move the generator to the front or rear as required. Slotted holes are provided in the generator to engine mounting bracket for this purpose.

(2) ADJUST FAN BELTS. Loosen bolt in generator end of slotted braces at each end of generator. Place one end of a bar thirty inches long between the generator and the engine and pry the generator away from the engine to tighten the fan belts. Hold pressure against the bar while tightening the bolts in the slotted braces. CAUTION: The belts must be tightened until they can be depressed ½-inch midway between the pulleys. It may be necessary to loosen the rear pivot bolt under the generator to move the generator away from the engine.

b. Fan Belt Replacement.

(1) REMOVE FAN BELTS. Remove the twenty-five bolts that attach the engine compartment cover and remove the cover. Loosen the bolt in the generator end of the slotted braces at each end of the generator. Place one end of a long bar between the center partition and the generator and pry the generator against the engine. Lift fan belts off generator, crankshaft, and fan pulleys.

(2) INSTALL FAN BELTS. Place two new belts over the fan and onto the fan pulley, crankshaft, and generator pulleys, and adjust the belt tension, as instructed in the previous paragraph. NOTE: The belts for the left and right engines are not interchangeable. Be sure that the arrows on the belts point toward the direction of travel.

87. THERMOSTAT.

a. Description. A bellows-type thermostat is located in the water outlet passage in the cylinder head. The valve in the thermostat starts to open at 156 F to 165 F and is fully opened at 185 F.

b. Thermostat Removal. Drain radiator. Remove the two cap screws which attach the water outlet to the thermostat housing. Remove the two bolts which attach the water outlet brace to the intake manifold and lift the outlet up to remove the thermostat and gasket.

c. Thermostat Installation. Place new thermostat in opening in cylinder head. Place a new outlet gasket in position and place the

MEDIUM ARMORED CAR T17E1

outlet connection in position and install the two cap screws. Attach the outlet brace to the intake manifold with the two bolts. Tighten the four bolts securely.

88. WATER PUMP.

a. Description. The water pump is a ball bearing type that requires no lubrication in service as it is packed and sealed at the time of manufacture.

b. Water Pump Removal.

(1) REMOVE HOSE. Drain the radiator and remove the two bolts that attach the water inlet to the pump body.

(2) LOOSEN FAN BELTS. Loosen the generator brace bolts and move the generator so as to loosen the fan belts.

(3) REMOVE PUMP PULLEY. Remove the four cap screws which attach the fan blade assembly and pump pulley to the drive flange and remove the fan and pulley.

(4) REMOVE PUMP. Remove the four cap screws that attach the pump to the cylinder block and remove the pump.

c. Water Pump Installation.

(1) INSTALL PUMP. Using a new gasket, place the pump in position on the cylinder block and install the four cap screws.

(2) INSTALL FAN BLADES. Install fan blade and pump pulley on the drive flange and install the four cap screws.

(3) ADJUST FAN BELTS. Adjust fan belts according to instructions given in paragraph 86.

(4) Install and tighten water inlet, using a new gasket, and refill radiator.

89. RADIATOR (fig. 67).

a. Description. Heavy duty construction radiators are used having ten and one-half fins per inch. They are supported in steel anchorage and bolted to the radiator mounting plates which are welded to the hull.

b. Radiator Removal.

(1) Remove the twenty-five retaining bolts from each engine compartment cover and remove the covers.

(2) Remove radiator filler cap from radiator.

(3) Drain cooling system by removing drain plugs from bottom of hull.

(4) Loosen the clamp and remove air cleaner inlet tube from top of carburetor.

COOLING SYSTEM

Figure 67 — Radiator Replacement

(5) Loosen hose clamp at radiator inlet and water outlet; remove hose (fig. 39).

(6) Disconnect both engine primary ignition leads from ignition filter box.

(7) Disconnect lower end of right engine oil filter inlet line from engine; disconnect filter outlet line from bottom of filter. Remove the four attaching bolts and remove the filter. NOTE: The two top bolts also attach the engine compartment junction box to fire wall.

(8) Disconnect all engine and transmission oil cooler oil inlet and outlet lines from the oil cooler. Wire the oil lines to the engine to hold them out of the way (fig. 68). NOTE: Mark the lines to identify from which connection they were removed.

(9) Disconnect Hydrovac vacuum hose from fitting on the left engine intake manifold.

(10) Disconnect the leads from the four terminals on the top of filter and generator regulator box. Remove the four filter and generator regulator box-to-hull bracket bolts and remove the box. NOTE: Mark the

TM 9-741
89

MEDIUM ARMORED CAR T17E1

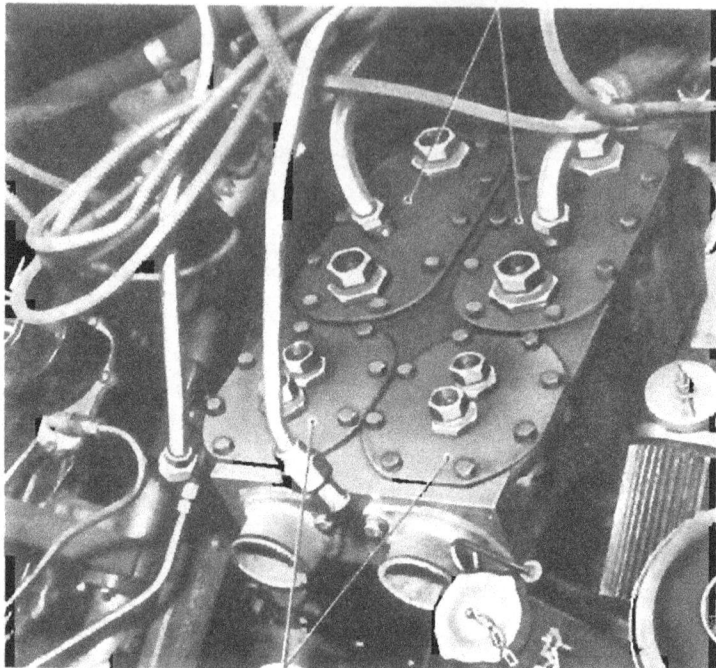

Figure 68 — Oil Cooler

generator regulator box and wires so that they can be installed correctly, or damage to unit will result (fig. 123).

(11) Remove the two bolts which attach forward fire extinguisher nozzle to bracket. Unscrew nozzle and remove bracket bolts.

(12) Disconnect the line from center fire extinguisher nozzle and remove the nozzle bracket-to-fire wall bolts and nozzles.

(13) Disconnect throttle rods at cross shaft end.

(14) Disconnect transmission throttle control rods from cross shaft.

(15) Disconnect throttle cross shaft brackets from engines and swing the cross shaft out of the way of the radiator.

(16) Disconnect temperature indicator lead from left engine thermostat housing fitting and remove clip; disconnect oil pressure gage lead from gage unit on left engine.

COOLING SYSTEM

(17) Disconnect right jettison and main fuel tank vent lines at each side of fire wall and at main fuel tank filler neck; remove main tank to fire wall fitting vent line.

(18) Remove the three bolts which attach the engine compartment cover center support to hull, also the bolts which attach it to fire wall, and remove the support.

(19) Remove the bolts which attach the rear fire wall panel to center panel and to hull. Remove panel.

(20) Remove the bolts which attach center fire wall panel to front panel and swing the center panel up on the hull out of the way.

(21) Remove the front fire wall panel-to-hull bolts and remove panel.

(22) Loosen radiator to cooler hose clamps at radiator end.

(23) Remove the two bolts which attach each water inlet connector to water pump.

(24) Remove the three oil cooler bracket bolts and remove the oil cooler by pulling it toward the rear until the hoses are off the radiators (fig. 68).

(25) Remove the eight bolts which retain the air inlet grille and remove the grille with a hoist.

(26) Loosen the radiator drain hose clamp and disconnect the drain hose from the radiator.

(27) Slide a piece of 2- x 4-inch beam down along the fire wall side of the engine and onto the top of flywheel housing to support the radiator. Remove the six bolts which retain the radiator and work the radiator up out of the engine compartment. Support the radiator on the 2- x 4-inch beam to avoid damaging the core (fig. 67).

c. Radiator Core Installation.

(1) Due to the serious consequences that might result from leaks in the oil, fuel, or cooling system, all connections in these systems are sealed with permatex. In reassembling, care should be taken to use a small amount of permatex on all joints and fittings which indicate the use of permatex in production. CAUTION: The left and right radiators are not interchangeable.

(2) Place a piece of 2- x 4-inch beam down along the fire wall side of engine and over the flywheel housing. Slide the radiator down on the 2- x 4-inch beam to avoid damaging the radiator. Hold the cables, lines and throttle cross shaft out of the way as the radiator is lowered into position (fig. 67). Install the six radiator retaining bolts.

(3) Permatex the radiator drain hose fitting. Install the hose and tighten clamp securely.

MEDIUM ARMORED CAR T17E1

(4) Attach throttle cross shaft brackets to the engines; hook up the cross shaft to carburetor throttle rods, and cross shaft to transmission rods.

(5) Attach forward fire extinguisher bracket, screw extinguisher nozzle on line and bolt to bracket.

(6) Bolt the terminal and generator regulator box to hull and attach the leads to the four terminals. CAUTION: Be sure to install the leads according to the marking made when removing the unit or serious damage will result (fig. 123).

(7) Permatex the oil cooler to radiator outlet hoses and attach them securely to cooler. Put a light coat of permatex on the radiator outlet and in the cooler hose. Slide the cooler into position. Raise the engine end of cooler and work the two lower hoses over the radiator outlet (fig. 68).

(8) Install the three cooler bracket bolts and tighten the cooler-to-radiator hose clamps securely.

(9) Attach inlet connectors to water pumps using new gaskets.

(10) Attach the transmission and engine oil lines to cooler.

(11) Install the fire wall front panel and bolt it loosely to radiator support center panel.

(12) Swing the fire wall center panel down into position and bolt it loosely to front panel.

(13) Install the fire wall rear panel and bolt it loosely to the center panel.

(14) Install engine compartment cover center support and bolt it to hull and fire wall. Tighten all fire wall panel bolts securely.

(15) Install right engine oil filter and engine compartment terminal box to fire wall panel.

(16) Connect the oil filter lines.

(17) Connect right jettison and main fuel tank vent lines to fitting at front of fire wall.

(18) Install main fuel tank vent line under wire on left side of fire wall center panel and attach it to main tank filler neck and fitting at division panel.

(19) Attach all wiring clips to fire wall.

(20) Connect temperature gage terminal to fitting on thermostat housing and oil gage terminal to oil gage unit on left side of engine block.

(21) Connect fire wall fire extinguisher nozzles to lines and attach the nozzle brackets to fire wall.

(22) Attach Hydrovac line to fitting on intake manifold.

TM 9-741
89-90

COOLING SYSTEM

(23) Install the water outlet hose between radiator and water outlet. Tighten clamps securely.

(24) Attach ignition wires to terminal on bottom of filter box.

(25) Install carburetor air cleaner inlet tube on carburetor (fig. 39).

(26) Place the air inlet grille on the hull and bolt it in position with the eight bolts.

(27) Install the radiator drain plug and fill cooling system with a suitable cooling solution for the temperature expected. Check cooling systems for leaks.

(28) Install radiator filler caps and tighten securely.

(29) Install engine compartment covers.

90. OIL COOLER.

a. Description.

(1) The engines, transmissions, and gear reduction units operate in an engine compartment which is completely enclosed and subjected under some conditions to quite high temperatures. To provide adequate cooling of the lubricant, an oil cooler assembly is provided (fig. 68).

(2) The oil cooler assembly consists of an oil cooler case located between the two engine assemblies. In this case four core assemblies are used for cooling the oils from two engines and two transmission and gear reduction units.

(3) The cooling agent is the water from the engine water circulating systems. The cool water from the bottom of each radiator core is piped directly to the oil cooler assembly and from the cooler is drawn to the water pump inlets by the pumps.

(4) The two water circulation systems are kept separate in the oil cooler case by a partition running lengthwise in the case. One engine oil and one transmission oil cooler core are placed on each side of this partition so that the oils from each system are cooled by the water from its own cooling system. This permits operation of one engine assembly without affecting the other.

(5) The oils from each unit are circulated by pumps in each of the units involved; the engine oil pumps for the engine oil coolers, and the transmission oil pumps for the transmission and gear reduction units. Each oil cooler core unit is supplied with two connections providing an inlet and return to the unit involved.

(6) The engine oiling system is protected against failure, due to obstructions in the cooler or oil lines, by a bypass valve on each engine (fig. 69). The pressure line from the pump is directed to one side of

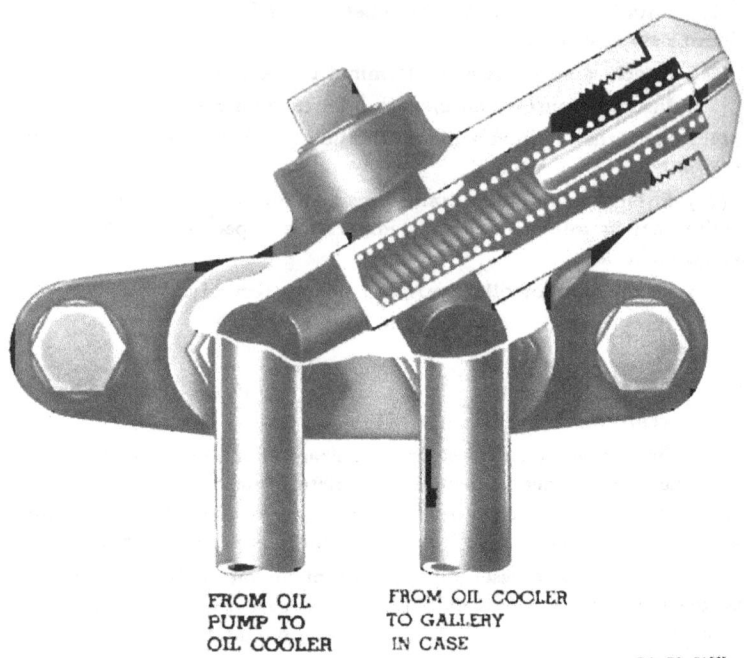

FROM OIL PUMP TO OIL COOLER

FROM OIL COOLER TO GALLERY IN CASE

RA PD 56283

Figure 69 — Oil Cooler Bypass Valve Assembly

this bypass valve and directly to the oil cooler. The oil returns from the cooler to the other fitting on the bypass valve and to the engine oil lines.

(7) In case of obstructions in the oil cooler system, the pump pressure forces the bypass valve off its seat and oil can flow directly to the return line and to the engine oil lines.

b. Oil Cooler Cleaning.

(1) Flush the entire cooling system at least twice a year, preferably before and after using antifreeze.

(2) Due to the large area in the oil cooler assembly, it is quite probable that this unit will not flush out as thoroughly as the rest of the cooling system. Therefore, the four oil cooler units should be removed from the housing for cleaning.

(3) Remove the two engine compartment covers.

(4) Remove both radiator filler caps and drain cooling systems by removing drain plugs from bottom of hull.

COOLING SYSTEM

(5) Disconnect the oil cooler lines from fittings on cooler units.

(6) Remove cooler unit-to-cooler case attaching bolts and remove the cooler units.

(7) Clean the cooler units externally with radiator cleaning solution and a stiff brush.

(8) Clean the cooler units internally with a strong cleaning solvent. After cleaning the units internally, it is advisable to let them stand filled with solvent for several hours and then flush them out again. After cleaning the units, they must be flushed out with light engine oil.

(9) Clean all rust and scale from the inside of the cooler case.

(10) Clean the gasket seats on the cooler case and the cooler units.

(11) Put a light coat of permatex on the gasket seats; install new gaskets; replace the cooler units and bolt them securely.

(12) Attach all oil cooler lines and tighten securely.

(13) Permatex the radiator drain plugs and install them in hull and fill cooling systems.

(14) Install radiator filler caps.

(15) Run the engines and check for oil or water leaks.

(16) Install the engine compartment cover.

c. Oil Cooler Removal (fig. 68).

(1) Remove the twenty-five retaining bolts from each engine compartment cover and remove the covers.

(2) Remove radiator filler caps from both radiators.

(3) Drain cooling systems by removing drain plugs from bottom of hull.

(4) Disconnect both engine primary ignition leads from ignition filter box.

(5) Disconnect lower end of right engine oil filter inlet line from engine. Disconnect filter outlet line from bottom of filter. Remove the four attaching bolts and remove the filter. NOTE: The two top bolts also attach the engine compartment junction box to fire wall.

(6) Disconnect all engine and transmission oil cooler oil inlet and outlet lines from the oil cooler. Wire the oil lines to the engine to hold them out of the way. NOTE: Mark the lines to identify from which connection they were removed.

(7) Disconnect Hydrovac vacuum hose from fitting on the left engine intake manifold.

(8) Disconnect the line from center fire extinguisher nozzle and remove the nozzle bracket-to-fire wall bolts and nozzles.

MEDIUM ARMORED CAR T17E1

(9) Disconnect temperature indicator lead from left engine thermostat housing fitting and remove clip; disconnect oil pressure gage lead from gage unit on left engine.

(10) Disconnect right jettison and main fuel tank vent lines at each side of fire wall and at main fuel tank filler neck; remove main tank-to-fire wall fitting vent line.

(11) Remove the three bolts which attach the engine compartment cover center support to hull, also the bolts which attach it to fire wall, and remove support.

(12) Remove the bolts which attach the rear fire wall panel to center panel and to hull; remove panel.

(13) Remove the bolts which attach center fire wall panel to front panel and swing the center panel up on the hull out of the way.

(14) Remove the front fire wall panel-to-hull bolts and remove panel.

(15) Loosen radiator-to-cooler hose clamps at radiator end.

(16) Remove the two bolts which attach each water inlet connector to water pump.

(17) Remove the three oil cooler bracket bolts and remove the oil cooler by pulling it toward the rear until the hoses are off the radiator (fig. 68).

d. Oil Cooler Installation.

(1) Due to the serious consequences that might result from leaks in the oil, fuel, or cooling system, all connections in these systems are sealed with permatex. In reassembling, care should be taken to use a small amount of permatex on all joints and fittings which indicate the use of permatex in production.

(2) Permatex the oil cooler to radiator outlet hoses and attach them securely to cooler. Put a light coat of permatex on the radiator outlet and in the cooler hose. Slide the cooler into position. Raise the engine end of cooler and work the two lower hoses over the radiator outlet (fig. 68).

(3) Install the three cooler bracket bolts and tighten the cooler-to-radiator hose clamps securely.

(4) Attach inlet connectors to water pumps, using new gaskets.

(5) Attach the transmission and engine oil lines to cooler.

(6) Install the fire wall front panel and bolt it loosely to radiator support center panel.

(7) Swing the fire wall center panel down into position and bolt it loosely to front panel.

(8) Install the fire wall rear panel and bolt it loosely to the center panel.

COOLING SYSTEM

(9) Install engine compartment cover center support and bolt it to hull and fire wall. Tighten all fire wall panel bolts securely.

(10) Install right engine oil filter and engine compartment terminal box to fire wall panel.

(11) Connect the oil filter lines.

(12) Connect right jettison and main fuel tank vent lines to fitting at front of fire wall.

(13) Install main fuel tank vent line under wires on left side of fire wall center panel and attach it to main tank filler neck and fitting at division panel.

(14) Attach all wiring clips to fire wall.

(15) Connect temperature gage terminal to fitting on thermostat housing and oil gage terminal to oil gage unit on left side of engine block.

(16) Connect fire wall fire extinguisher nozzles to lines and attach the nozzle brackets to fire wall.

(17) Attach Hydrovac line to fitting on intake manifold.

(18) Attach ignition wires to terminal on bottom of filter box.

(19) Install the radiator drain plug and fill cooling systems with a suitable cooling solution for the temperature expected. Check cooling systems for leaks.

(20) Install radiator filler caps and tighten securely.

(21) Install engine compartment covers.

TM 9-741
91-94

MEDIUM ARMORED CAR T17E1
Section XVII
EXHAUST SYSTEM

	Paragraph
Description	91
Trouble shooting	92
Muffler replacement	93
Exhaust pipe replacement	94

91. DESCRIPTION (fig. 5).

a. Exhaust from each engine is taken from the exhaust manifold through a flexible connection to the exhaust pipe and muffler. A bracket and strap hold each muffler in place on the outside of the hull.

92. TROUBLE SHOOTING.

a. **Noisy Muffler.**

Probable Remedy	Probable Cause
Open seams in muffler.	Replace muffler (par. 93).
Corroded metal.	Replace muffler (par. 93).
Loose baffle.	Replace muffler (par. 93).
Loose flange.	Tighten or replace gasket (par. 94).
Dented or damaged.	Replace (par. 93).

93. MUFFLER REPLACEMENT.

a. **Removal Procedure** (fig. 30). Remove nut and lock washer from one end of muffler strap. Remove the four bolts and lock washers from the flange at top of muffler and lift off muffler.

b. **Installation Procedure.** Place a new flange gasket between the muffler and exhaust pipe and install the four bolts and lockwashers. Place muffler clamp screw through hole in bracket and install the lock washer and nut. Tighten the nut and bolts securely.

94. EXHAUST PIPE REPLACEMENT.

a. **Removal Procedure** (fig. 39). Remove the clamp end of muffler strap. Remove the three bolts that attach the support to the hull. Loosen the clamp that holds the flexible connection to the exhaust pipe and drive out the exhaust pipe. Loosen the set screw that retains the exhaust pipe to the support and drive the support off the exhaust pipe.

b. **Installation Procedure.** Place exhaust pipe through opening in hull and into position in flexible connection. Place support over end of exhaust pipe and install the three bolts in the support. Tighten the set screw in the support and the clamp screw on the flexible connection. Place muffler clamp screw through hole in bracket and install the lock washer and nut.

Section XVIII

FLUID COUPLING

	Paragraph
Introduction	95

95. INTRODUCTION.

a. A fluid coupling is used in conjunction with the hydra-matic drive to transmit power from the engine to the transmission. The fluid coupling is operated by centrifugal force through the use of oil which is forced into the coupling by the front pump in the transmission.

b. The coupling needs no adjustments or maintenance other than keeping the transmission oil level between the high and low mark on the oil measuring rod.

c. No repairs should be attempted by the using arm personnel.

TM 9-741
96-97

MEDIUM ARMORED CAR T17E1

Section XIX

TRANSMISSION

	Paragraph
Description	96
Trouble shooting	97
Adjustment of servo bands	98
Hydraulic throttle control	99
Manual shift control	100
Removal of hydra-matic transmission	101

96. DESCRIPTION.

a. General. The hydra-matic drive is a method of converting torque and transmitting power from the engines to the wheels of the vehicle. With hydra-matic drive, there is no gear shifting. The gear ratio in which the vehicle is operating at any time is selected automatically by the mechanism itself in accordance with the performance demands made upon the vehicle by the driver and by road conditions. The gear selected always provides maximum efficiency under any combination of conditions. The hydra-matic drive consists of a hydraulic coupling combined with an automatic hydraulic-controlled transmission having four forward speeds and reverse. The hydra-matic drive simplifies driving. The only control relating to the transmission is the lever in driving compartment, which is used to select neutral, reverse, or one of the two forward speed ranges.

b. Towing to Start. It is possible to start the engines by pushing or towing the vehicle. To do this, the vehicle should be towed in neutral until a speed of 15 to 20 miles per hour is reached. Then the control lever should be moved to the "DRIVE" position (never to "LOW") and the engines will ordinarily start within a few seconds.

c. Parking on Hills. When the vehicle is stopped on a hill or other steep incline, the vehicle can be locked in gear by shutting off the engines and placing the lever in reverse. This locks the transmission in two gears simultaneously, making it impossible to turn the wheels until the lever is shifted out of reverse.

97. TROUBLE SHOOTING.

a. Transmission Jumps Out of Reverse Gear.

Probable Cause	Probable Remedy
Rod connecting manual control lever to transmission adjusted too short.	Readjust manual control linkage (par. 100).

TM 9-741

TRANSMISSION

Probable Cause	Probable Remedy
Reverse pawl bottoming on reverse gear and preventing roller from contacting stop.	Report to ordnance personnel.
Reverse shifter lever binding in notch of inside manual control lever and preventing roller from contacting stop.	Report to ordnance personnel.
Weld on reverse shifter lever of reverse gear anchor assembly broken.	Report to ordnance personnel.
Reverse pawl plunger spring broken.	Report to ordnance personnel.

h. Transmission Downshifts from Third to Second or Third to First When Throttle is Opened at Speed Above 14 Miles per Hour and Downshifts at Speeds Above 14 Miles per Hour with Throttle Closed and Transfer Case in High Ratio.

Rods connecting manual control lever to transmission adjusted too long, mispositioning the manual control valve in the valve body assembly.	To check, before making any adjustments, move outer shifter lever out of detent slightly toward neutral. If this corrects the trouble, readjust manual control linkage (par. 100).
Low governor pressure resulting from low capacity front oil pump.	Check front oil pump pressure with oil gage and fixture KM-J1467M6. With transmission oil warm, move control lever from the neutral position into one of the forward ranges. Oil pressure should be from 55 to 85 pounds at engine idle. If not, report to ordnance personnel.
Leaks around governor, valve body assembly, or oil pump.	Report to ordnance personnel.
Pressure regulator valve stuck in open position.	Report to ordnance personnel.
Leaky rear servo.	Report to ordnance personnel.
Governor valves sticking.	Report to ordnance personnel.
Throttle valve stuck in an open position.	Report to ordnance personnel.
T-valve stuck in an open position.	Report to ordnance personnel.

TM 9-741
97

MEDIUM ARMORED CAR T17E1

c. Transmission Shifts Back and Forth Between Fourth and Second on Acceleration.

Probable Cause	Probable Remedy
Rods connecting manual control to transmission out of adjustment.	To check, before making any adjustments, move outer shifter lever out of detent toward neutral. If this corrects the trouble, adjustment is incorrect and all manual control rods should be readjusted (par. 100).

d. Engine Speeds Up and Fails to Drive Vehicle During Acceleration with Shift Lever in "Low" Range, Transmission in Second Gear.

Rod connecting manual control lever to transmission adjusted too long.	Adjust manual control linkage to correct position (par. 100).
No oil pressure to front clutch because of leak in valve body assembly causing drop in line pressure of front clutch.	Report to ordnance personnel.
Drilled hole in case to oil delivery sleeve partially obstructed.	Report to ordnance personnel.
Oil delivery sleeve not correctly located on dowel.	Report to ordnance personnel.
Oil delivery sleeve ring for front clutch broken.	Report to ordnance personnel.

e. In "Drive" Range Vehicle Shifts From First to Second Properly but Second to Third to Fourth Shifts Come at Too High Speeds or Not at All; Throttle Downshifts Fourth to Second Rather Than Fourth to Third.

Rods connecting outer shifter lever to transmission adjusted too long, mispositioning the manual valve, "starving" the governor of input pressure.	Readjust manual control linkage to correct position (par. 100).
Valves in governor sticking closed.	Report to ordnance personnel.

f. Vehicle Creeps Excessively with Accelerator Closed and Shift Lever in a Driving Position.

154

TRANSMISSION

Probable Cause	Probable Remedy
Engines idle too fast.	Carburetor idle adjustments improperly adjusted or hydraulic throttle mechanical linkage fails to return to its stops (pars. 77 and 99).
Idle adjustments incorrect.	Adjust each carburetor to approximately 600 revolutions per minute and then readjust mechanical linkage (pars. 77 and 99).
Weak or broken return springs.	Replace spring.
Restriction in hydraulic throttle lines resulting from a sharp bend in the tubing or hoses.	Straighten or replace the hydraulic lines affected. Refer to ordnance personnel.
Intermediate shaft bushing seizing on main shaft.	Report to ordnance personnel.

g. When Accelerating Through the Shifts with Manual Lever in "Drive" Range the Shifts All Come at Too High a Speed.

The throttle controls are not properly adjusted. Worn throttle control linkage.	To check for the cause of this condition, accelerate with wide open throttle from a standing start with transmission case in high range. The first to second, and second to third, and third to fourth upshifts occur at approximately 9, 20, and 43 miles per hour respectively, the transmission is normal and the difficulty is in the throttle linkage. Make complete throttle linkage adjustment according to paragraph 99.
Loss of governor pressure. If the governor pressure is low, then it will require higher vehicle speeds to increase governor pressure before the shifter valves can be shifted.	Report to ordnance personnel.
Throttle valve stuck in wide open position.	Report to ordnance personnel.

TM 9-741

MEDIUM ARMORED CAR T17E1

h. When Accelerating Through the Shifts, Part Throttle, in "Drive" Range the Shift from Second to Third Occurs Too Soon After the First to Second Shift. Likewise the Third to Fourth Shift Occurs Too Soon After Second to Third.

Probable Cause	Probable Remedy
The throttle controls are not properly adjusted.	Properly adjust the throttle linkage controls.
Governor G2 valve stuck open.	Report to ordnance personnel.
Second to third regulator plug stuck in closed position.	Report to ordnance personnel.

i. When Driving in "Drive" Range at Speeds Above 12 Miles per Hour, Transmission Will Throttle Downshift Before Hitting the Detent Plug in Valve Body, Transfer Case in High Ratio. NOTE: The Transmission Should Remain in Fourth Gear When the Throttle is Opened Just to the Detent Plug.

Detent plug stuck closed in outer valve body.	Report to ordnance personnel.

j. Transmission Will Not Throttle Downshift From Fourth to Third. Usually This Occurs at Speeds Above 12 Miles per Hour.

Stop on hydraulic throttle control will not let carburetor throttle valves open.	Adjust hydraulic throttle controls (par. 99). Make sure throttle levers on side of transmission cases are tight on shaft, and linkage allows transmission throttle arm to move to the extreme forward position.
Detent plug stuck open in outer valve body.	Report to ordnance personnel.
Drilled passage for T-valve pressure to third to fourth shifter valve plugged.	Report to ordnance personnel.
Fourth to third regulator plug stuck closed.	Report to ordnance personnel.

k. Transmission Does Not Respond to Shift Lever Position or Will Only Move Forward, or Move Forward Only After Accelerator is Opened Excessively.

Outer shifter lever on side of transmission loose on shaft.	Properly position the outer shifter lever on the shaft and tighten in place.

TM 9-741

TRANSMISSION

Probable Cause	Probable Remedy
The pin which picks up the manual valve in the valve body is not positioned in the groove in the manual valve.	Report to ordnance personnel.
Vanes in torus damaged, laying over and not permitting full drive to be taken through the fluid coupling.	Report to ordnance personnel.
Obstruction between end of servo and piston not permitting band to be held tight to the drum.	Report to ordnance personnel.
Servo bands adjusted too loosely.	Adjust servo bands (par. 98).

l. **Transmission Fails to Drive Vehicle When Control Lever is Moved into Either "Drive" or "Low" Range. NOTE:** In General this Condition is Caused by Failure of One or Both Bands to be Applied.

Low oil level.	Be sure transmission oil level is up to full mark (par. 24).
Front or rear band extremely loose — not applied.	Inspect band adjustment lock nuts for tightness. Lock nuts may have become loose; if so, adjust both bands and lock nuts (par. 98).
Front oil pump failure.	With oil gage and fixture **KM-J 1467 M6** assembled into place in rear of side cover and governor sleeve, check front pump pressure (oil warm). Move control lever into one of the forward ranges. Oil pressures should show 55 to 85 pounds at engine idle. If oil pressure is not 55 to 85 pounds at engine idle, report to ordnance personnel.
Pressure regulator valve stuck closed.	Report to ordnance personnel.
Manual valve out of proper position.	Make manual control adjustment. See paragraph 100 or report to ordnance personnel.

TM 9-741
97

MEDIUM ARMORED CAR T17E1

Probable Cause	Probable Remedy
Excessive oil leakage due to loose valve body assembly, loose or leaky pressure regulator body or valve, or leaky front servo.	Report to ordnance personnel.
Broken servo band on front or rear drum.	Report to ordnance personnel.

m. **Transmission Drives Vehicle in First Gear but Acts as Though in Neutral After Shifting Out of First, or First to Second Shift is Normal but Acts as Though in Neutral After Shifting Out of Second. When Shifting Out of First, if Transmission Acts as Though in Neutral, then there is No Oil Pressure to Front Clutch. If First and Second Shifts are Normal, but Act as Though in Neutral after Shifting Out of Second, Then There Is No Oil Pressure to Rear Clutch.**

Oil leakage or restriction in the passages.	Report to ordnance personnel

n. **Transmission Slips in First, Third, and Reverse, but Operates Properly in Second and Fourth.**

Front band not holding due to improper adjustment, loss of oil pressure, or restriction.	Adjust front band to proper setting (par. 98). If trouble still exists, report to ordnance personnel.

o. **Transmission Slips in First and Second but Operates Properly in Third and Fourth and Reverse.**

Rear band not holding due to improper adjustment or loss of pressure or restriction.	Adjust the rear servo band to proper setting (par. 98). If trouble still exists report to ordnance personnel.

p. **Transmission Slips or Chatters on Throttle Downshift Third to Second, Open Throttle, or Third to First.**

Front band not holding due to improper adjustment, or loss of pressure or restriction.	Adjust front band to proper setting (par. 98). If front band still slips, report to ordnance personnel.
Check if front band slips, move shift lever into the reverse position, apply brakes and open throttle. If engine speeds up excessively, front band is not holding.	Report to ordnance personnel.

TRANSMISSION

Probable Cause	Probable Remedy
Rear band not holding due to improper adjustment or momentarily trapping oil in the accumulator section of the servo.	Adjust rear band to proper setting (par. 98). If rear band still slips, report to ordnance personnel.

q. Noise on Third to First Downshift, Throttle Closed.

Check valve not properly seated in rear servo accumulator body, or it may be broken.	Report to ordnance personnel.
Stuck or broken ring in accumulator piston.	Report to ordnance personnel.
Compensator valve in valve body assembly stuck open, holding check valve in rear servo off seat.	Report to ordnance personnel.

r. Transmission Will Not Force Shift (Vehicle Started in Motion) with Full Throttle Second to Third, or Third to Fourth. Throttle Downshift Fourth to Third Occurs before Throttle Is Wide Open. Light Throttle Shifts and First to Second Forced Shift Occur at Proper Speeds.

Leakage in governor input line caused by omitting governor sleeve plug.	Report to ordnance personnel.
Center governor pipe loose fit at either end.	Report to ordnance personnel.
Governor loose on pump flange.	Report to ordnance personnel.
Low pump pressure.	Report to ordnance personnel.
Valve body loose on transmission case.	Report to ordnance personnel.

s. Transmission Misses Second Gear on Upshift.

Double transition valve stuck inward.	Report to ordnance personnel.
First to second shifter valve stuck open.	Report to ordnance personnel.
First to second governor plug stuck open.	Report to ordnance personnel.
First to second regulator plug stuck in an open position.	Report to ordnance personnel.

TM 9-741

MEDIUM ARMORED CAR T17E1

t. Vehicle Starts in Second Gear "Low" Range.

Probable Cause	Probable Remedy
G1 governor valve stuck open.	Report to ordnance personnel.
First to second shifter valve stuck open or first to second governor plug stuck open.	Report to ordnance personnel.

u. Transmission Misses Third Gear, Shifting from Second to Fourth Gear.

Double transition valve stuck or compensator auxiliary plug stop pin broken.	Report to ordnance personnel.
The second to third governor plug stuck.	Report to ordnance personnel.
The second to third shifter valve stuck.	Report to ordnance personnel.

v. Transmission Starts in Second but Makes All Other Shifts.

First to second shifter valve stuck open.	Report to ordnance personnel.
G1 governor valve stuck open in governor assembly.	Report to ordnance personnel.
The first to second governor plug stuck open.	Report to ordnance personnel.
The first to second regulator plug stuck in an open position.	Report to ordnance personnel.

w. Transmission Has Only First and Second Gears, or Only Third and Fourth Gears.

Only first and second gear.

Second to third shifter valve stuck closed.	Report to ordnance personnel.
Second to third governor plug stuck closed.	Report to ordnance personnel.
Second to third regulator plug stuck open.	Report to ordnance personnel.
G2 governor valve stuck closed.	Report to ordnance personnel.

Only third and fourth gear.

Second to third shifter valve stuck open.	Report to ordnance personnel.
Second to third governor plug stuck open.	Report to ordnance personnel.

TM 9-741
97

TRANSMISSION

x. Transmission Shifts Through First, Second, and Third Gears Normally, but Has No Fourth Gear.

Probable Cause	Probable Remedy
Third to fourth shifter valve stuck closed.	Report to ordnance personnel.
Third to fourth governor plug stuck closed.	Report to ordnance personnel.
Third to fourth regulator plug stuck open.	Report to ordnance personnel.

y. Transmission Upshifts with Light Throttle, Occurs at Speeds When They Should Occur with Wide Open Throttle. Band Application When Shift Lever is Moved from "Neutral" into "Drive" or "Low" Range with Vehicle Standing Is Very Violent, or Transmission Will Downshift from Third to Second at Approximately 10 Miles per Hour with Throttle Closed.

Throttle valve stuck open in valve body assembly.	Report to ordnance personnel.

z. Shifts Very Severe at Light Throttle but Normal at Greater Throttle Openings.

New clutch plates not broken in.	Report to ordnance personnel.
Stuck throttle valve in outer valve body.	Report to ordnance personnel.
Compensator valve stuck in outer valve body.	Report to ordnance personnel.

aa. Vehicle Fails to Move for the First Few Minutes After Starting Engine Regardless of Shift Lever Position Unless Throttle Is Wide Open. In General, This Condition Is Caused by the Oil from the Fluid Coupling Draining Back Excessively into the Transmission. To Check for this Condition Allow the Vehicle to Stand for Length of Time Necessary to Bring About This Condition and Then Check Oil Level in Transmission. If Level Is Considerably Above the Full Mark, There Is an Excessive Leakage Caused by:

Broken oil seal ring in front cover assembly.	Report to ordnance personnel.
Ball check in driven torus stuck open.	Report to ordnance personnel.

bb. Vehicle Moves Forward with Control Lever in Reverse. NOTE: In Severe Cold Weather, If Vehicle Has Been Allowed to Stand Out Over Night, This May Be Noticeable for Just an Instant. If the

161

MEDIUM ARMORED CAR T17E1

Condition Exists Even After the Transmission Is Warm, the Drive Is Being Taken Through the Rear Unit.

Probable Cause	Probable Remedy
Bent clutch release spring.	Report to ordnance personnel.
Clutch plates not sliding freely on rear unit hub.	Report to ordnance personnel.
Scored bronze washer under rear unit hub.	Report to ordnance personnel.
Clutch pistons sticking in cylinder bores in rear unit.	Report to ordnance personnel.

cc. **Shift Becomes Erratic Following Fourth to Third Downshift.**

Throttle valve stuck in open position.	Report to ordnance personnel.

dd. **Excessive Drag Is Noticed When Driving Vehicle in Reverse. Vehicle Does Not Roll Freely When Coasting in Reverse. When Coasting in Reverse, Vehicle Comes to an Abrupt Stop. Rear Band Is Not Released or Is Applied at Slow Speed While Car Is Under Motion.**

Faulty front oil pump.	Report to ordnance personnel.
Compensator valve auxiliary plug pin broken.	Report to ordnance personnel.
Pressure regulator body bolts loose, or pressure regulator valve stuck open, dropping line pressure.	Report to ordnance personnel.
Valve body assembly attaching screws or end plate loose.	Report to ordnance personnel.
Shuttle valve in pressure regulator body stuck open to rear pump.	Report to ordnance personnel.
Leak causing drop in oil pressure.	Report to ordnance personnel.

ee. **Vehicle Fails to Reach Maximum Speed. Throttle Fails to Open Wide Open.**

Improperly adjusted stop screw at hydraulic throttle slave cylinder.	Adjust stop screw (par. 99).
Air in the lines. This will be evidenced by a spongy pedal feel at part throttle.	Bleed system (par. 99).
Badly worn and leaking piston cup in either the master or slave cylinder.	Refer to ordnance personnel.
Carburetor throttle rods too short.	Adjust linkage (par. 99).

TM 9-741
97-98

TRANSMISSION

ff. Excessive Loss of Hydraulic Throttle Fluid Requiring an Abnormal Amount of Fluid to Keep the System Filled.

Probable Cause	Probable Remedy
Master cylinder shaft packing gland leaks.	Refer to ordnance personnel.
Welch plug in master cylinder opposite the shaft packing gland leaks.	Refer to ordnance personnel.
Master cylinder housing gasket leaks.	Tighten or replace gasket.
Slave cylinder breather loose.	Tighten, or refer to ordnance personnel.
Slave cylinder cover gasket leaks.	Refer to ordnance personnel.
Slave cylinder shaft seal leaks.	Refer to ordnance personnel.
Hydraulic hose or pipes leaking at fittings.	Tighten fittings, or refer to ordnance personnel.

98. ADJUSTMENT OF SERVO BANDS.

a. The front and rear servo bands should be adjusted every 3,000 miles, or each time the transmission oil is changed.

b. Adjustment Procedure.

(1) REMOVE SERVICE PLATES (fig. 31). Remove the service plates from the bottom of hull.

(2) DRAIN TRANSMISSION OIL. Remove the drain plug from transmission oil pans and drain oil.

(3) REMOVE TRANSMISSION OIL PANS (fig. 70). Remove the oil pan retaining screws and remove the pan.

(4) REMOVE SIDE COVER. Remove the manual control shift lever with detent ball and spring and remove side cover.

(5) ADJUST FRONT SERVO BAND (fig. 71). Remove lock nut from front servo adjusting screw. Place tool KM-J 1459 over a cap screw on front planetary unit. Tighten front servo adjusting screw just to the point where the front drum cannot be turned in either direction KM-J 1459. Mark servo adjusting screw for identification position. Back off the adjusting screw (7 turns). Install and tighten lock nut, holding the adjusting screw to prevent turning. NOTE: It is important in the operation of the unit that the adjusting screw is backed off the exact number of turns (7) and that only tool KM-J 1459, or its equivalent in length, be used to turn drum.

(6) ADJUST REAR SERVO BAND (fig. 72). Loosen lock nut on rear servo adjusting screw. Place the servo gage KM-J 1460 over the rear

MEDIUM ARMORED CAR T17E1

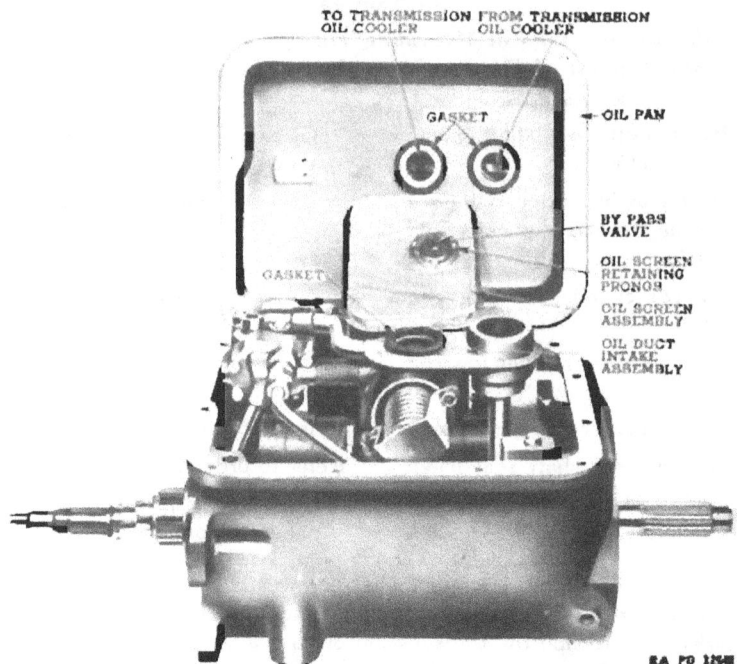

Figure 70 — Transmission Oil Pan Replacement

servo assembly, engaging one end over the outer edge of the accumulator body. Use a small pry bar to force servo lever over against the servo gage (fig. 72) and turn the adjusting screw until the servo lever holds in position, just touching the outer edge of servo gage. NOTE: A pry bar is used in the above operation to relieve strain on the adjusting screw while making this adjustment. Tighten the rear servo adjusting screw lock nut.

(7) CHECK SHIFTER SHAFT SEAL (fig. 73). Check cork gasket on manual control shaft to insure a tight oil seal.

(8) CHECK POSITION OF GOVERNOR SLEEVE PLUG (fig. 73). Check to make sure governor sleeve plug is in place in governor valve body.

(9) INSTALL GASKET AND SIDE COVER (fig. 82). Install new side cover gasket and side cover. Attach in place with cap screws and washers. NOTE: The bottom row of cap screws must be provided with copper washers to seal against oil leak. On the balance, use lock washers.

(10) INSTALL OIL PAN (fig. 70). Install the oil pan, using a new gasket. Tighten the screws securely. NOTE: Check the oil pressure gage plug in rear of oil pan. This must be tight and gasket in place.

TRANSMISSION

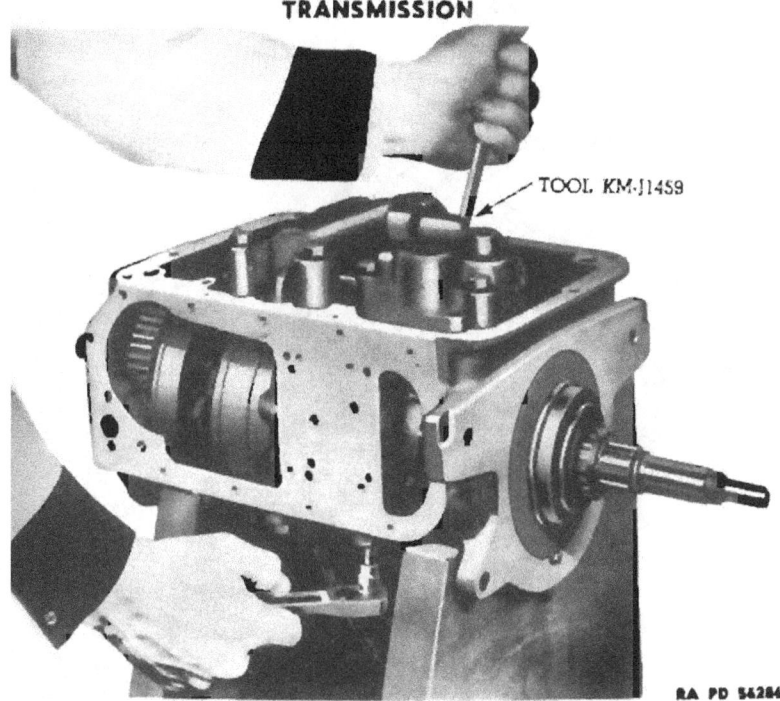

Figure 71 — Adjusting Front Band

(11) INSTALL OUTER SHIFT LEVER (fig. 82). Install outer manual shift lever, spring, and detent ball on splined shaft and tighten clamp screw.

(12) INSTALL THROTTLE LEVER (fig. 82).

(13) REFILL TRANSMISSION. Refill transmission with the correct lubricant, as specified in paragraph 24.

(14) INSTALL SERVICE PLATE. Bolt the transmission service plate securely to bottom of hull.

99. HYDRAULIC THROTTLE CONTROL.

a. Description (figs. 74 and 80). The foot throttle control is hydraulically operated and consists of a foot accelerator pedal, master cylinder, fluid reservoir, slave cylinder, hydraulic lines and the necessary linkage. The foot accelerator pedal is attached to the master cylinder and the fluid reservoir is the same reservoir that is used in the brake system. The slave cylinder is mounted on a bracket which is attached to the flywheel housing of the right-hand engine. A set of control rods is used to connect the left-hand engine, the carburetors and the transmissions, which must be synchronized.

MEDIUM ARMORED CAR T17E1

Figure 72 — Adjusting Rear Band

b. Operation. Pressure applied to the foot accelerator pedal forces the piston in the master cylinder down, exerting hydraulic pressure on the piston in the slave cylinder, which moves the slave cylinder linkage and opens the throttles and controls the shifting of the gears in the transmission.

c. Control Rod Adjustments (figs. 74, 75, 76, 77, 78, 79 and 80).

(1) Retract the set screws in the slave cylinder idler lever, cross shaft accelerator lever, and slave cylinder support bracket (1, 2, and 3, fig. 74) until they do not protrude beyond the levers and bracket.

(2) Disconnect clevises at lower end of throttle rods and swivel at upper end of transmission throttle control rods for both left- and right-hand engines (8 and 9, fig. 74).

(3) Adjust idling speed of each engine separately to approximately 600 revolutions per minute. This is a fast idle.

TM 9-741
99

TRANSMISSION

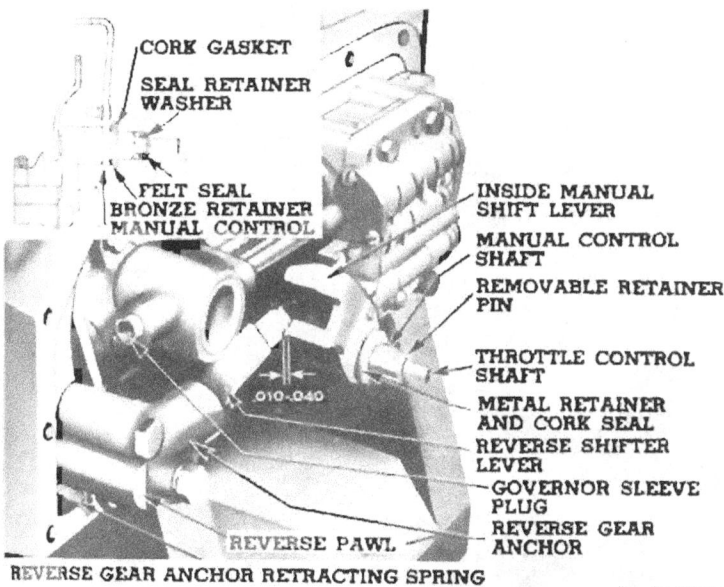

Figure 73 — *Assembly of Reverse Pawl and Reverse Gear Shift Mechanism*

(4) Place the throttle control cross shaft lever checking gage (which is furnished with tool kit) on the slave cylinder idler lever by engaging the gage pins in the hole and slot provided in the lever, as shown in figure 75.

(5) Rotate gage and lever until the end of the gage rests on the top of the exhaust manifold flange.

(6) With the lever in this position, and holding the throttle rod down so the carburetor throttle lever is in the idle position, adjust the length of the throttle rod by rotating the clevis at the lower end until the clevis pin may enter freely into the hole in the slave cylinder idler lever (fig. 75). Install clevis pin with new cotter pin and lock clevis nut. Remove gage.

(7) Holding transmission throttle control rod for right-hand engine (9, fig. 74) down so that the throttle valve lever (10, fig. 74) is at the limit of its travel in the closed throttle position, adjust the length of the rod by rotating the swivel on the upper end of the rod until the swivel pin may enter freely into the hole in the slave cylinder idler lever (fig. 76).

167

TM 9-741
99

MEDIUM ARMORED CAR T17E1

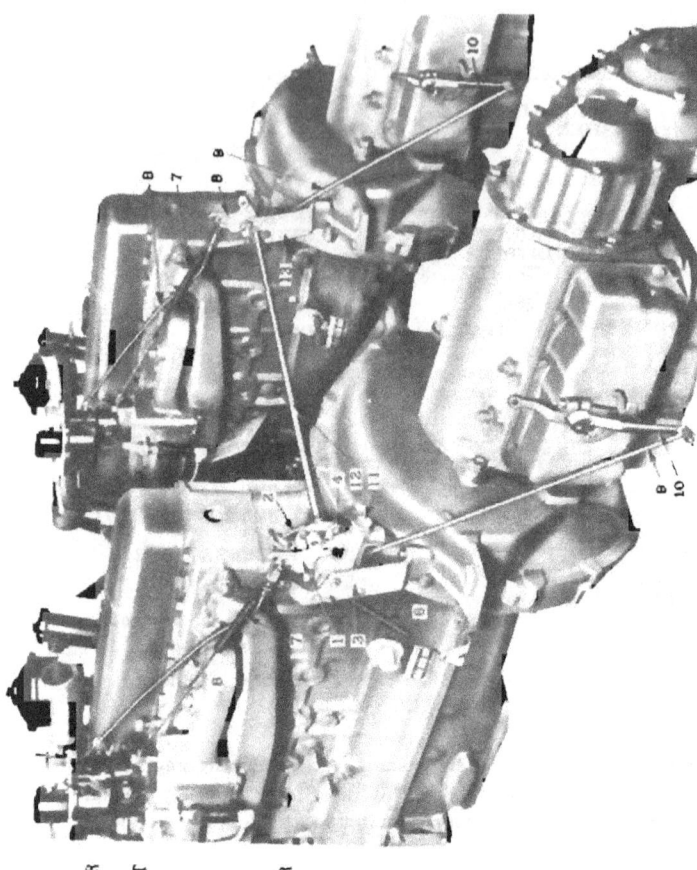

1. SLAVE CYLINDER IDLER LEVER
2. CROSS SHAFT ACCELERATOR LEVER
3. SLAVE CYLINDER SUPPORT BRACKET
4. CROSS SHAFT COUPLING LEVER
5. CROSS SHAFT OPERATING LEVER
6. SLAVE CYLINDER COUPLING LEVER RETURN SPRING
7. THROTTLE ROD RETURN SPRINGS
8. THROTTLE RODS
9. TRANSMISSION THROTTLE CONTROL RODS
10. THROTTLE VALVE LEVERS
11. SLAVE CYLINDER
12. CROSS SHAFT
13. CROSS SHAFT SUPPORT BRACKET

Figure 74 — Hydraulic Throttle Control Mechanical Linkage

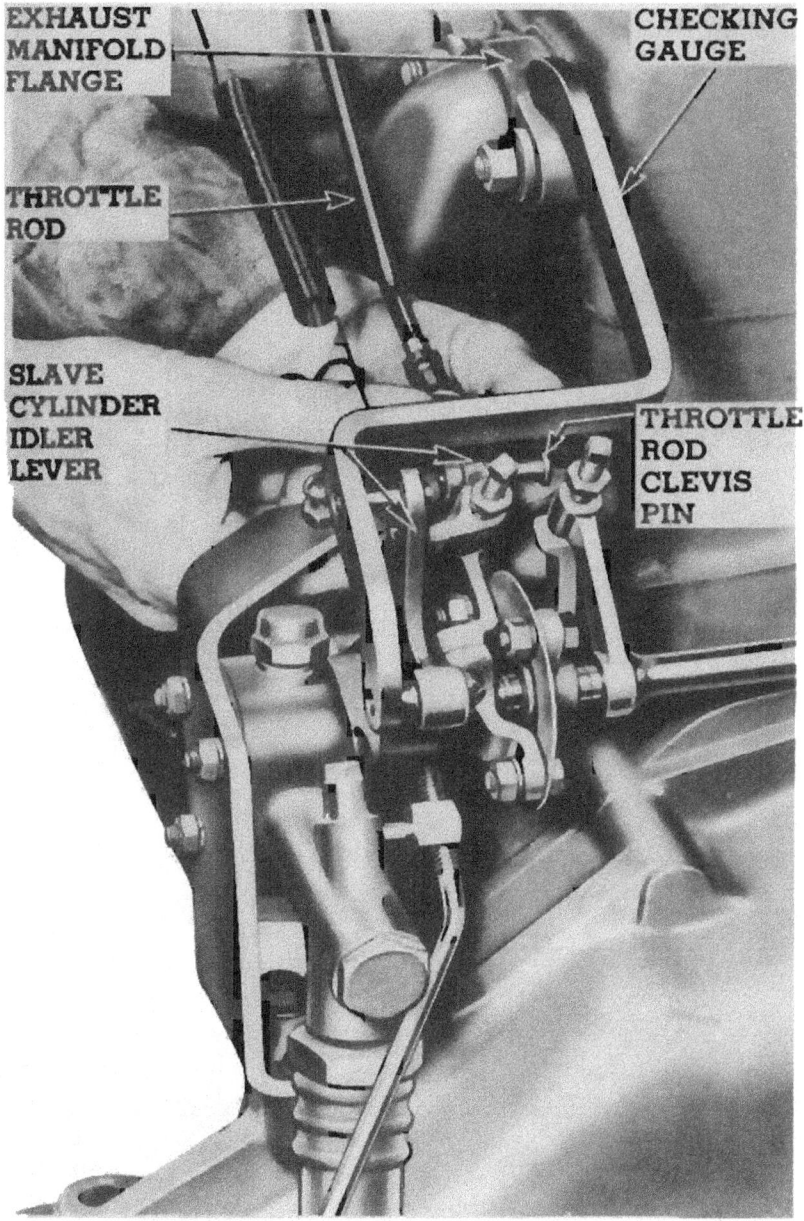

Figure 75 — Adjusting Carburetor Throttle Rod

MEDIUM ARMORED CAR T17E1

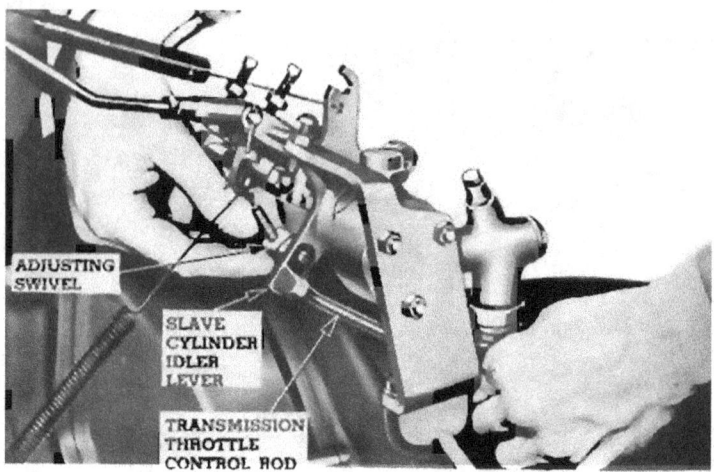

Figure 76 — Adjusting Transmission Throttle Rod

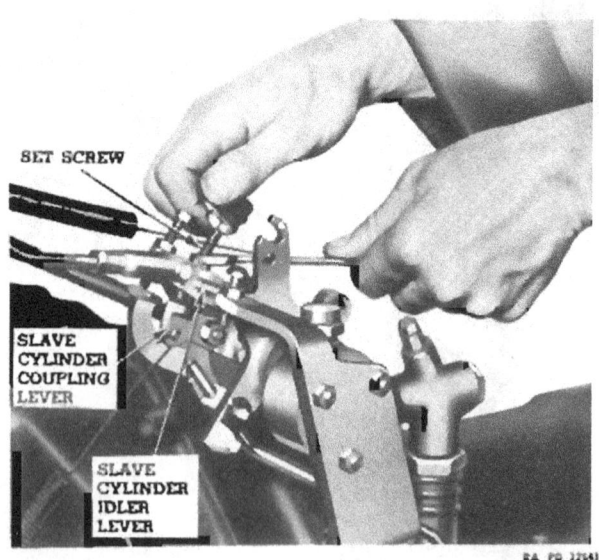

Figure 77 — Adjusting Set Screw on Slave Cylinder Idler Lever

TRANSMISSION

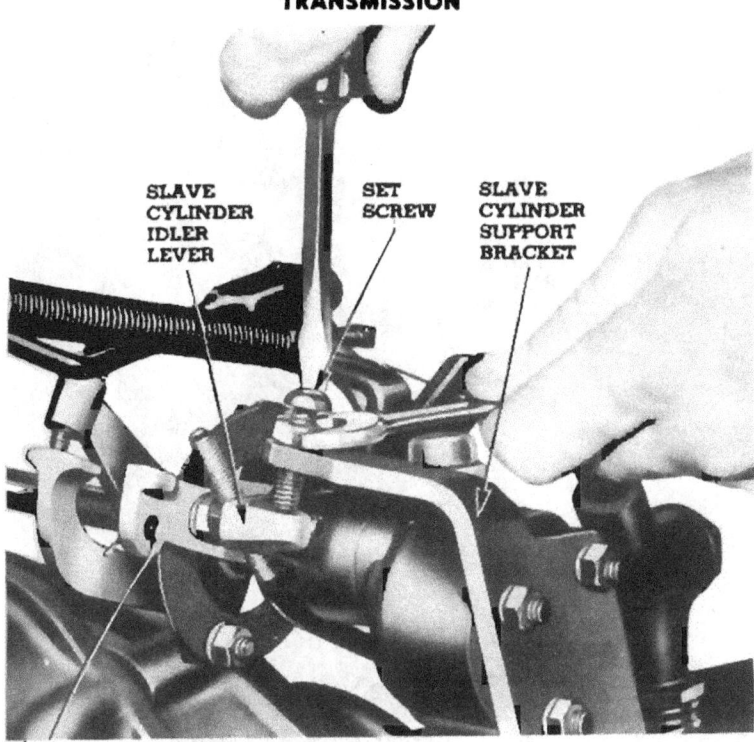

Figure 78 — Adjusting Set Screw on Slave Cylinder Support Bracket

(8) Turn set screw in slave cylinder idler lever until it touches the contact pad in the slave cylinder coupling lever, then back off set screw one full turn and tighten lock nut (fig. 77).

(9) Place the throttle control cross shaft lever checking gage on the cross shaft operating lever at the left-hand engine (5, fig. 74), by engaging the gage pins in the hole and slot provided in the lever.

(10) Adjust the lengths of the throttle rod and transmission throttle control rod for the left-hand engine by repeating the procedure detailed in (6) and (7) above.

(11) Turn set screw on the cross shaft accelerator lever (2, fig. 74) until it touches the contact pad on the cross shaft coupling lever (fig. 74), then back off set screw one full turn and tighten lock nut.

(12) Rotate cross shaft coupling (cross shaft coupling lever and slave cylinder coupling lever) toward open throttle position until the throttle valve lever on the transmission is at the limit of its travel. In this position,

171

MEDIUM ARMORED CAR T17E1

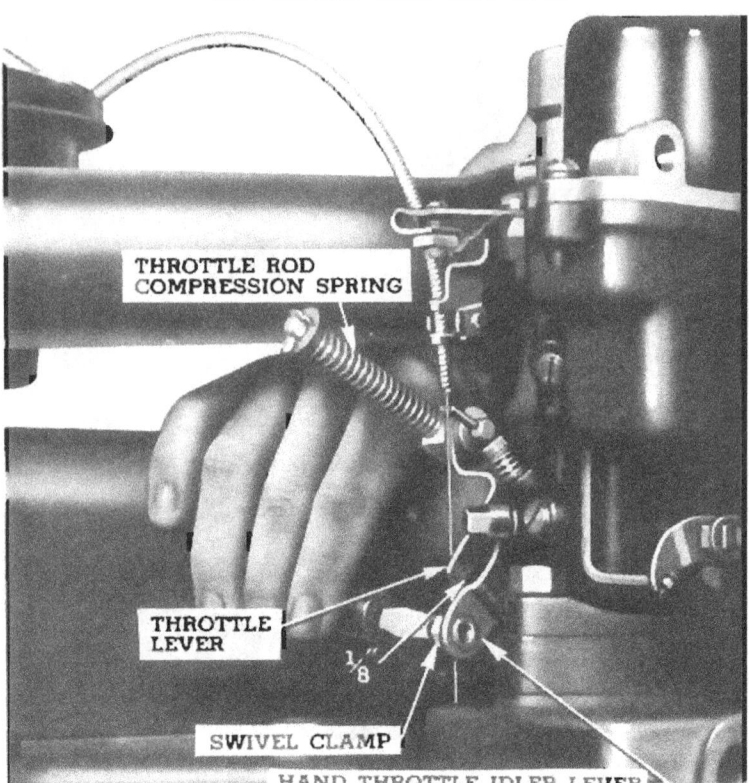

Figure 79 — Hand Throttle Adjustment

turn set screw on the slave cylinder support bracket until it touches the contact pad on the slave cylinder idler lever and tighten lo 'k nut (fig. 78). CAUTION: Do not back off this set screw. Set screw must touch contact pad.

(13) With hand throttle levers in driver's compartment in closed position, adjust swivel clamp on hand throttle control wire at carburetor to provide 1/8-inch clearance between the hand throttle idler lever and throttle lever on carburetor on each engine (fig. 79).

d. **Hydraulic System** (fig. 80).

(1) GENERAL. Whenever air enters the hydraulic system it will be necessary to bleed the system. Air will enter the system from any one of the following causes:

TRANSMISSION

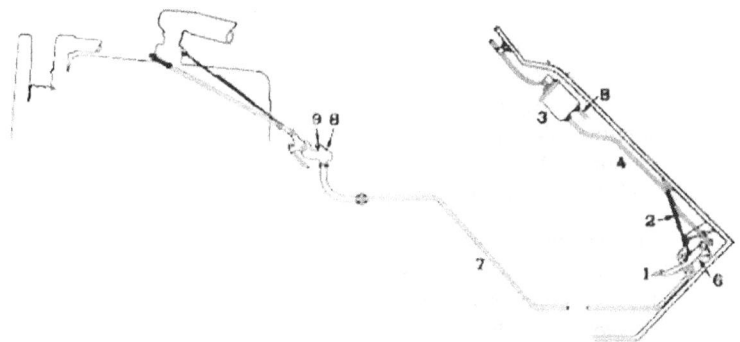

1. ACCELERATOR PEDAL
2. PEDAL PULL BACK SPRING
3. OIL RESERVOIR
4. RESERVOIR LINE TO HYDRAULIC THROTTLE MASTER CYLINDER
5. RESERVOIR LINE TO BRAKE MAIN CYLINDER
6. HYDRAULIC THROTTLE MAIN CYLINDER
7. HYDRAULIC LINE MASTER CYLINDER TO SLAVE CYLINDER
8. HYDRAULIC THROTTLE SLAVE CYLINDER
9. SLAVE CYLINDER BLEEDER VALVE

RA PD 32539

Figure 80 — Hydraulic Throttle Control System

(a) Whenever a line has been disconnected.

(b) Broken line.

(c) Leaky connection.

(d) If oil in reservoir goes below half full.

(e) Weak or broken pull-back springs.

(f) Worn piston in slave cylinder.

(2) BLEED SYSTEM. Air in the system is usually indicated by a spongy pedal, especially at part throttle, and the system can be bled as follows: Remove the filler plug in the reservoir and fill the reservoir with genuine brake fluid. Open bleeder valve on slave cylinder (9, fig. 80), attach one end of bleeding hose to valve and place the other end in a jar with about an inch of clean brake fluid. Pump the accelerator very slowly between each downward thrust of the pedal to allow the fluid to feed, by gravity, past the valve until all the air has been expelled from the system, as indicated by the air bubbles ceasing to rise in the bleeder jar. Close bleeder valve and disconnect the hose. Fill reservoir to a point one inch below the top of the reservoir.

e. **Master Cylinder Replacement.**

(1) REMOVE CYLINDER. Disconnect the pedal pull-back spring. Disconnect the oil pipe that leads to the slave cylinder. Remove the four

mounting bolts that attach the cylinder to bracket and remove the cylinder with pedal. Loosen clamp screw on pedal and pull pedal off the shaft.

(2) INSTALL CYLINDER. Place cylinder in position on bracket and install the four mounting bolts. Tighten bolts securely. Connect the two oil lines and the pedal pull-back spring. Place pedal on the shaft in the proper position and tighten clamp screw. Bleed the system as outlined in paragraph 99 d (2) above.

f. **Slave Cylinder Replacement.**

(1) REMOVE SLAVE CYLINDER. Disconnect the pull-back spring. Disconnect the hydraulic oil hose. Remove the three nuts and lock washers that attach the slave cylinder to the bracket. Loosen the clamp screw on the control lever and remove the lever.

(2) INSTALL SLAVE CYLINDER. Install the control lever in the same position it was in before removal. Tighten clamp screw. Place cylinder in position on bracket and install the three lock washers and nuts. Connect hydraulic hose and pull-back spring. Bleed system as outlined in paragraph 99 d (2) above. Adjust control linkage as instructed in subparagraph c.

100. MANUAL SHIFT CONTROL.

a. **Operation** (figs. 81 and 82). The manual shift levers on the side of the transmissions are controlled by a lever located on the side of the hull, at the left of the driver, which is connected to the shift levers by a mechanical linkage. A flexible cable connects the hand lever to the end of a lever located on the front side of the bulkhead. The lever extends through a hole in the bulkhead and rotates a cross shaft mounted on the rear slide of the bulkhead. Two additional levers on the cross shaft operate the transmission manual shift levers through connecting rods which are adjustable for length. Details of the transmission manual control lever and rod are shown in figure 82. The hole through the bulkhead is closed by a sliding spring-loaded seal, as shown in figure 82.

b. **Adjustment.** It is important that the manual shift linkage be properly adjusted in order that the movement of the hand lever in the driver's compartment may provide the proper movement of the manual control lever on the transmission.

c. **Procedure.**

(1) DISCONNECT CONTROL RODS. Disconnect both transmission manual control rods at the transmission end, and disconnect the flexible cable at the manual control lever. Pull the transmission outer manual shift lever on left transmission assembly forward until the detent engages in the neutral position.

TM 9-741

TRANSMISSION

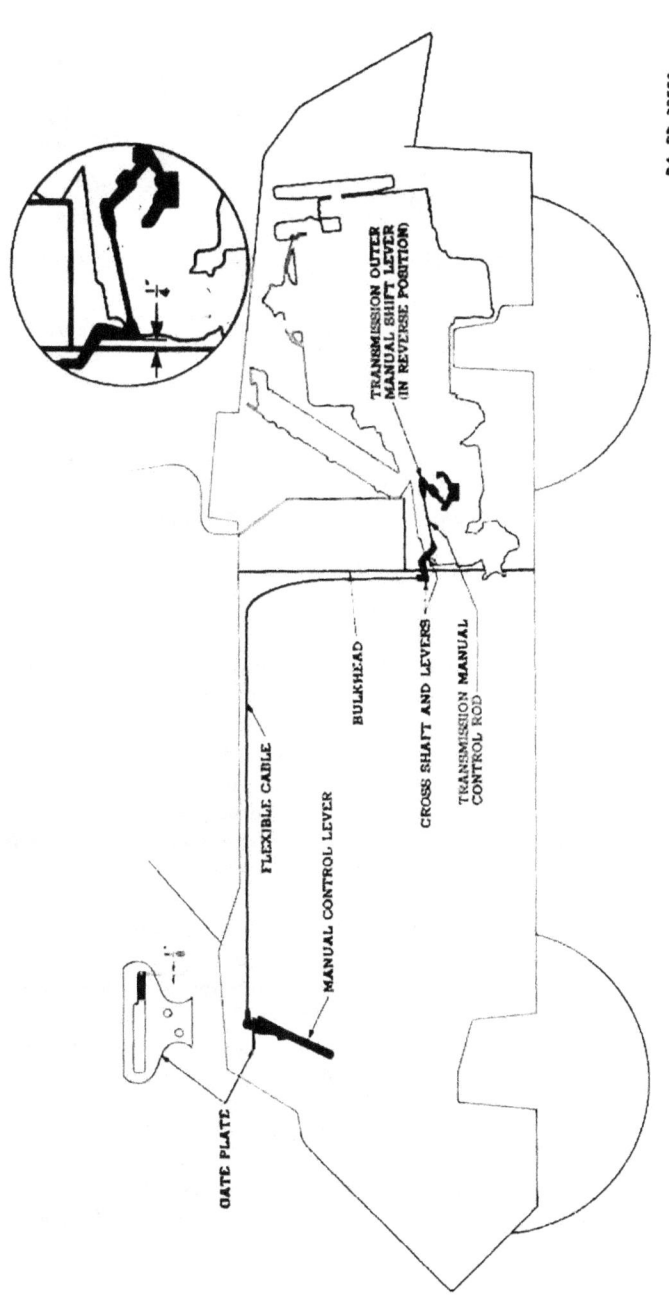

Figure 81 — Transmission Manual Control Linkage and Adjustment

TM 9-741
100

MEDIUM ARMORED CAR T17E1

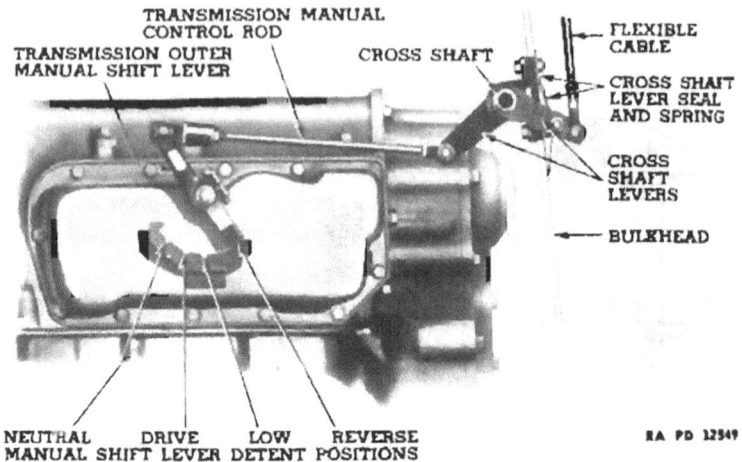

Figure 82 — Transmission Manual Control Levers

(2) ADJUST CONTROL LEVERS. Rotate the cross shaft until the control rod levers are spaced 1/4 inch away from the bulkhead. See inset on figure 81.

(3) ADJUST CLEVIS. Adjust clevis on manual control rod until the clevis pin will engage freely into the hole in the transmission outer manual shift lever. Lock the clevis pin with new cotter pin.

(4) SHIFT OUTER MANUAL SHIFT LEVERS. Shift the outer manual shift levers on both transmissions to the rear limit of their travel, so that both transmissions are fully engaged in reverse gear (fig. 81). CAUTION: It may be necessary to turn one or both engines by the fan blades to properly engage the reverse pawls into the external teeth of the reverse gear.

(5) ADJUST MANUAL CONTROL CLEVIS, RIGHT SIDE. Adjust the manual control rod clevis on the right rod until the clevis pin will engage freely into the hole in the right transmission outer manual shift lever. Lock the clevis pin with new cotter pin.

(6) POSITION TRANSMISSION CONTROL LEVER. Position the transmission control lever in driver's compartment so that there is 1/8-inch clearance between the lever and the rear end of the slot in the gate plate, as shown in figure 81. Without changing the position of the cross shaft and levers, adjust clevis pin until it will engage freely in the hole in the hand lever. If enough adjustment cannot be obtained from this clevis, further adjustment is available at the clevis at bulkhead. Lock clevis pin with new cotter pin.

176

TRANSMISSION

(7) CHECK OPERATION OF MANUAL SHIFT. Check the operation of manual shift linkage by moving hand lever to various shift positions and check the engagement of the detents in the transmission manual control lever. The engagement of the detent can be detected by a hesitation or "catch" in the movement of the hand lever.

101. REMOVAL OF HYDRA-MATIC TRANSMISSION.

a. General. In order to replace the hydra-matic transmission, the complete engine, transmission, and gear reduction must be removed from the vehicle. Instructions covering the replacement of this assembly are given in paragraph 63.

b. Removal of Transmission from Engine. Removal of the transmission from the engine involves partial disassembly of the reduction case and fluid coupling. These operations should not be performed without authority and instructions from ordnance personnel.

MEDIUM ARMORED CAR T17E1

Section XX

TRANSMISSION GEAR REDUCTION ASSEMBLY

	Paragraph
Description	102
Removal of gear reduction assembly	103
Removal of gear reduction assembly from transmission	104

102. DESCRIPTION (figs. 35 and 37).

a. General. A gear reduction assembly is attached to each of the transmission assemblies to transmit the drive from the transmissions to the propeller shafts leading to the transfer case.

b. Purpose. This unit has a gear reduction of 1.52:1 and the method of attachment affords a means of bringing the propeller shafts closer together, avoiding excess angularity at the shaft universal joints. This is accomplished by mounting the gear reduction assembly at an angle to the transmission, so the drive shafts come out of the gear reduction case at a point below and toward the center of the vehicle from the engine center lines.

c. Operation. The drive to the gear reduction assembly is accomplished by means of a splined transmission driven shaft, extending into the splined hub of the gear reduction drive gear. The power is transmitted by helical cut teeth from the drive gear to an idler gear and then to a driven gear, which in turn is connected to the transfer case propeller shaft flange.

d. Lubrication. Lubrication for the unit is provided from the transmission assembly through the drive gear ball bearing. Return of the oil is accomplished by an oil pump at the bottom of the reduction case assembly which returns the oil to the transmission. A drain plug is provided at the bottom of each assembly.

103. REMOVAL OF GEAR REDUCTION ASSEMBLY.

a. Reference. In order to replace the gear reduction assembly, the complete engine, transmission and gear reduction assembly must be removed from the vehicle as a unit. Instructions covering the replacement of this unit are given in paragraph 63.

104. REMOVAL OF GEAR REDUCTION ASSEMBLY FROM TRANSMISSION.

a. Reference. Removal of the gear reduction assembly from the transmission involves partial disassembly of the gear reduction assembly. This operation should not be performed without authority and instructions from the ordnance personnel.

TM 9-741
105-106

Section XXI

PROPELLER SHAFTS AND UNIVERSAL JOINTS

	Paragraph
Description—propeller shafts	105
Description—universal joints	106
Trouble shooting	107
Propeller shaft assembly removal (gear case to transfer case)	108
Propeller shaft assembly service (gear case to transfer case)	109
Propeller shaft assembly installation (gear case to transfer case)	110
Propeller shaft assembly removal (transfer case to axle)	111
Propeller shaft assembly service (transfer case to axle)	112
Propeller shaft assembly installation (transfer case to axle)	113

105. DESCRIPTION — PROPELLER SHAFTS.

a. Four propeller shafts are used on the Medium Armored Car T17E1 to provide drive lines between the different units. Each of these shafts is fitted with two universal joints.

b. Two solid type shafts which are alike and interchangeable are used between the two transmission gear reduction units and the transfer case.

c. Two tubular shafts are used between the transfer case and axles. These two shafts are constructed the same except for length. The front shaft is longer due to the transfer case location in relation to axle positions.

d. Each of the four shafts is fitted with a spline at one end and a joint yoke at the other (fig. 83). The two universal joints for each shaft are assembled to these two ends. The four shafts are fitted with a slip joint and a permanent joint to make up the drive shaft assemblies.

106. DESCRIPTION — UNIVERSAL JOINTS.

a. In addition to the slip and permanent type joints, there are two different methods of construction used. The joints used between the transmissions and transfer case use U-bolts to attach the trunnions to the yoke (fig. 84). The one between the transfer case and axle uses a flange yoke (fig. 83).

b. One slip type joint is used on each shaft to allow for slight changing in the distance between units due to road and load conditions. The slip joints on the two transmissions to transfer case shafts go at the transmission end while the slip joints on the transfer case to axle propeller shafts should be at the transfer case ends.

179

TM 9-741
106

MEDIUM ARMORED CAR T17E1

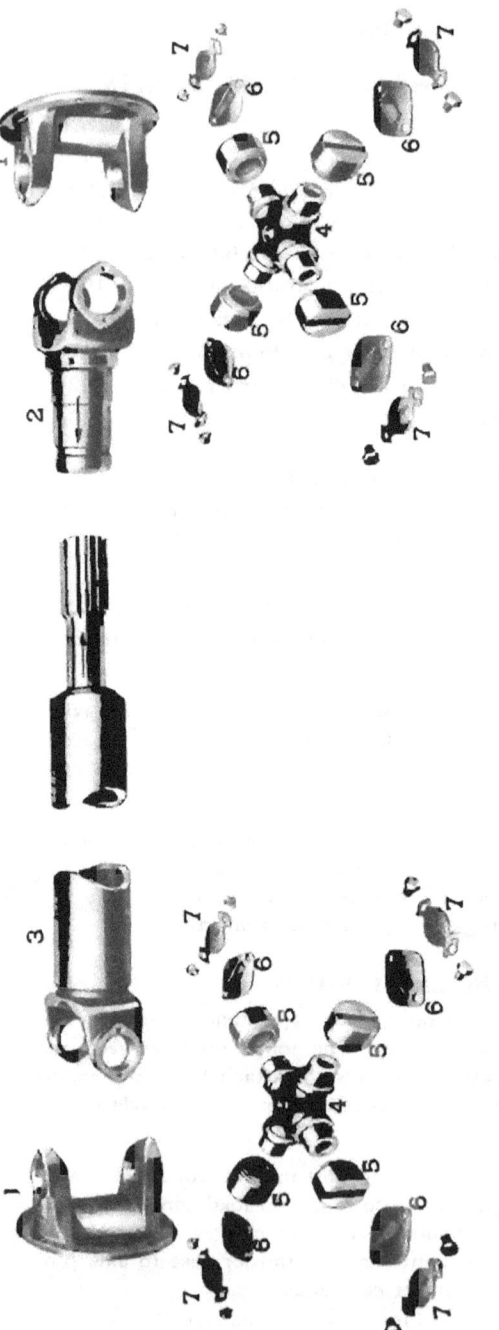

1. CIRCULAR FLANGE YOKES
2. SLEEVE YOKE
3. WELDED YOKE AND PROPELLER SHAFT
4. YOKE TRUNNIONS
5. TRUNNION BEARINGS WITH NEEDLES
6. BEARING CAPS
7. LOCK STRAPS AND BOLTS

RA PD 32490

Figure 83 — Universal Joint (Flange Type)

PROPELLER SHAFTS AND UNIVERSAL JOINTS

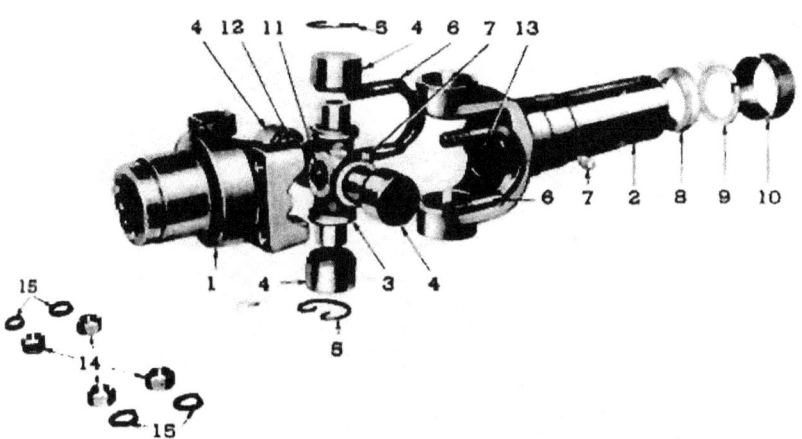

1. UNIVERSAL JOINT YOKE
2. UNIVERSAL JOINT SLEEVE YOKE
3. YOKE TRUNNION
4. TRUNNION BEARING
5. TRUNNION BEARING LOCK RING
6. "U" BOLT
7. LUBRICATION FITTING
8. SLEEVE YOKE GASKET
9. SLEEVE YOKE WASHER
10. SLEEVE YOKE CAP
11. RELIEF VALVE
12. TRUNNION BEARING ROLLERS
13. AIR VENT
14. "U" BOLT RETAINING NUT
15. LOCKWASHER

RA PD 32491

Figure 84 — Universal Joint ("U" Bolt Type)

c. Both types of joints use the needle bearings and trunnion yoke principle. The trunnion yokes are drilled to lubricate their four bearings from one central fitting. In order to lubricate the joints, a special adapter which is furnished in the tool kit must be attached to the lubrication gun. A special relief valve is mounted on the trunnion yoke, opposite the lubrication fitting to prevent overlubrication and damage to the trunnion bearing seals. A lubrication fitting is installed in the sleeve yoke to lubricate the splines. A plug is pressed into the joint end of the sleeve yoke to retain the lubricant and keep dirt from entering the splines. A small hole is drilled in the plug to relieve trapped air. A cork seal and screw cap is used at the propeller shaft end of the sleeve yoke to retain the lubricant and prevent dirt from entering at this point (fig. 83).

107. TROUBLE SHOOTING.

a. Vibration.

Probable Cause	Probable Remedy
Bent propeller shaft.	Replace propeller shaft (par. 108).
Universal joint improperly assembled to shaft.	It is important that the slip joint be installed on the spline of the shaft so that the arrows line up (fig. 88).

MEDIUM ARMORED CAR T17E1

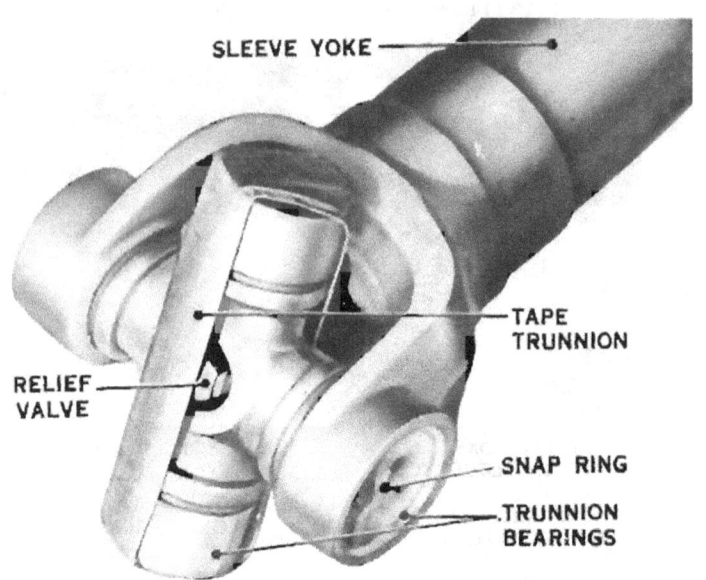

Figure 85 — Trunnion Bearings Held in Place with Tape

Probable Cause	Probable Remedy
Universal joints badly worn.	Replace or overhaul universal joints (par. 109).
b. Excessive Backlash.	
Worn universal joints.	Replace or overhaul universal joints (par. 109).
Universal joint U-clamps or flange bolts loose.	Tighten clamp bolt nuts or flange bolt nuts. Replace worn parts (par. 109).

108. PROPELLER SHAFT ASSEMBLY REMOVAL (GEAR CASE TO TRANSFER CASE).

a. The two propeller shafts between the transmission gear reduction cases and the transfer case can be reached by removing a panel from the floor of the turret basket.

b. Disconnect the two universal joints from the yokes by removing the nuts and washers from the U-bolts which attach the trunnion bearings to the yokes. Remove the U-bolts (fig. 84).

TM 9-741
108-109

PROPELLER SHAFTS AND UNIVERSAL JOINTS

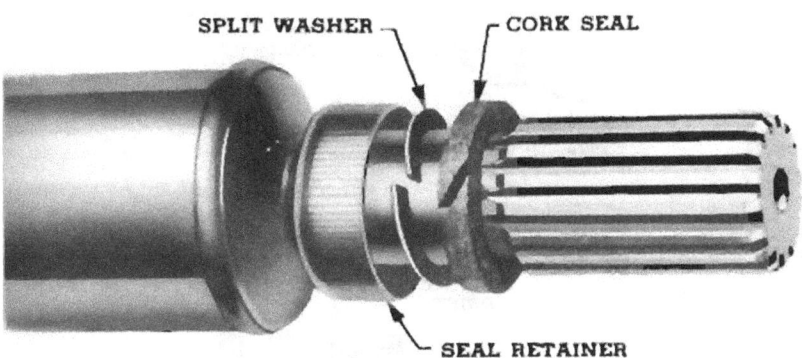

Figure 86 — Universal Slip Joint Oil Seal

c. Hold the trunnion bearings in place, slide the slip joint onto the spline, and put a piece of tape over the trunnion bearings (fig. 85). This will keep dirt out of the bearings and hold them on the trunnion.

d. Tape the trunnion on the front joint as explained above. Remove the propeller shaft and joints.

109. PROPELLER SHAFT ASSEMBLY SERVICE (GEAR CASE TO TRANSFER CASE).

a. Disassembly.

(1) The slip joint may be disassembled from the propeller shaft by unscrewing the seal retainer with a chain grip wrench and pulling the universal joint assembly off the splines on the propeller shaft.

(2) When necessary to replace a cork seal in the grease retainer, push it out of the retainer, and as it is split (fig. 86), it may be removed from the propeller shaft.

(3) Remove the two loose trunnion bearings which were taped in place when splitting the universal.

(4) Remove the lubrication fitting.

(5) Remove the two trunnion bearing lock rings from the ends of the yoke.

(6) Clamp the yoke in a bench vise and, using a brass drift and hammer, drive on one of the trunnion bearings until the trunnion just strikes the lower yoke (fig. 87). This will drive the other trunnion bearing almost out. It can be carefully removed with a pair of pliers.

183

TM 9-741
109

MEDIUM ARMORED CAR T17E1

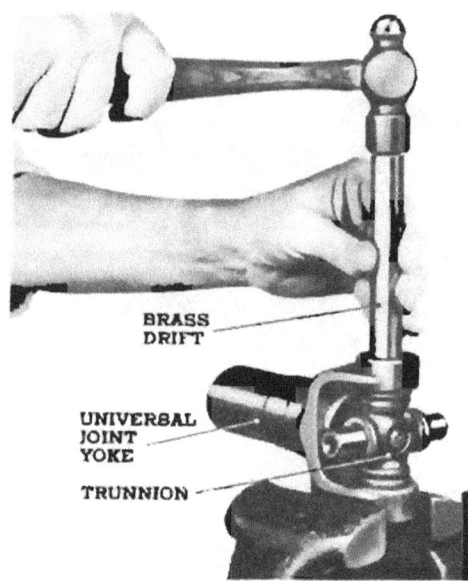

Figure 87 — Trunnion Removal

(7) Turn the yoke over, clamp the other side in the vise, and drive on the end of the trunnion pin until the other bearing is removed.

(8) Slide the trunnion into one side of the yoke, tip the other side away from the yoke, and lift it out of place.

b. Cleaning and Inspection.

(1) Wash all parts in SOLVENT, dry-cleaning.

(2) Inspect the needle bearings, bearing cages, trunnion cork seals, and splines for wear.

(3) Replace all damaged parts.

c. Reassembly.

(1) Assemble the 27 needle bearings to each bearing cage, using general purpose grease to hold them in place.

(2) Install the trunnion in the yoke.

(3) Install new cork seals on the trunnion.

(4) Start one trunnion bearing into the yoke and start the trunnion into the bearing to hold the needle bearings in place.

(5) Using the bench vise as a press, force the bearing in flush with the yoke.

TM 9-741
109-110

PROPELLER SHAFTS AND UNIVERSAL JOINTS

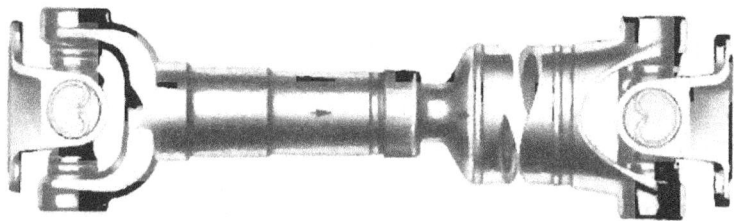

RA PD 32493

Figure 88 — Universal Joint Yoke Alining Marks

(6) Start the other trunnion bearing into the yoke and enter the trunnion pin into it to hold the needles in place and press the bearing in flush with the yoke.

(7) Drive or press one trunnion bearing in far enough to install the lock ring.

(8) Drive or press the other bearing in and install the lock ring.

(9) Install the lubrication fitting.

(10) Hold the joint in one hand and strike the yoke a light blow on each side to seat the trunnion bearings against the lock rings.

(11) Check to see that the bearings are seated against the lock rings as this provides normal clearance and prevents overheating.

(12) Install the slip joint onto the propeller shaft, making sure that the yoke of the slip joint is in the same plane with the yoke of the joint welded in the opposite end of the shaft. Arrows are used to indicate the correct position (fig. 88). Install and tighten the oil seal retainer with the chain grip wrench.

(13) Install the four loose trunnion bearings to the trunnions and tape them to hold them in place.

110. PROPELLER SHAFT ASSEMBLY INSTALLATION (GEAR CASE TO TRANSFER CASE).

a. Place the assembly in the hull through the opening in bottom of turret basket.

b. Remove the tape from the front joint trunnion bearings, place the bearings in their seats on the flange, and install the U-clamps, washers, and nuts.

c. Remove the tape from the slip joint trunnions, place the joint in position against the yoke, and install the U-clamps, washers, and nuts. Tighten the nuts securely.

MEDIUM ARMORED CAR T17E1

111. PROPELLER SHAFT ASSEMBLY REMOVAL (TRANSFER CASE TO AXLE).

a. The propeller shafts between the transfer case and axles can be removed from under the vehicle.

b. Remove the four bolts which attach the flange yoke to flange at each universal and remove the propeller shaft and joints as an assembly.

112. PROPELLER SHAFT ASSEMBLY SERVICE (TRANSFER CASE TO AXLE).

a. Disassembly.

(1) Remove the slip joint by unscrewing the seal retainer with a chain grip wrench and pulling the universal joint assembly off the spline of the propeller shaft.

(2) When necessary to replace a cork seal in the grease retainer, push it out of the retainer, and as it is split (fig. 86), it may be removed from the propeller shaft.

(3) Bend down the locking lugs on the lock plate with a screwdriver (7, fig. 83).

(4) Remove the two bolts from each bearing cap.

(5) Support the yoke; then push on the exposed face of one bearing until the opposite bearing assembly comes out. (The bearings are a slip fit in the yoke.)

(6) Push on the exposed end of the trunnion until the other bearing is forced out of the yoke.

(7) Repeat the above operation to remove the other trunnion bearings, slide the trunnion to one side, and tip the free end away from the yoke to remove it.

(8) Wash all parts in SOLVENT, dry-cleaning, and inspect them for wear or damage.

b. Reassembly.

(1) Assemble the needle bearings to the cups, using a little general purpose grease to hold them in place.

(2) Install new seals on the trunnion, if necessary.

(3) Lubricate all parts with LUBRICANT, gear, universal, SAE 90; install the trunnion in one yoke.

(4) Place the bearings in the yoke, being sure that the needle bearings stay in place until the trunnion is in position.

(5) Turn the bearings until the slot in the bearing cup lines up with the key in the cups. Install the caps, lock plates, and bolts. Lock the bolts by bending the tang of the lock plate up against the bolts.

PROPELLER SHAFTS AND UNIVERSAL JOINTS

(6) Install the other yoke over the trunnion and install the bearings as explained above.

(7) Assemble the other joint in the same manner.

113. PROPELLER SHAFT ASSEMBLY INSTALLATION (TRANSFER CASE TO AXLE).

a. Place the propeller shaft in position under the vehicle.

b. Install the four bolts, washers, and nuts at each flange. Tighten the nuts securely.

TM 9-741
114-115

MEDIUM ARMORED CAR T17E1

Section XXII

TRANSFER CASE

	Paragraph
Description	114
Trouble shooting	115
Transfer case removal	116
Transfer case installation	117

114. DESCRIPTION (figs. 89 and 90).

   a. General. The transfer case and brake assembly used is an auxiliary unit located ahead of the bulkhead.

   b. Purpose. The transfer case is essentially a two-speed transmission unit which provides the means of connecting the two power lines with both the front and rear axles.

   c. Shifting. A shifter mechanism engages and disengages the drive to the front axle. An additional shift mechanism selects the high and low speeds. Each of these controls is operated by a lever in the driver's compartment within reach of the operator. The shift lever for high and low ratios also incorporates a selector lever so that the sliding gears in the transfer case can be shifted selectively. With the selector lever in the center position, both sliding gears are shifted. With the selector lever moved to one side or the other, only the sliding gear selected will be shifted. This permits operation of one engine only, if desired.

   d. Drive Connections. Two drive shafts enter the transfer case, one on each side from the two transmission reduction assemblies (fig. 89).

   e. Parking Brake (fig. 90). At the front of each of the transfer case drive shafts is a brake drum and brake band. This comprises the parking brake mechanism described in paragraph 132.

115. TROUBLE SHOOTING.

   a. Slips Out of Gear.

Probable Cause	Probable Remedy
Shift lock spring weak or broken.	Refer to ordnance personnel.
Shift lock ball sticking.	Refer to ordnance personnel.
Misalinement of front axle drive shaft.	Refer to ordnance personnel.

   b. Leaks Lubricant.

Damaged bearing cap gaskets.	Refer to ordnance personnel.
Bearing caps loose.	Tighten caps.
Damaged drive flange seals.	Refer to ordnance personnel.

TM 9-741
115

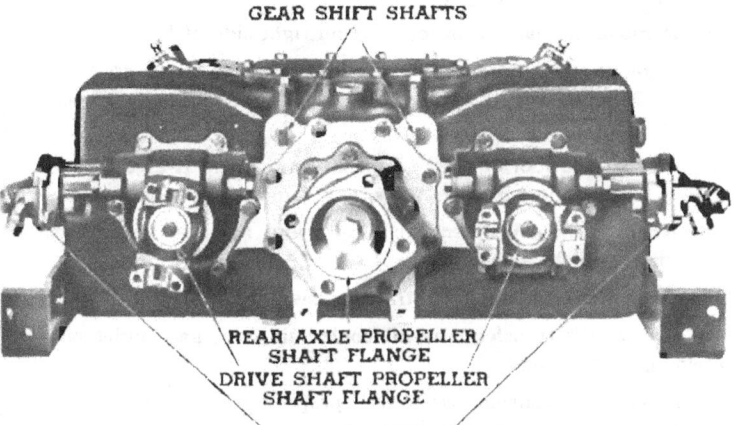

Figure 89 — Transfer Case — Engine End

Figure 90 — Transfer Case — Brake End

Probable Cause	Probable Remedy
Leak at drive flange retaining nuts.	Remove cotter pin and tighten nut. (Make sure serrated washer fits into splines.)

c. Noisy.

Probable Cause	Probable Remedy
Insufficient lubricant.	Fill to correct level.
Damaged gears or bearings.	Refer to ordnance personnel.
Damaged speedometer gears.	Refer to ordnance personnel.

MEDIUM ARMORED CAR T17E1

116. TRANSFER CASE REMOVAL.

a. Remove dunnage from tool box on right side of hull.

b. Remove the three tool-box-to-hull bracket bolts at hull front bracket. Remove the three tool-box-to-hull bracket bolts at box rear bracket.

c. Support the tool box with a jack, remove the twelve tool box to jettison tank support bracket bolts, and remove the tool box.

d. Remove the seventeen cap screws and remove the hull transfer case removal plate and gasket.

e. Drain the lubricant from transfer case.

f. Disconnect transfer case oil cooler air inlet and outlet pipes by loosening hose clamps.

g. Disconnect transfer case to axle propeller shaft joints at the transfer case ends and remove flanges.

h. Remove the two turret basket floor plates. Turn the turret until the gear reduction case to transfer case universal joints can be reached through the opening.

i. Disconnect the joints at transfer case by removing the U-bolts and remove flanges. NOTE: Tape the trunnion bearings to hold them in place and to prevent dirt from getting into the bearings.

j. Remove propeller shaft brake shroud top, front, and rear plates from left side of transfer case.

k. Remove the clevis pin from each end of the front axle shift pull rod and remove the rod.

l. Loosen the two transfer case cross shaft clamp bolts, remove the two cross shaft support-to-hull bolts, and slide the support end levers off end of shaft.

m. Remove the lock wire from the left brake band top adjusting screw and remove the screw.

n. Remove the transfer case oil pump inlet pipe and elbow at left side of transfer case.

o. Remove parking brake pull back spring and unhook the ball end of parking brake cable from parking brake equalizer.

p. Disconnect speedometer cable at transfer case.

q. Remove the four transfer case bracket-to-hull attaching bolts and remove the shims between front end of bracket and hull cross member on left side.

r. Remove propeller shaft brake shroud top, front, rear end side plates from right side.

TRANSFER CASE

s. Remove the transfer case oil pump inlet pipe and elbow at right side.

t. Remove the four transfer case bracket-to-hull attaching bolts and remove the shims between front end of bracket and hull cross member on right side.

u. Loosen the transfer case propeller shaft seal retaining nut at front and rear of transfer case with spanner wrench.

v. Remove the four clutch head screws which attach the turret slip ring to basket floor and block the slip ring up to provide clearance for the transfer case.

w. Slide the transfer case to the right until it reaches the hull side flange. Raise the transfer case slightly to clear the hull flange and slide to the right until the transfer case bracket rests on the hull.

x. Place a roller jack under the end of transfer case and raise until the case clears the hull. Slide out until the case bracket rests firmly on the jack. CAUTION: Support the transfer case so that the oil lines along the bottom will not be damaged.

y. Work the transfer case and jack to the right and remove the propeller shaft oil seals.

z. When the case is about two-thirds of the way out of the hull opening, block the outer end to support it and place the jack securely under the center so that the case will balance on the jack.

117. TRANSFER CASE INSTALLATION.

a. Place the transfer case on a roller jack and start the left end of case into the opening on right side of hull with brakes forward (fig. 91).

b. Install the two propeller shaft flange oil seals. Support the case and place the jack under the transfer case support at right end of case.

c. Work the case into hull and into position.

d. Install the eight transfer case bracket-to-hull attaching bolts and replace the shims between front end of brackets and hull cross member.

e. Replace the four clutch head screws which attach the turret slip ring to basket floor.

f. Connect the transfer case propeller shaft seal retaining nut at front and rear of transfer case. NOTE: Use spanner wrench.

g. Replace transfer case oil pump inlet pipe and elbow at right side. NOTE: Permatex threads of elbow.

h. Replace propeller shaft brake shroud top, front, rear and side plates on right side.

TM 9-741
MEDIUM ARMORED CAR T17E1

Figure 91 — Transfer Case Replacement

i. Rotate basket and connect speedometer cable at transfer case.

j. Replace parking brake pull-back spring and connect the ball end of parking brake cable on parking brake equalizer.

k. Install the transfer case inlet elbow and oil pump pipe at left side of case. NOTE: Permatex threads of elbow.

l. Replace the left brake band adjusting screw and adjust bands to 0.020-inch clearance; then install screw lock wire.

m. Slide the cross shaft support and levers on cross shaft. Install two cross shaft support hull bolts and tighten lever clamp bolts.

n. Replace front axle shift pull rod and install clevis pins and cotter keys in each end.

o. Replace propeller shaft brake shroud top, front, and rear plates on left side of transfer case.

p. Install drive flanges and connect the U-joints at transfer case by tightening U-bolts.

q. Install drive flanges and connect transfer case to axle propeller joints at transfer case ends.

TRANSFER CASE

r. Connect transfer case oil cooler air inlet and outlet hoses by tightening hose clamps.

s. Fill the transfer case to correct level with lubricant. (See lubrication instructions.)

t. Install the turret basket floor plates.

u. Install the plate at right side of hull using a new gasket. Tighten up the seventeen cap screws.

v. Raise the tool box into position and bolt it to the jettison tank brackets and the hull brackets.

w. Install the dunnage in tool box.

x. Lubricate all U-joints. (See lubrication instructions.)

TM 9-741
118-119

MEDIUM ARMORED CAR T17E1

Section XXIII

FRONT AXLE

	Paragraph
Description	118
Trouble shooting	119
Front axle drive flange gasket replacement	120
Front axle shaft replacement	121
Tie rod and tie rod yoke replacement	122
Adjust toe-in	123

118. DESCRIPTION.

a. The front axle assembly on this vehicle is a front wheel driving unit, consisting of a banjo type axle housing and detachable axle housing outer ends which are used for mounting the steering knuckle assemblies. The front wheels are driven by the hypoid drive unit through the axle shafts which are equipped with universal joints enclosed in the steering knuckle assemblies.

119. TROUBLE SHOOTING.

a. Lubricant Leaks.

Probable Cause	Probable Remedy
Leak at housing outer end seal or end seal gasket.	Refer to ordnance personnel.
Leak at differential carrier to axle housing.	Refer to ordnance personnel.
Leak at hypoid drive gear thrust bearing retainer.	Refer to ordnance personnel.
Leak at hypoid pinion thrust bearing retainer.	Refer to ordnance personnel.
Leak at housing filler plug or drain plug.	Tighten plug or replace gasket.
Leak at axle drive flange plate.	Replace gasket (par. 120).
Leak at axle drive flange.	Tighten axle drive flange bolts or replace terneplate drive flange gasket (par. 120).

b. Hard Steering.

Lack of lubrication.	Lubricate tie rod yoke bolts, steering connecting rod, and steering gear.

TM 9-741
119

FRONT AXLE

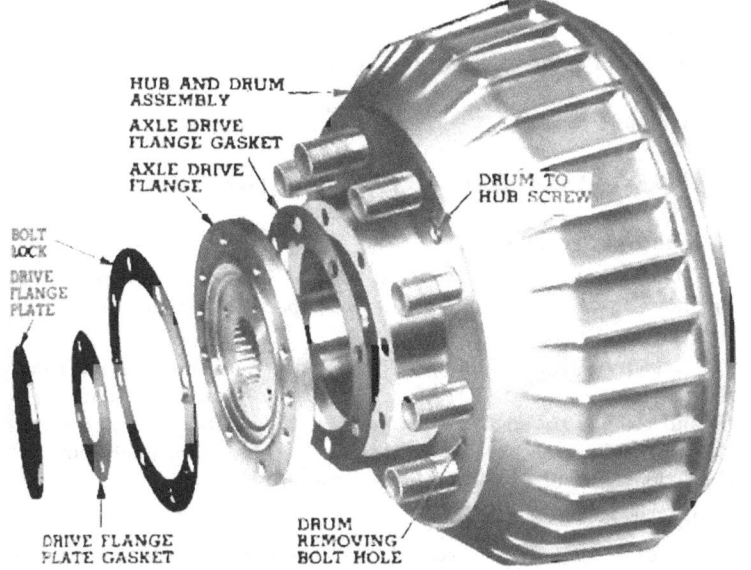

Figure 92 — Drive Flange and Hub and Drum Assembly

Probable Cause	Probable Remedy
Tires improperly inflated.	Inflate front tires to 70 pounds and rear tires to 80 pounds.
Tight steering knuckle support trunnion bearings.	Refer to ordnance personnel.
Improper toe-in.	Adjust toe-in (par. 123).
Faulty steering mechanism.	Refer to ordnance personnel.

c. Front Wheel Shimmy.

Improper tire inflation.	Inflate front tires to 70 pounds.
Loose wheels on hubs.	Tighten wheel hub nuts.
Improper toe-in.	Adjust toe-in (par. 123).
Loose front wheel bearings.	Refer to ordnance personnel.
Steering knuckle support trunnion bearings loose.	Refer to ordnance personnel.

d. Wandering.

Axle shifted due to loose or worn radius rod ball studs.	Refer to ordnance personnel.

195

MEDIUM ARMORED CAR T17E1

Probable Cause	Probable Remedy
Loose front wheel bearings.	Refer to ordnance personnel.
Improper toe-in caused by bent tie rod.	Refer to ordnance personnel.
Front axle noise.	See paragraph 125.

120. FRONT AXLE DRIVE FLANGE GASKET REPLACEMENT (fig. 92).

a. Removal Procedure.

(1) DRAIN OIL FROM BANJO HOUSING. Remove drain plug from banjo housing and drain oil into a drain pan. Install drain plug and gasket when oil is drained.

(2) REMOVE DRIVE FLANGE. Remove the four bolts, drive flange plate, and gasket. Bend the axle shaft bolt lock lugs away from the bolts and remove the eight bolts and bolt lock. Install two of the bolts in the tapped holes of the flange, turning the bolts alternately until the flange is loose. Remove the flange and gasket.

b. Installation Procedure.

(1) INSTALL DRIVE FLANGE. Place new flange gasket over flange and push flange in place, lining up the bolt holes in the hub, gasket, and flange. Place a new bolt lock in position and install the eight bolts and tighten them securely. Bend tangs on lock against bolt heads. Install flange plate and gasket, and tighten the four bolts.

(2) REFILL WITH LUBRICANT. Remove filler plug in housing cover and fill with LUBRICANT, gear, universal, seasonal grade, to proper level and install filler plug and gasket.

121. FRONT AXLE SHAFT REPLACEMENT.

a. Removal Procedure.

(1) DRAIN OIL FROM BANJO HOUSING. Remove the drain plug from banjo housing and drain oil into a drain pan. Install drain plug and gasket when oil is drained.

(2) RAISE FRONT OF VEHICLE. Jack up front end of vehicle and support on suitable blocks. Remove the ten wheel nuts and remove the wheel.

(3) REMOVE DRIVE FLANGE. Remove the four bolts that attach the drive flange plate and gasket. Bend the axle shaft bolt lock tangs away from the bolts and remove the eight bolts and bolt lock. Install two of the bolts in the tapped holes of the flange, turning the bolts alternately until the flange is loose. Remove the flange and gasket.

(4) REMOVE HUB AND DRUM ASSEMBLY (fig. 93). Bend tang away

TM 9-741
121

FRONT AXLE

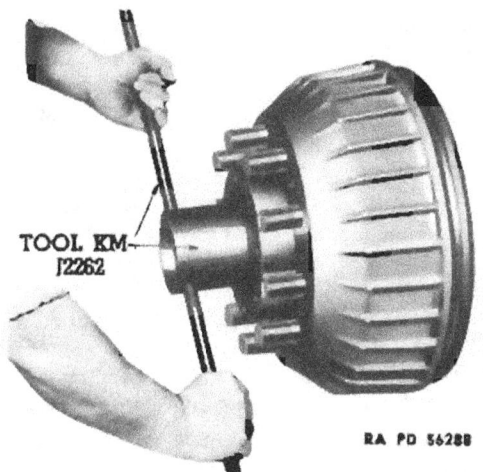

Figure 93 — Adjusting Wheel Bearings

from wheel bearing lock nut and remove lock nut, lock, adjusting nut and washer, using wheel bearing adjusting nut wrench KM-J2262. Lift the drum assembly off the axle housing.

(5) DISCONNECT BRAKE HOSE. Loosen hose clip on upper trunnion. Loosen bolt at hose to housing connector. Disconnect the brake hose at the T-connector at top of brake flange plate.

(6) REMOVE AXLE SHAFT. Remove the ten nuts that attach the oil deflector, flange plate assembly, and spindle to the knuckle support. Remove the oil deflector, flange plate assembly and spindle, and pull out axle shaft.

b. Installation Procedure.

(1) INSTALL AXLE SHAFT. Hand pack universal joint with LUBRICANT, gear, universal, seasonal grade, and rub a coating of lubricant on the outer axle end, bushing contact surfaces, and thrust washer surfaces. Thread axle shaft into housing with extreme care until it touches the differential case; then "break" the outer end of the universal joint down as far as it will go to be used as a lever to raise the inner end of the shaft until it lines up and meshes in the splines in the differential case. When splines are lined up, push shaft in until the joint contacts the thrust washer. Straighten outer end of shaft to line it up with inner end.

(2) INSTALL SPINDLE. Place new gasket in position on steering knuckle support; then thread spindle over axle and place in position

MEDIUM ARMORED CAR T17E1

with outer keyway up. Place brake flange plate assembly and oil deflector in position and install the bolts and lock washers. Tighten the bolts securely.

(3) INSTALL HUB AND DRUM ASSEMBLY. Install wheel hub and drum assembly on spindle and install outer bearing cone and roller assembly, rotating hub to aline bearings. Install washer and inner adjusting nut.

(4) ADJUST WHEEL BEARING (fig. 93). Using wheel bearing wrench KM-J2262, tighten inner adjusting nut "wrench tight," then back it off 1/8 turn (45 deg. minimum). Install new adjusting nut lock and check the alinement of one of the notches with one of the short lugs on the lock. If lug and notch do not line up, back off nut to line up with nearest lug; then bend lug into notch of nut. Rotate hub to see that bearings are seated properly; then install lock nut, tightening lock nut securely. Bend tang into notch of lock nut. Check assembly for free rotation.

(5) INSTALL DRIVE FLANGE. Place new flange gasket over flange and push flange in place, lining up the bolt holes in the hub, gasket, and flange. Place a new bolt lock in position and install the eight bolts and tighten them securely. Bend tangs on lock against bolt heads. Install drive flange plate and gasket. Tighten the four bolts securely.

(6) CONNECT BRAKE HOSE. Connect brake hose at T-connection at top of brake flange plate, tighten to housing flange connector, and clip at upper trunnion.

(7) INSTALL WHEELS. Place wheels on hub and install the ten wheel nuts. Tighten nuts securely.

(8) LOWER FRONT OF VEHICLE. Raise front of vehicle and remove the blocking; then lower vehicle and remove the jack.

(9) BLEED BRAKES. Bleed brakes as instructed in paragraph 133.

(10) REFILL WITH LUBRICANT. Remove filler plug in housing cover and fill with LUBRICANT, gear, universal, seasonal grade, to proper level and install filler plug and gasket.

122. TIE ROD AND TIE ROD YOKE REPLACEMENT.

a. Removal Procedure. Remove the cotter pins from the castellated nuts on the bottom of the tie rod yoke bolts and remove the nuts, bolts, and tie rod. Loosen the outer clamp bolt on left yoke. Remove the inner bolt and tie rod lock; then screw the yoke off of rod. Loosen both clamp bolts on right yoke and screw the yoke off of rod.

b. Installation Procedure. Screw both yokes on rod about $3\frac{3}{16}$ inches. Install tie rod lock and bolt in left yoke and tighten bolts securely. Place tie rod in position on vehicle with locked yoke at left of vehicle. Install the left tie rod yoke bolt and nut, tighten nut down snug,

TM 9-741
122-123

FRONT AXLE

Figure 94 — Checking Toe-In

RA PD 32390

and back off ¼ to ½ turn to line up slot in nut with hole in bolt and install cotter pin. Install bolt in right tie rod yoke, but do not install nut until after the toe-in has been adjusted.

123. ADJUST TOE-IN.

a. Procedure. Jack up front of vehicle so that tires clear the ground and set wheels in straight ahead position. Rotate wheels and with a pointed instrument, scribe a line in the center of the tire tread around the whole circumference of both tires. Lower vehicle and remove jack. Use a plumb bob, hold the plumb line against scribed line on tire at the center line of the axle, and mark a point on the floor where the bob touches the floor at the front and rear of each tire (fig. 94). Measure the distance between marks on the floor, the front measurement should be between 0 and $\frac{3}{16}$ inch less than the rear measurement. To adjust to proper distance, remove the tie rod bolt from the right yoke and screw the yoke in to increase the distance or out to decrease the distance; then turn steering wheel to right or left to line up tie rod bolt hole and install bolt. Replumb the points again and repeat the adjusting operation until the measurement is correct. When the measurement is correct, tighten the right yoke clamp bolts securely and install the castellated nut on the tie rod bolt, adjusting it in the same manner as the left tie rod nut (par. 122).

199

MEDIUM ARMORED CAR T17E1

Section XXIV

REAR AXLE

	Paragraph
Description	124
Trouble shooting	125
Axle shaft replacement	126

124. DESCRIPTION.

a. The rear axle is of the full-floating type, designed so that the load is carried by the rear axle housing. This type of construction permits replacing the axle shaft without raising the vehicle.

125. TROUBLE SHOOTING.

a. Axle Noisy on Drive.

Probable Cause	Probable Remedy
Hypoid drive pinion and drive gear lash too tight.	Refer to ordnance personnel.
Rear side of hypoid pinion double row thrust bearing rough.	Refer to ordnance personnel.

b. Axle Noisy on Coast.

Probable Cause	Probable Remedy
Excessive lash between hypoid pinion and drive gear.	Refer to ordnance personnel.
Front side of hypoid pinion double row thrust bearing rough.	Refer to ordnance personnel.
End play in hypoid pinion double row thrust bearing.	Refer to ordnance personnel.
End play in hypoid drive gear double row thrust bearing.	Refer to ordnance personnel.

c. Axle Noisy on Both Drive and Coast.

Probable Cause	Probable Remedy
Hypoid drive pinion too deep in drive gear.	Refer to ordnance personnel.
Worn or damaged hypoid pinion bearings or drive gear thrust bearing.	Refer to ordnance personnel.
Loose or worn wheel bearings.	Adjust bearings as instructed in paragraph *153*.

d. Backlash.

Probable Cause	Probable Remedy
Axle drive flange loose.	Replace drive flange gasket and bolt lock, and tighten bolts securely (par. 126).

TM 9-741
125-126

REAR AXLE

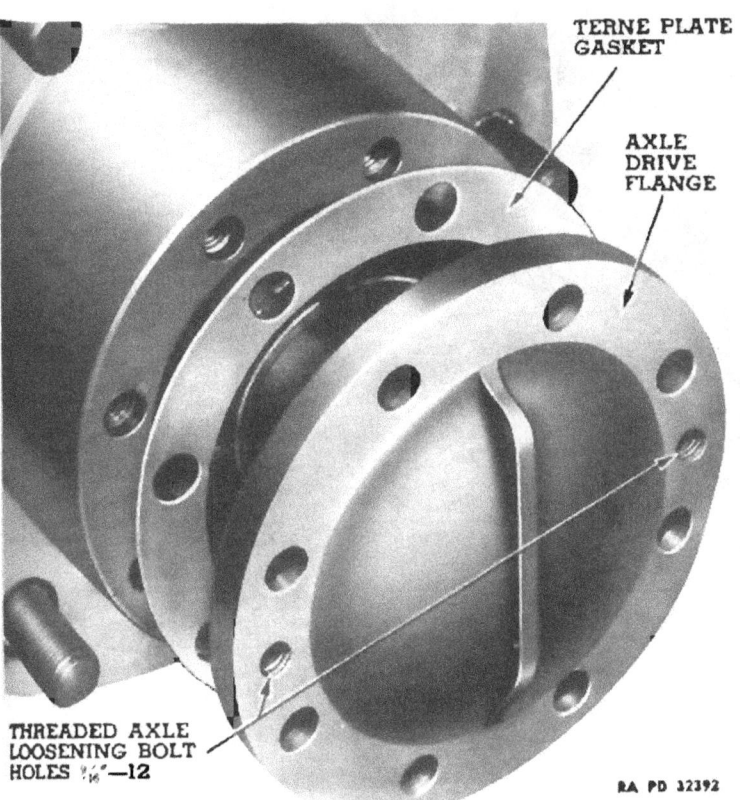

Figure 95 — Axle Shaft Removal

Probable Cause	Probable Remedy
Worn differential side gear and pinion thrust washers.	Refer to ordnance personnel.
Excessive herringbone drive gear to pinion lash.	Refer to ordnance personnel.
Worn universal joints.	Replace worn universal joint parts (par. 112).

126. AXLE SHAFT REPLACEMENT.

a. Removal Procedure.

(1) REMOVE LUBRICANT FROM BANJO HOUSING. Remove drain plug from banjo housing and drain oil into a drain pan. Install drain plug and gasket when oil is drained.

MEDIUM ARMORED CAR T17E1

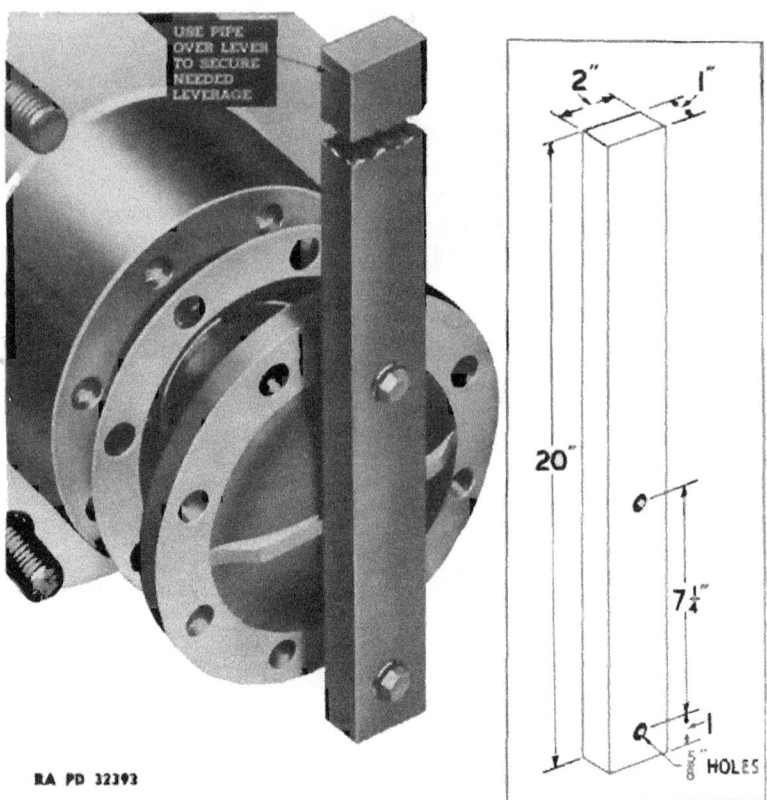

Figure 96 — Axle Shaft Installation

(2) REMOVE AXLE SHAFT (fig. 95). Bend the axle shaft bolt lock tangs away from the bolts and remove the eight bolts and lock. Insert two of the bolts in the tapped holes in the flange and screw them in alternately until the shaft is loose. Pull out the shaft.

b. **Installation Procedure.**

"(1) INSTALL AXLE SHAFT. Fasten a lever to the axle shaft flange with two bolts. This lever can be made from a piece of steel 1 inch thick, 2 inches wide, and about 20 inches long (fig. 96). Install new flange gasket over shaft and push shaft in housing, using care not to damage splines on end of shaft or in differential case. When shaft contacts case, use lever to lift inside end of shaft and turn shaft slightly to line up the splines. Push shaft in case until flange is against hub. Remove the lever and line up the holes in the hub, gasket and axle shaft flange.

REAR AXLE

Install a new bolt lock and install the eight bolts, tighten bolts securely and lock them by bending tangs against the bolts.

(2) REFILL WITH LUBRICANT. Remove the rear axle filler plug and fill with LUBRICANT, gear, universal, seasonal grade. Install filler plug and gasket.

TM 9-741
127-128

MEDIUM ARMORED CAR T17E1

Section XXV

BRAKE SYSTEM

	Paragraph
Description	127
Operation	128
Hydraulic system	129
Trouble shooting	130
Brake adjustment	131
Parking brake adjustment	132
Bleeding hydraulic system	133
Brake shoe replacement	134
Wheel cylinder	135
Main cylinder	136
Hydrovac system	137
Brake lines	138

127. DESCRIPTION (fig. 97).

a. The service or foot brake system is the hydraulic type operating two shoes at each wheel. In addition to the hydraulic system, a Hydrovac unit utilizes engine vacuum to assist the operator in application of brakes. Units used in this system include brake shoes, main cylinder, main cylinder reserve tank, wheel cylinder, Hydrovac unit (power cylinder, relay valve, and hydraulic slave cylinder), vacuum check valve, vacuum reserve tanks, vacuum and hydraulic lines. Two hand brake drums and linings are located at the front of the transfer case as described later in this section. A hand brake lever operates the brakes through connecting cable and linkage.

128. OPERATION.

a. Application or operation of the brakes is dependent upon the function of two systems: hydraulic system and vacuum. These two systems are connected in such a manner that hydraulic pressure actuates a relay valve which controls vacuum and atmospheric pressure in the power cylinder. Movement of the power cylinder pistons, controlled by vacuum and atmospheric pressure, forces fluid from the hydraulic slave cylinder into the brake lines and wheel cylinder. Two complete brake systems are used, one for the front wheels and the other for the rear wheels. The main cylinder assembly consists of two cylinders, one of which operates through a Hydrovac unit to the wheel cylinders at the front wheels, and the other operates through a second separate

BRAKE SYSTEM

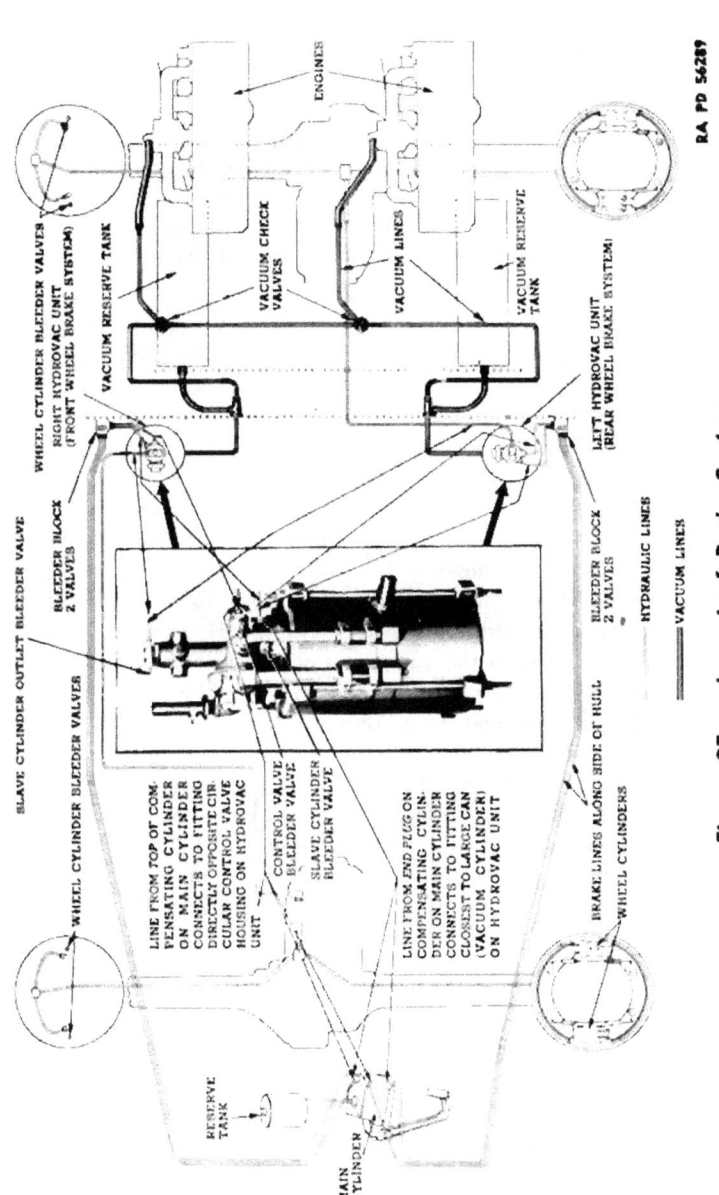

Figure 97 — Layout of Brake System

Hydrovac unit to the wheel cylinders at the rear wheels. Figure 97 shows a layout of the brake system.

129. HYDRAULIC SYSTEM.

a. **General.** The hydraulic system, which is the actual brake operating system, consists of a master cylinder and brake pedal assembly connected to the vacuum boosters and wheel cylinders by hydraulic lines or tubing. Briefly, the hydraulic system employs the principle of manually compressing a column of fluid with which the wheel cylinders are operated and, in turn, actuate the brake shoes.

b. **Sequence of Operation.** As pressure is applied to the brake pedal and is transmitted from the push rod to the pistons in the main cylinder, each primary cup closes the compensating port and fluid is forced into the compensating cylinders, out the unrestricted outlet at the top of the cylinder to the Hydrovac, and thence into the wheel cylinder. The function of the Hydrovac in this sequence is explained later in this section. This pressure forces the pistons in the wheel cylinder outward, expanding the brake shoes against the drums. As the pedal is further depressed, higher pressure is built up within the hydraulic system, causing the brake shoes to exert greater force against the brake drums. In the event the Hydrovac is inoperative the fluid then leaves the compensating cylinder through the holes in the check valve cage, around the lip of the rubber valve cup, through the outlet in the end plug, through the Hydrovac slave cylinder (though it does not operate), and thence into the wheel cylinders. As the pedal is released, the hydraulic pressure is relieved and the brake shoe retracting springs draw the shoes together, pressing the wheel cylinder pistons inward and forcing the fluid out of the wheel cylinders back into the lines toward the Hydrovac and main cylinder. The piston return springs in the main cylinder return the pistons to the pedal stop faster than the brake fluid is forced back into the lines, creating a slight vacuum in that part of the cylinder ahead of the piston. This vacuum causes a small amount of fluid to flow through the holes in the piston head, past the lip of the primary cup and into the forward part of the cylinder. This action keeps the cylinder filled with fluid at all times, ready for the next brake application. As fluid is drawn from the space behind the piston head, it is replenished from the reservoir through the inlet port. When the piston is in a full released position, the primary cup clears the compensating port, allowing excess fluid to flow from the cylinder into the reservoir as the brake shoe retracting springs force the fluid out of the wheel cylinder.

BRAKE SYSTEM

130. TROUBLE SHOOTING.

a. Brake Pedal.

Probable Cause	Probable Remedy
Brake pedal "spongy."	Air in lines, bleed brakes (par. 133).

b. All Brakes Drag.

Probable Cause	Probable Remedy
Mineral oil in system.	Thoroughly wash out all lines and cylinder, and replace all rubber parts. Refer to ordnance personnel.
Dirt in main cylinder.	Replace main cylinder (par. 136).

c. One Brake Drags.

Probable Cause	Probable Remedy
Loose wheel bearing.	Adjust wheel bearing (par. 153).
Weak retractor spring.	Replace spring (par. 134).
Brake shoes adjusted too close to drum.	Readjust brakes according to instructions in paragraph 131.

d. Loose Brakes.

Probable Cause	Probable Remedy
Normal lining wear.	Readjust brakes according to instructions in paragraph 131.
Brake lining worn out.	Replace shoes and readjust (par. 134).
Fluid low in main cylinder and reservoir.	Fill main cylinder and reservoir; bleed all brake lines (par. 136).

e. Brakes Uneven.

Probable Cause	Probable Remedy
Oil on lining.	Replace wheel cylinder and install new shoes (pars. 134 and 135).
Tires improperly inflated.	Inflate tires to 70 pounds front and 80 pounds rear.
Spring center bolt sheared and spring shifted on axle.	Refer to ordnance personnel.

f. Excessive Pedal Pressure, Poor Brakes.

Probable Cause	Probable Remedy
Oil on linings.	Replace wheel cylinder and install new shoes (pars. 134 and 135).
Full area of lining not contacting drum.	Sand shoes so linings contact drums properly.
Scored brake drum.	Replace drum. If lining is badly scored, it should also be replaced (par. 134).

MEDIUM ARMORED CAR T17E1

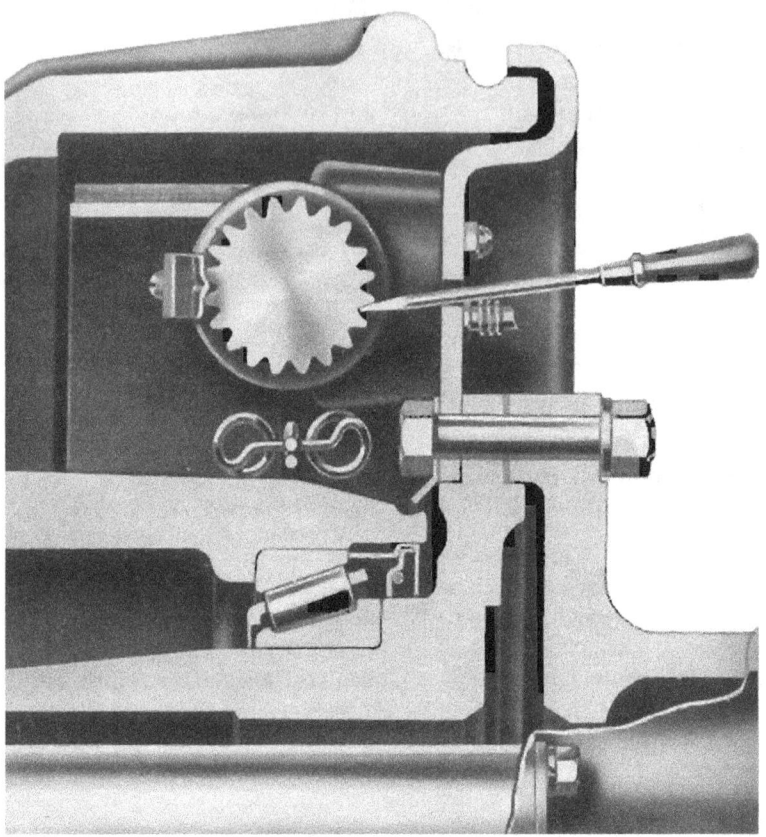

Figure 98 — Adjusting Wheel Cylinder

g. Hydrovac System Power Brakes Fail to Operate.

Probable Cause	Probable Remedy
Vacuum line leaks.	Find location of leak and correct leak or replace line.
Vacuum valve sticking.	Refer to ordnance personnel.
Lack of lubrication in vacuum cylinder.	Lubricate vacuum cylinder (par. 24).
Worn parts in Hydrovac unit.	Replace unit (par. 137).

131. BRAKE ADJUSTMENT.

a. General. The brakes can be adjusted without the removal of the wheels as the brake flange plates have openings with spring snap covers through which the adjustment may be reached.

TM 9-741
131

BRAKE SYSTEM

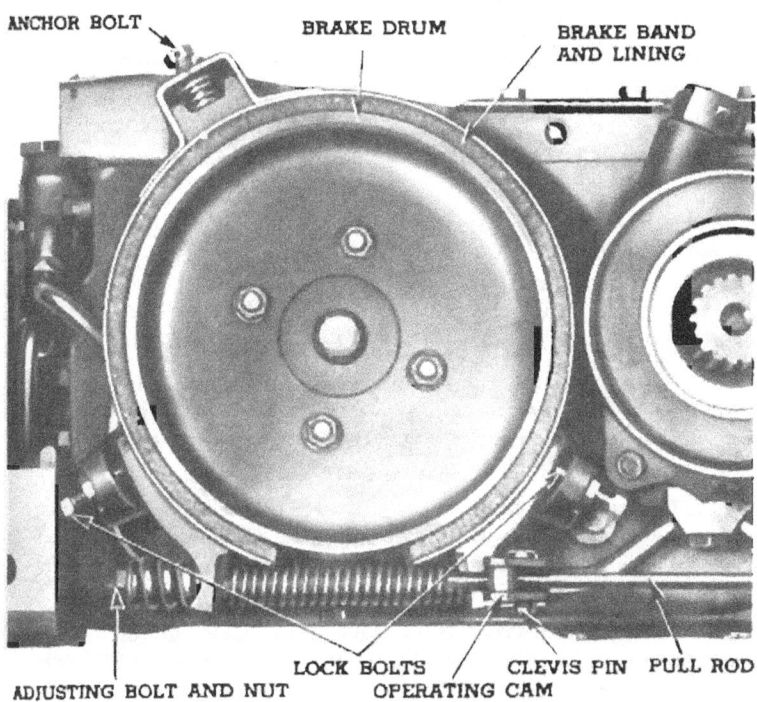

Figure 99 — Parking Brake

b. Procedure.

(1) REMOVE ADJUSTING COVER (fig. 98). Remove the adjusting hole cover from the flange plate, insert a screwdriver through the adjusting hole, and engage the teeth on the adjusting screw of the wheel cylinder.

(2) TURN ADJUSTING SCREW. Turn the adjusting screw until the shoe is snug in the drum, or until the adjusting screw can be turned no more. Turn the adjusting screw back six notches. Each notch backed off will be indicated by a faint click of the adjusting screw lock spring as the screw is turned. This backing-off of the adjusting screw moves the brake shoe away from the drum to insure proper running clearance of the shoe in the drum.

(3) INSTALL COVER. Install the adjusting hole cover.

(4) ADJUST SHOES. Adjust both shoes on all wheels according to the above procedure.

MEDIUM ARMORED CAR T17E1

132. PARKING BRAKE ADJUSTMENT.

a. **General** (fig. 99). The parking brake adjustment should be checked each time the hydraulic service brakes are adjusted. Use the same procedure for both parking brakes.

b. **Procedure.**

(1) RELEASE BRAKE LEVER. Set hand brake lever in fully released position.

(2) ROTATE TURRET. Rotate the turret until the trap door in the bottom of the basket is over the parking brakes. Working through this opening, break the short hose connection at the transfer case oil cooler, and remove the top, front, and end panels of the air duct over the parking brakes.

(3) ADJUST ANCHOR BOLT. Release the lock wire from the head of the anchor bolt near the top of the brake band and turn down the bolt until a 0.020-inch feeler gage will just pass between the brake band and the drum at the anchor bolt position.

(4) TIGHTEN ADJUSTING NUT. Tighten the self-locking adjusting nut on the adjusting bolt at the bottom of the brake until a 0.020-inch feeler gage will just pass between the two ends of the brake lining and drum.

(5) TIGHTEN ADJUSTING BOLT. Loosen the lock nuts on the two brake band lock bolts at each side of the adjusting bolt and tighten the bolts until a 0.020-inch feeler gage will just pass between the brake band and the drum at these locations.

(6) RECHECK CLEARANCE. Recheck the 0.020-inch clearance at the end of the brake bands and readjust the adjusting bolt at the bottom, if necessary. Check the brake band to drum clearance around the entire surface of the drum with a 0.020-inch feeler gage. Readjust the adjusting bolt, lock bolts, and anchor bolt, if necessary to obtain a minimum of 0.020-inch clearance between the brake lining and drum at any point. CAUTION: It is very important that this clearance is maintained. At the high speed at which the parking brake drums revolve under operating conditions, even a slight drag of the linings would result in overheating and rapid wear of the linings, as well as serious overheating of adjacent parts, with possible damage resulting.

(7) TIGHTEN LOCK NUTS. Tighten the lock nuts on the lock bolts and install a new locking wire on the anchor bolt, using care, so that the adjustment is not disturbed.

(8) LUBRICATE CONTROL LINKAGE. Lubricate all frictional surfaces of the brake control linkage and anchor bolts with oil.

TM 9-741
133

BRAKE SYSTEM

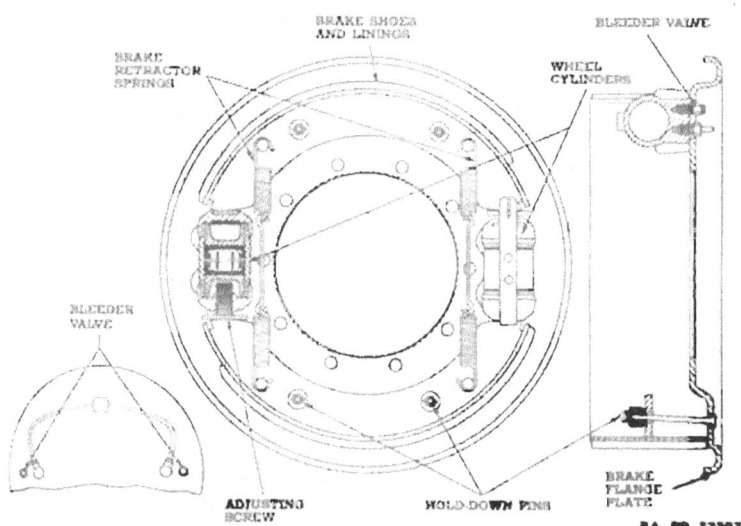

Figure 100 — Brake Construction at Wheel

133. BLEEDING HYDRAULIC SYSTEM (fig. 97).

a. General. The hydraulic brake system must be bled whenever a line has been disconnected, or when a leak has allowed air to enter the system. A leak in the system may sometimes be noticed by a "spongy" brake pedal. Air trapped in the system is compressible and does not permit all pressure applied to the brake pedal to be transmitted to the brake shoes. The system must be absolutely free from air at all times. If the disconnected pipe or leak affected only one of the two complete brake systems, that system only needs to be bled. Make sure that the main cylinder reservoir tank is nearly full of fluid. During all bleeding operations, the reservoir tank must be kept at least one-quarter full of hydraulic brake fluid. NOTE: Before removing the filler plug from the reservoir, clean all dirt and foreign matter from around the plug so that none will fall into the reservoir. Bleed at each of the bleeder valves listed below and in the sequence listed by performing the following operations.

b. Procedure.

(1) REMOVE BLEEDER VALVE SCREW (fig. 100). Remove bleeder valve screw and attach brake bleeder hose KM-J2252 in place of the screw. Place the free end of the tube in a glass jar partially filled with brake fluid so that the end of the tube is below the surface of the fluid in jar.

MEDIUM ARMORED CAR T17E1

(2) UNSCREW BLEEDER VALVE. Unscrew the bleeder valve ½ to ¾ turn.

(3) DEPRESS BRAKE PEDAL. Depress the brake pedal and allow it to return slowly. Continuing this pumping action forces the fluid through the lines and out of the bleeder tube, carrying with it any air in that portion of the system as may be noted by the bubbles from end of the bleeder tube.

(4) TIGHTEN BLEEDER VALVE. When bubbles cease to appear at the end of the bleeder tube and the stream is a solid fluid mass, tighten the bleeder valve and remove the bleeder tube.

(5) INSTALL VALVE SCREW. Install and tighten the bleeder valve screw.

c. **Bleeding Sequence.** Referring to the schematic diagram in figure 97, perform the above bleeding procedure at each of the following points in the order listed:

(1) FRONT WHEEL BLEEDING SEQUENCE. Bleed at each of the two valves at the bleeder block on the right side of the hull near the top and just ahead of the bulkhead.

(2) BLEED SLAVE CYLINDER. Bleed at the slave cylinder bleeder valve on the right-hand Hydrovac assembly.

(3) BLEED CONTROL VALVE. Bleed the right-hand Hydrovac assembly at the control valve bleeder.

(4) BLEED SLAVE CYLINDER OUTLET VALVE. Bleed at the slave cylinder outlet bleeder valve at the connector block on outlet (upper) end of the right-hand Hydrovac assembly.

(5) BLEED LEFT FRONT WHEEL (fig. 100). Bleed at each of the two right front wheel cylinder bleeder valves at the back of front brake flange plate.

(6) BLEED RIGHT FRONT WHEEL (fig. 100). Bleed at each of the two right front wheel cylinder bleeder valves at the back of front brake flange plate.

(7) REAR WHEEL BLEEDING SEQUENCE. Bleed at each of the two valves at the bleeder block on the left side of the hull near the top and just ahead of the bulkhead.

(8) BLEED SLAVE CYLINDER. Bleed at the slave cylinder bleeder valve on left-hand Hydrovac assembly.

(9) BLEED CONTROL VALVE. Bleed at the control valve bleeder valve on left-hand Hydrovac assembly.

(10) BLEED SLAVE CYLINDER OUTLET VALVE. Bleed at the slave cylinder outlet bleeder valve at the connector block on outlet (upper) end of left-hand Hydrovac assembly.

TM 9-741
133-134

BRAKE SYSTEM

(11) BLEED LEFT REAR WHEEL (fig. 100). Bleed at each of the two left rear wheel cylinder bleeder valves at the back of left rear brake flange plate.

(12) BLEED RIGHT REAR WHEEL (fig. 100). Bleed at each of the two right rear wheel cylinder bleeder valves at back of right rear brake flange plate. After bleeding operations have been completed at each point listed above, fill the reservoir tank approximately full of new, clean FLUID, brake, hydraulic, and replace the filler plug. NOTE: Fluid withdrawn in the bleeding operation must not be used again.

d. **Hydraulic Brake Fluids.** As there are several general classifications of hydraulic brake fluids available, care should be taken to make certain that the fluid being used will not injure the brake parts. Some brake fluids may have a rather severe action on the rubber parts, causing them to become sticky, preventing proper piston action or, due to expansion of the rubber parts, causing them to lose their sealing qualities. Other types of fluid may cause vapor lock or, due to extreme thinness, leak past the rubber cups in the wheel cylinders and saturate the brake linings. In the event that improper fluid has entered the system, it will be necessary to drain the entire system. Thoroughly flush out the system with clean alcohol, 188 proof, or a hydraulic brake cleaning fluid. Replace all rubber parts of the system. Refill with proper FLUID, brake, hydraulic.

134. BRAKE SHOE REPLACEMENT.

a. **General.** Brake shoes which have become unserviceable should be replaced in the following manner:

b. **Procedure.**

(1) LOOSEN WHEELS. Loosen the ten wheel nuts on each wheel two turns.

(2) RAISE VEHICLE. Place vehicle on level spot, jack up vehicle, and place on jack stands or suitable blocks.

(3) REMOVE WHEELS. Remove stud nuts and take off all wheels.

(4) REMOVE DRUMS (fig. 101). Remove the three drum retaining screws, install three screws in tapped holes in drum, and tighten alternately until drum is removed.

(5) INSPECTION. After removal of the brake drums and before disassembly of the shoes from the flange plate, all linings should be inspected for wear, improper alinement causing uneven wear, and oil or grease on linings. If any of these conditions exist, it will be necessary to replace the shoes. If in checking the lining it is noticed that they have the appearance of being glazed, this is a normal condition with the hard type lining used. Do not use a wire brush or any abrasive on the

TM 9-741
134

MEDIUM ARMORED CAR T17E1

Figure 101 — Removing Brake Drum

lining to destroy this glazed surface as it is essential for proper operation. Shoes should be changed in sets; both shoes on both front wheels, or both shoes on both rear wheels.

Figure 102 — Wheel Cylinder Clamp Installed

(6) INSTALL WHEEL CYLINDER CLAMP (fig. 102). Install wheel cylinder clamp KM-J2280 to keep the wheel cylinder pistons in place and prevent leakage of brake fluid while replacing shoes.

BRAKE SYSTEM

(7) REMOVE RETRACTING SPRINGS (fig. 103). Remove brake shoe retracting springs by pulling the inner ends of the double springs together using special brake spring pliers KM-KMO142, and remove the connecting link.

(8) REMOVE BRAKE SHOES (fig. 104). Remove the cotter pins, retaining nuts, steel cup, and springs from the four hold down pins. The brake shoe assemblies now can be slipped off the hold down pins.

(9) INSTALL SHOES. Replace the brake shoe assemblies on the hold down pins with the short end of the lining on the leading end of the shoe. NOTE: Care must be taken to install the brake shoe and lining assembly in the proper position on the brake flange plate. To determine the leading end, point a finger at the flange plate and rotate it in a circular path the same as the brake drum would rotate with the vehicle moving forward. That end of each shoe to which the finger points first in its rotation is the leading end of the shoe. If the wheel cylinders have not been removed, the adjustment screw end of the wheel cylinders will bear against the leading end of the shoe.

(10) REASSEMBLE COMPONENT PARTS. Reassemble steel cups, springs, and retaining nuts, using new cotter pins. Replace brake shoe retracting springs and connecting links, using spring pliers KM-KMO142.

(11) REMOVE CYLINDER CLAMPS (fig. 104). Remove wheel cylinder clamps.

(12) INSTALL DRUMS. Install drums and tighten securely.

(13) ADJUST BRAKES. Adjust brakes as previously outlined in paragraph 131.

(14) INSTALL WHEELS. Install wheels and tighten retaining nuts securely. Remove jack and blocks.

135. WHEEL CYLINDER.

a. General (fig. 100). At each wheel, two brake shoes are actuated by two wheel cylinders, each containing two pistons which bear against end covers which cover the end of the shoes and expand them outward against the drum when fluid under pressure is introduced between the pistons. Rubber piston cups at the head of each piston seal the cylinder and prevent leakage of fluid past the pistons. The rubber cups are held against the pistons by a coil spring between them which engages in suitable retainers. One end cup in each cylinder is provided with an adjusting screw that bears against the shoe and provides a means of adjusting the brake shoes to the proper drum clearance, since the shoes are held in contact with the pistons (even when at rest) by retractor springs.

TM 9-741
135

MEDIUM ARMORED CAR T17E1

Figure 103 — Removing Retracting Spring

b. Procedure.

(1) REMOVE WHEEL CYLINDERS. In order to remove the hydraulic brake wheel cylinders, the vehicle must be jacked up, the wheel and brake drum removed (par. 134), and the hydraulic lines leading to the wheel cylinder disconnected at the back of the flange plate. Remove the brake retractor springs by pulling the inner ends of the springs together with special pliers KM-KMO142, and by removing the connecting link. Remove the four cap screws per cylinder from the back side of the brake flange plate and the wheel cylinders may be removed.

(2) INSTALL WHEEL CYLINDER. Reassemble the cylinder to the flange plate, making sure that the bleeder valve on the back of the cylinder is toward the outer edge of the flange plate, and the adjustment screw end of the cylinder is toward the leading end of the brake shoe or the end having the lining cut short. If the brake shoes have been removed, see paragraph 134, for the installation of the shoes. If it is found that the end covers have been assembled backwards, they may be exchanged end for end.

(3) REASSEMBLE COMPONENT PARTS. Connect the brake shoe retracting springs and the hydraulic lines at the back of the flange plate. Replace the brake drum and wheel. NOTE: After a wheel cylinder has been removed and replaced. **The entire hydraulic brake system must be bled** (par. 133).

TM 9-741
136

BRAKE SYSTEM

Figure 104 — Removing Brake Shoe

136. MAIN CYLINDER.

a. General. The function of the main cylinder is to displace fluid from a central source into lines and into wheel cylinders. The main cylinder includes two complete hydraulic cylinders, side by side, and connected by a compensating cylinder. The main cylinder is mounted on the steering gear housing.

b. Procedure. Replace main cylinder as a unit.

(1) REMOVE MAIN CYLINDER. Disconnect the four brake lines from the compensating cylinder and the tube from the reserve tank and allow the fluid from these lines to drain into a suitable container. Disconnect the stop lamp switch wires and the brake pedal pull back spring. Dismount the main cylinder assembly from the steering gear housing by removing the three attaching bolts. Remove link bolts and nut which attach the main cylinder push rods to top of brake pedal. Drive pin out of pedal shaft collar and pedal shaft and remove brake pedal. Remove stop lamp switch.

(2) INSTALL NEW MAIN CYLINDER. Install stop lamp switch and brake pedal. Install the cylinder on the steering gear housing with the three attaching bolts. Connect the pipe from the reserve tank, the

217

MEDIUM ARMORED CAR T17E1

brake lines from the compensating cylinder to the Hydrovac units, the brake pedal pull-back spring, and the stop lamp wires. NOTE: It is very important that the proper brake lines be attached to the correct outlets on the compensating cylinder. Referring to figure 97, note that the lines from the outlet on each Hydrovac unit which is nearest the large vacuum cylinder, connect to the end plugs on the compensating cylinder of the brake main cylinder. Also, the line from each Hydrovac unit attaching to the outlet which is **directly opposite** the circular control valve housing (having removable end plate) connects to the outlets at the top of the compensating cylinder on the brake main cylinder assembly. After all connections have been made to the main cylinder, the main cylinder and reservoir must be filled with new **FLUID, brake, hydraulic**, and the hydraulic brake system bled in accordance with instructions given in paragraph 133.

137. HYDROVAC SYSTEM.

a. **General.** (fig. 105). The Hydrovac power brake system includes two auxiliary power units that apply additional force to the hydraulic brake system and which are controlled automatically by the normal operation of the brake pedal. This greatly increases the driver's ability to stop quickly from high speeds or on steep grades. One Hydrovac unit is used in the front wheel brake system and one in the rear wheel system (fig. 97). The function and servicing of the Hydrovac unit is identical for each system. If one of the Hydrovac units is not functioning properly, it should be replaced as a unit.

b. **Procedure.** The following are the steps to use in replacing the Hydrovac assembly.

(1) REMOVE HYDROVAC. Rotate turret until one of the large panel openings in the basket floor is opposite the Hydrovac unit to be removed. Working through this opening, disconnect the three hydraulic lines from the upper end of the Hydrovac assembly.

(2) DISCONNECT VACUUM HOSE. Disconnect the vacuum hose leading through the bulkhead at the center of the power cylinder.

(3) DISCONNECT CONTROL VALVE PIPE. Disconnect the line at the control valve which leads to the air cleaner above the Hydrovac assembly.

(4) LOOSEN MOUNTING BRACKET NUTS. Loosen the lock nuts and screw the anchor nuts down on the upper ends of the two power cylinder clamp studs which attach the Hydrovac to the upper mounting brackets on the hull. This will allow the Hydrovac assembly to be tipped away from the hull and lifted up to disengage the lower ends of the two studs from the holes in the lower mounting brackets. Remove the Hydrovac.

BRAKE SYSTEM

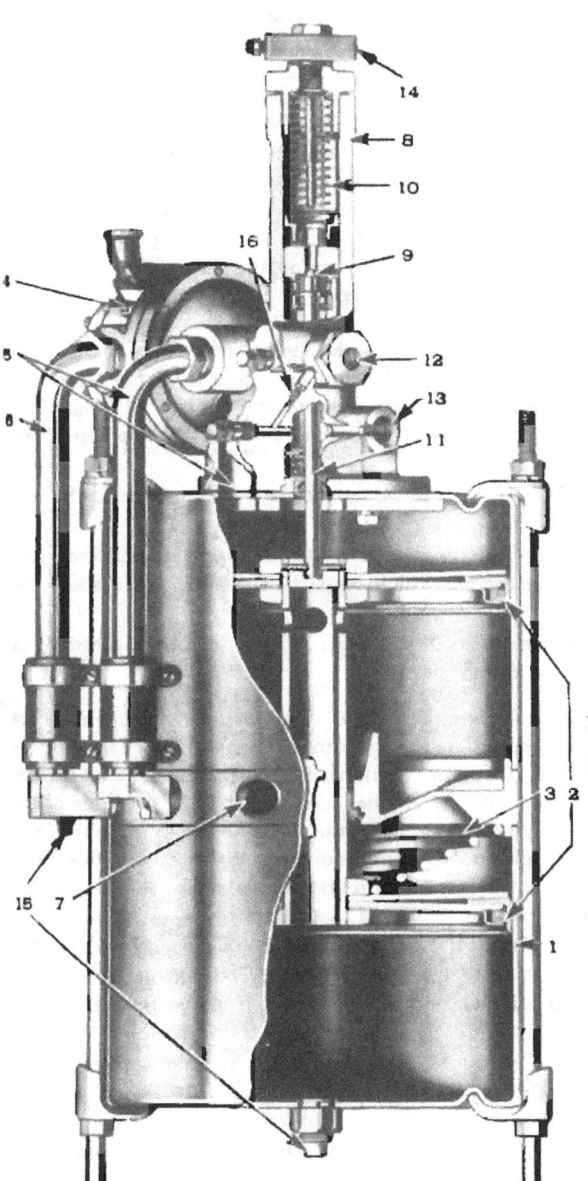

1. VACUUM POWER CYLINDER
2. CYLINDER PISTONS
3. PISTON RETURN SPRING
4. CONTROL VALVE HOUSING
5. VACUUM LINE
6. AIR AND VACUUM LINE
7. CONNECTION FOR LINES FROM VACUUM SOURCE
8. SLAVE CYLINDER BODY
9. VALVE
10. PISTON RETURN SPRING
11. POWER CYLINDER PUSH ROD
12. CONTROL VALVE HYDRAULIC LINE CONNECTION FROM MAIN CYLINDER
13. SLAVE CYLINDER LINE CONNECTION FROM MAIN CYLINDER
14. LINE CONNECTION TO WHEEL CYLINDERS
15. LUBRICATION PLUGS
16. SURGE RELIEF VALVE

RA PD 32402

Figure 105 — Hydrovac Assembly

TM 9-741

MEDIUM ARMORED CAR T17E1

(5) INSTALL NEW HYDROVAC. Rotate turret until one of the large panel openings in the basket floor is opposite the corner of the hull in which the Hydrovac unit is to be installed. Working through this opening, insert the Hydrovac on an angle, bottom end first, and insert the unthreaded end of the long studs in the holes in lower mounting brackets welded to the hull; at the same time enter the vacuum hose on fitting at center of Hydrovac. Straighten the Hydrovac and raise it so that the upper end of the long studs protrude through the holes in the upper mounting brackets on the hull. Holding the assembly in this position, tighten the retaining nuts on the upper end of the long studs, and tighten the upper lock nuts.

(6) CONNECT HYDRAULIC LINES. Connect the hydraulic line from the end plug of the main cylinder to the lower connection on the slave cylinder (13, fig. 105). Connect the hydraulic line from the top of the compensating port on the main cylinder to the control (12, fig. 105). Connect the hydraulic line from the wheel cylinders to the top of the slave cylinder (14, fig. 105).

(7) CONNECT VACUUM HOSE. Connect the vacuum hose to the connector on the center plate.

(8) CONNECT CONTROL VALVE TO AIR CLEANER PIPE. Reconnect the line at the control valve which leads to the air cleaner above the Hydrovac assembly.

(9) BLEED SYSTEM. Bleed the entire brake system according to instructions given in paragraph 133.

c. Vacuum Check Valve (fig. 106). The vacuum check valves are mounted on the bulkhead in the radiator compartment. The purpose of these valves is to trap the vacuum in the Hydrovac and reserve tanks so that in the event the engines stall, several applications of the Hydrovac braking system may be made after the engines stop. Ordinarily the valves require no attention; however, if the valves stick, there will be evidence of no reserve vacuum and the valve should be removed and cleaned. The repair procedure is as follows:

(1) REMOVE VACUUM VALVE. Remove the eight cap screws that attach the radiator grille to the hull and remove the grille. Disconnect the vacuum lines from each end of the valve. Remove the valve from the support fittings.

(2) DISASSEMBLE VACUUM CHECK VALVE (fig. 106). Remove the four screws that attach the two halves of the valve, pull the two halves apart, and remove the valve and gasket.

(3) CLEAN AND INSPECT PARTS. Wash all parts thoroughly in SOLVENT, dry-cleaning, and inspect the valve and valve seat for wear or other damage. Replace the damaged parts.

TM 9-741
137-138

BRAKE SYSTEM

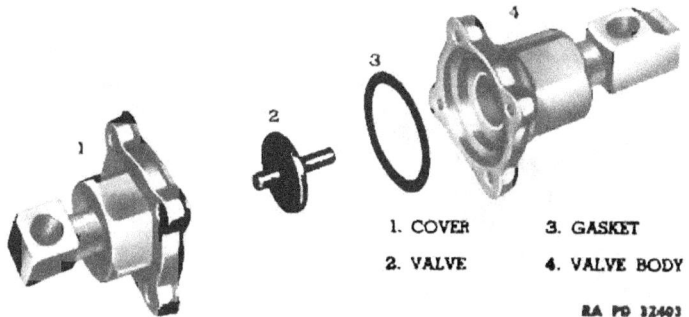

1. COVER
2. VALVE
3. GASKET
4. VALVE BODY

RA PD 32403

Figure 106 — Vacuum Check Valve

(4) ASSEMBLE VALVE. Install the valve and a new gasket. Place the cover in position and install the four screws. Tighten the screws securely.

(5) INSTALL VALVE ASSEMBLY. Mount the valve assembly in the support fittings so that the valve cover is toward the top. Connect the vacuum lines to the fittings at each end of the valve.

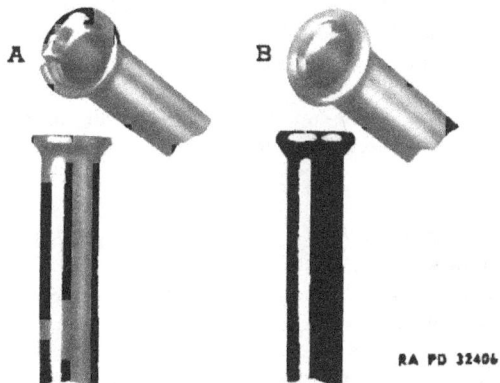

Figure 107 — Single and Double Lap Flares

138. BRAKE LINES.

a. General. The hydraulic brake tubing is a double layer flexible steel tubing, treated to resist corrosion and also stand up under the high pressures which are developed when applying the brakes. It is

MEDIUM ARMORED CAR T17E1

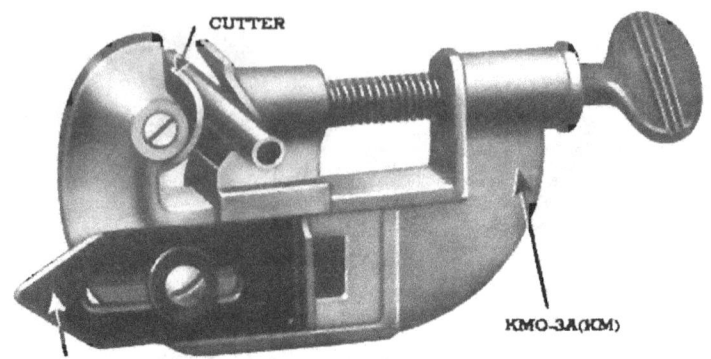

Figure 108 — Brake Tube Cutting Tool

important when making up hydraulic brake tubing that they have the proper flaring at the end of tubing for the compression couplings. Unless the tubing is properly flared, the couplings will leak and the brake will become ineffective. This safety steel tubing must be double-flap flared in order to produce a strong, leakproof joint. The brake tube flaring tool (fig. 109) is used to form the double-lap flare. Figure 107 shows two pieces of tubing, one with a single-lap flare (A), and the other with the double-lap flare (B). It will be noted that the single-lap flare splits the tubing while the one shown in B has a heavy, well-formed joint.

b. Procedure.

(1) CUTTING TUBING (fig. 108). Cut tube to correct length, using tube cutter KM-KMO3A.

(2) INSTALL FITTINGS. Place new compression fittings on the tubing with the hex ends toward each other. Dip the end of the tube to be flared into hydraulic fluid to lubricate it.

(3) PLACE TUBING IN DIE BLOCKS. Select the correct size die block for the tubing (the halves of the die blocks are marked for tubing size). Place the tube in one die block half, allowing approximately ¼ inch to project beyond the countersunk end of the block. Tap the tube into the block with a lead hammer. Place the remaining half of the die block over the tube and tap block halves firmly together.

(4) PLACE DIE BLOCK INTO TOOL BODY (fig. 109). Place the die block with tubing into tool body, seat block firmly against bottom of

TM 9-741
138

BRAKE SYSTEM

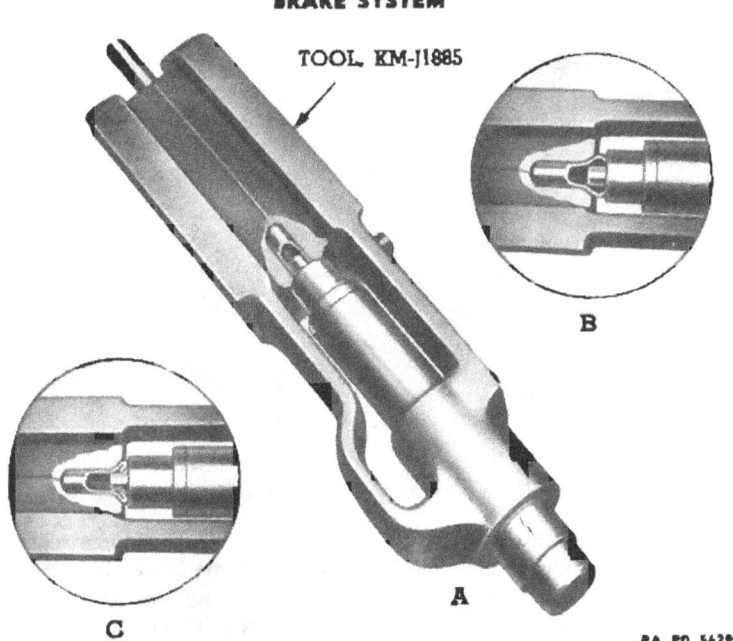

Figure 109 — Brake Tube Flaring Operations

body and against stop pin in the end. Tap block and tube into place with wood block and hammer.

(5) INSTALL ASSEMBLY IN VISE (fig. 110). Install body, die block, and tube assembly in a bench vise. The flange on the top of tool body should rest on top of the vise jaws and the stop pin in the side of body against the side of the jaws. Tighten vise jaws just enough to keep tube from slipping in die block.

(6) INSTALL FLARE-FORMING TOOL (fig. 109). Place the punch ram in tool body (open end toward die block); select the proper size flare-forming tool and place it in the open end of the punch ram with the concave end toward the die block.

(7) ADJUST TUBE FOR FLARING. Push ram and flaring tool toward tube and die block until the tool pilot engages the tubing. Tap end of ram until the guide mark (A, fig. 109) is in line with the end of body casting. This adjusts tubing to proper depth in die block.

(8) TIGHTEN VISE. Draw vise jaws together as tightly as possible.

(9) START DOUBLE FLARE. Strike the end of the punch ram with a hammer until the shoulder of the flare-forming tool strikes the die block. This upsets the end of the tube (B, fig. 109).

223

TM 9-741
138

MEDIUM ARMORED CAR T17E1

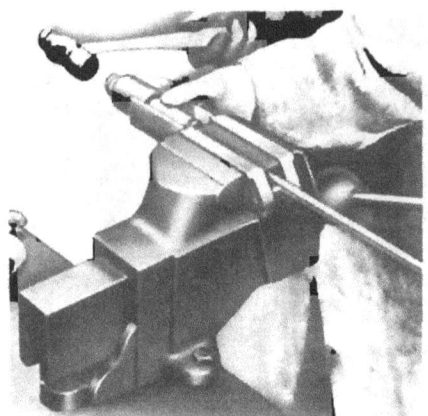

Figure 110 — Flaring Brake Tube

(10) FINISH FLARE. Pull ram back and reverse the flare-forming tool so that the tapered end of tool is toward the die block. Push the ram in until the flaring tool pilot engages the tubing, and strike the end of ram with a hammer. Slide ram away from tubing to observe the flaring operation. Engage ram in tube and strike with hammer until flare is formed (C, fig. 109).

(11) This method is employed on all tubing on the vehicle from $\frac{3}{16}$ inch to $\frac{1}{2}$ inch, using the correct die block and flare-forming tool for the tubing being worked.

(12) A clamp assembly is furnished with the tool for portable operation. However, the use of this clamp is not recommended for flaring tubes larger than $\frac{3}{8}$-inch.

c. **Hose Replacement.** To allow for axle movement, flexible brake hose is used at each of the front wheels and also between the main line and the line leading along each axle. The composition of the brake hose is such that it permits a great amount of flexing action without destroying the brake fluid seal, thus preventing air leaks or loss of fluid. When replacing a brake hose, care must be taken to make certain that the hose fitting is seating properly in the couplings, otherwise leaks will cause the brakes to become ineffective. This can be checked by applying pressure to the brake foot pedal and at the same time carefully inspecting the surface surrounding the hose joint for any signs of leaks. If even the smallest of bubbles appears, it will be necessary to tighten the connection further. If proper seal cannot be obtained by tightening the fitting, then replace the coupling (par. 138).

TM 9-741
139

Section XXVI

SPRINGS, RADIUS RODS, AND SHOCK ABSORBERS

                                                                        Paragraph

Description 139
Spring replacement 140
Radius rods 141
Shock absorbers 142

139. DESCRIPTION (fig. 111).

a. General. Both front and rear springs on this vehicle are of the semielliptic type, 50½ inch long and 3½ inch wide. Each spring has

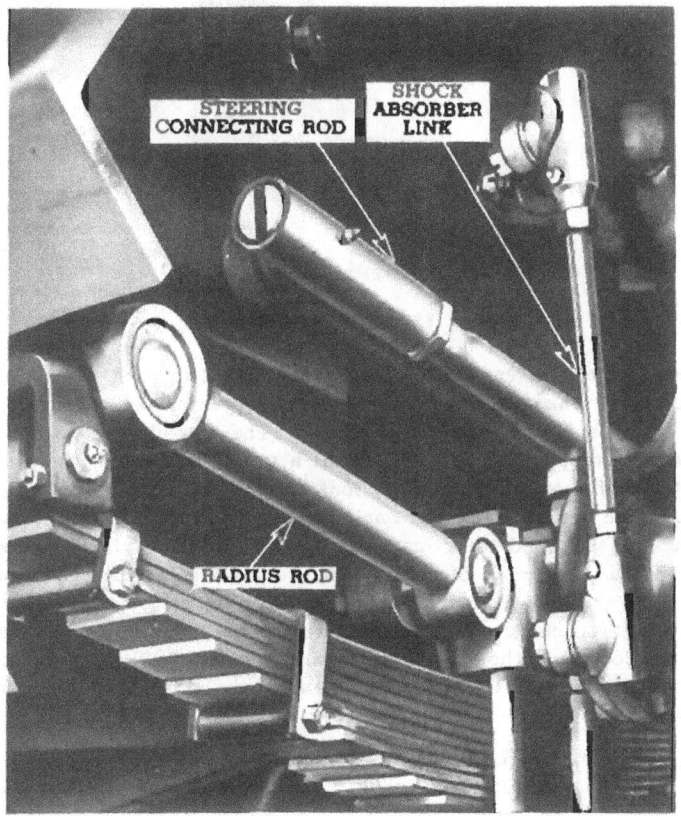

Figure 111 — Spring, Radius Rod and Shock Absorber Connections

MEDIUM ARMORED CAR T17E1

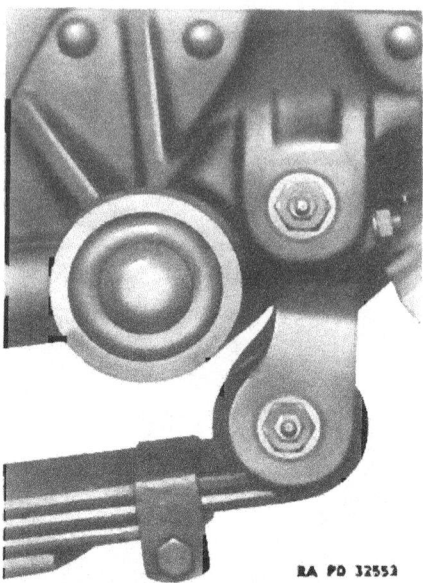

Figure 112 — Spring Shackle

eleven tapered leaves with four rebound clips, two each side of the center bolt. The top leaf has an overhung eye formed at each end into which a bronze bushing is pressed. The second leaf has a full thickness wrapped eye in each end as a safety measure in the event the top leaf should break. The front springs are attached to spring seats on the bottom of the front axle housing by straight bolts. The rear springs are attached to spring seats on the top of rear axle housing by U bolts. Both ends of each spring are shackled to spring hangers riveted to the hull. The springs are equipped with spring covers, and all springs, shackles, and shackle pins are interchangeable. Shackle pins are locked in the shackle with locking bolts. Lubrication fittings assembled in adapters on the outside end of the pins (fig. 112) admit lubricant to the bushings through drilled passages in the pin.

140. SPRING REPLACEMENT.

a. **General.** To remove a spring, raise that end of the vehicle until the tire is just ready to leave the ground and support with a suitable stand jack. If a rear spring is being removed, it will be necessary to disconnect the shock absorber link at the lower end.

TM 9-741
140

SPRINGS, RADIUS RODS, AND SHOCK ABSORBERS

b. Procedure.

(1) REMOVE SPRING. Disconnect the U-bolts (or straight bolts, as the case may be) and remove the lubrication fitting adapter plugs from front and rear spring shackle pins. Remove the spring shackle pin locking bolts from the lower shackle pins, and with special shackle pin and bushing puller KM-J2274, pull the front and rear lower shackle pins and remove spring. If it is desired to remove the shackles also, remove the lubrication fitting adapter plugs, and the shackle pin locking bolts from the upper shackle pins; then again use puller KM-J2274 to pull the pins. The upper shackle pins rotate in two solid bronze bushings in the shackle. The lower shackle pins rotate in a single split steel bushing in the spring eye. If any of these bushings are worn or scored, they should be replaced with new parts.

(2) REMOVE BUSHING. To remove the bushings, thread the puller shaft through the bushing and screw the bushing adapter on the end of puller shaft. Hold the puller shaft with one wrench and turn the nut on the puller shaft until the bushing is pulled out of the eye of the spring or shackle.

(3) INSTALL NEW BUSHINGS AND SPRING. To install a new bushing, remove sleeve from puller shaft; then thread the puller shaft through the eye of the spring or shackle and also through the bushing. Assemble bushing adapter to puller screw. Hold the puller shaft with one wrench and turn the nut on the puller shaft until the bushing is pulled into the eye of spring or shackle. When installing the two bronze bushings in the shackle, press one bushing in from each side until both are slightly below the face of the shackle. If the ends of the steel bushing protrude beyond the edges of the spring eye, they should be filed off flush. After the bushings are in place, ream them to size, using reamer KM-J2289. In replacing the springs and shackles, care must be taken to position the front springs and shackles properly. The front shackle of each front spring must be installed with the curved side of the shackle to the rear, regardless of the resulting location of the locking bolt. All other shackles should be installed with the locking bolt to the outside, away from the hull. Remove the lubrication fitting adapter from the spring pins and start them into the shackle with the tapped hole to the outside and the locking bolt grooves at the bottom. Drive the pin with a soft face hammer until the locking bolt groove in the pin alines with the bolt hole in shackle; then install locking bolt. When installing the front springs, if interference occurs between the front spring eye and the end of the radius rod, turn the spring end for end and the interference will be eliminated.

MEDIUM ARMORED CAR T17E1

141. RADIUS RODS (fig. 111).

a. **General.** The drive from the wheels is transmitted to the hull by four radius rods which link each end of each axle to the hull. The rods are tubular members having a sealed and lubricated ball stud at each end. The ball studs assemble to brackets provided on the axle housing and on the sides of the hull.

b. **Replace Radius Rod** (fig. 111). Since the ball stud assembly is sealed, it cannot be disassembled for replacement of individual parts; consequently, the complete radius rod is serviced as an assembly. The four radius rods are interchangeable. When installing a radius rod, it should be noted that one ball stud is longer than the other. NOTE: The longest ball stud should be assembled to the bracket on the hull. Installation of the radius rod will be facilitated if this end is mounted first. To engage the short ball stud in the bracket on the axle housing, it may be necessary to shift the axle assembly forward or backward slightly.

142. SHOCK ABSORBERS.

a. **Description.** Shock absorbers provide a means of dampening the spring vibrations as the wheels of the vehicle pass over irregularities in the road or ground. Both front and rear shock absorbers on this vehicle are of the hydraulic double-acting type and control or dampen the speed of both compression and rebound of the springs. Each shock absorber contains two pistons which are operated in tandem in opposed cylinders by a crank or cam between them. The cam is rotated by the shock absorber arm which is attached to the end of the camshaft and is connected to either the front or rear axle by a shock absorber link. The cylinders are filled with fluid which passes between the two cylinders through passageways as the axle moves up and down and operates the pistons.

b. **Maintenance.** The shock absorbers should be checked every 5,000 miles to see that they contain sufficient fluid. To fill shock absorbers, remove the filler plug and fill with FLUID, brake, hydraulic, to the bottom of the filler plug. If the shock absorbers leak fluid, they should be replaced. Apply grease gun to fittings every 1,000 miles and fill the ball housing with LUBRICANT, gear, universal.

c. **Checking Shock Absorber Operation.** After filling shock absorbers, disconnect the link by removing the end plug, ball seat, and pull link off of the ball. Move the arm up and down with a steady motion. If arm moves easily in either direction, the shock absorber should be replaced. Check the end play in the link for looseness and if loose, replace the link assembly.

TM 9-741
142

SPRINGS, RADIUS RODS, AND SHOCK ABSORBERS

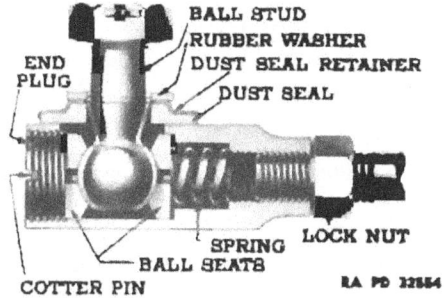

Figure 113 — Shock Absorber Link End

d. Shock Absorber Replacement.

(1) REMOVE SHOCK ABSORBER. Remove end plug and seat from shock absorber end of link and lift link off of ball on shock absorber arm. Remove the four bolts that attach the shock absorber to the hull.

(2) INSTALL SHOCK ABSORBER. Place shock absorber in position on hull and install the four mounting bolts. Tighten bolts securely. Place end of link over ball and install the seat and end plug. Screw end plug in until it is solid, then back it off until the slot lines up with the cotter pin hole. Install a cotter pin. Apply grease gun to fitting and fill the ball housing with LUBRICANT, gear, universal.

e. Shock Absorber Link Replacement (figs. 111 and 113).

(1) REMOVE LINK. Remove the cotter pin and end plug from each end of link and remove the outer half of the ball seat on each end. Lift the link off of both ball studs.

(2) INSTALL LINK. Place spring in end of link; then place ball seat on spring with spherical side of seat to outside. Place link on ball stud and install the other ball seat in end of link with spherical side of seat toward the ball stud. Install the end plug and screw it in until tight; then back it off until the slot lines up with cotter pin holes. Install cotter pin. Install other end of link in the same manner. Apply grease gun to fitting and fill the ball joint housing with LUBRICANT, gear, universal.

MEDIUM ARMORED CAR T17E1

Section XXVII

STEERING GEAR

	Paragraph
Description	143
Operation	144
Trouble shooting	145
Service operations	146
Steering gear motor	147
Steering connecting rod	148

143. DESCRIPTION.

a. The steering gear in this vehicle is of the recirculating ball type to which has been added a hydraulic power system. This hydraulic system augments the steering pressure which the driver creates by turning the steering wheel; therefore, normally easy steering is provided.

b. The steering gear ratio is 26.1 which is in the range of ordinary motor trucks and is sufficiently "fast" for safe handling at any speed the vehicle is capable of attaining.

c. The steering gear mounts to the left side of the hull with four large bolts. The pitman shaft extends through the hull for mounting the pitman arm and steering connecting rod.

144. OPERATION (fig. 114).

a. As the driver turns the steering wheel, the main shaft and worm turn in the ball nut. This causes the ball nut to move up or down the worm. This would normally turn the pitman shaft, causing the vehicle to turn. Due to the weight of the vehicle this would require considerable pressure. The pressure required to move the ball nut produces an end thrust on the main shaft. This shifts the main shaft endwise, slightly shifting the valve spool, opening a valve applying pressure to one side of the power cylinder piston. This aids the driver in turning the front wheels.

b. If the steering wheel is turned in the opposite direction, the same result is obtained only that main shaft pressure is applied in the opposite direction opening a valve to the other end of the power cylinder piston which assists in turning the vehicle in the opposite direction.

c. This design aids materially in steering stability as severe road shocks being transmitted to the steering gear open the valves, producing a counter pressure to hold the wheels in the plane determined by the steering wheel position.

STEERING GEAR

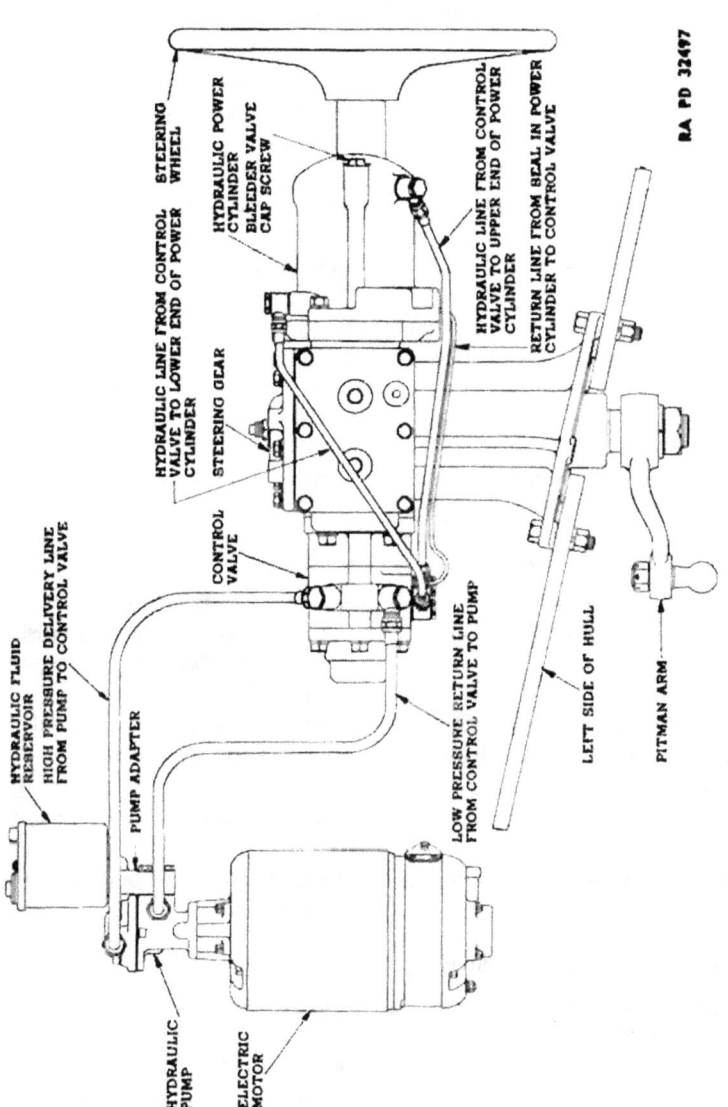

Figure 114 — Steering Gear Mounting

MEDIUM ARMORED CAR T17E1

145. TROUBLE SHOOTING.

a. Hard Steering.

Probable Cause	Probable Remedy
Lack of lubrication.	Lubricate steering gear, tie rod ends and steering connecting rod ball joints.
Tie rod end bolts too tight.	Readjust tie rod end bolts (par. 122).
Under-inflated tires.	Inflate front tires to 70 pounds and rear to 80 pounds.
Improper steering gear adjustment.	Report to ordnance personnel.
Hydraulic power cylinder not operating.	Check for oil in the reservoir, leaks in the system and operation of the electric motor and hydraulic pump.
Tight steering knuckle support bearings.	Report to ordnance personnel.

b. Loose Steering.

Probable Cause	Probable Remedy
Improper adjustment.	Report to ordnance personnel.
Loose ball joints.	Adjust plugs in steering connecting rod according to instructions (par. 122).
Worn steering arm bushings.	Report to ordnance personnel.
Loose or broken front wheel bearings.	Adjust bearings (par. 153).
Loose steering wheel.	Tighten steering wheel retaining nut (par. 146).
Steering gear loose at mounting.	Tighten steering gear mounting bolts (par. 146).

c. Motor Does Not Start.

Probable Cause	Probable Remedy
Steering gear motor switch not turned "ON."	Turn switch to "ON" position.
Circuit breaker "kicked out."	Reset.
Open circuit between switch and battery.	Tighten connections and/or replace line.
Internal short (motor smoking).	Replace (par. 147).
Defective switch.	Test and/or replace (par. 146 e).
Defective circuit breaker.	Test and/or replace (par. 146 e).

TM 9-741
145

STEERING GEAR

Probable Cause	Probable Remedy
Circuit breaker "Kicks out" continually.	Replace motor (par. 147).
Defective motor.	Replace motor (par. 147).

d. Slow Motor Speed.

Motor rotates, but is slow and hums.	Replace motor (par. 147).
Motor will not carry heavy load.	Replace motor (par. 147).
Sluggish motor operation.	Lubricate steering gear and check front axle trunnion bearings, etc., for binding. If no improvement, report or replace motor.
Tight armature bearings—excess heat at front and rear motor frames.	Disconnect hydraulic pump — rotate armature by hand—if rotation is tight, report or change motor.
Dragging armature—scratching or dragging sound.	Replace motor (par. 147).

e. Excessive Motor Speed.

Shunt field open. CAUTION: Pull starting switch up to "OFF" position immediately if this condition should develop to avoid damaging the motor.	Replace motor (par. 147).

f. Motor Vibrates.

Bent armature shaft — motor vibrates.	Replace motor (par. 147).
Worn armature bearings — grinding noise.	Replace motor (par. 147).
Armature out-of-balance.	Replace motor (par. 147).

g. Motor Noisy.

Bad bearings.	Replace motor (par. 147).
Armature dragging.	Replace motor (par. 147).
Whining—excessive speed.	Replace motor (par. 147).
Amplified motor noises.	If noise seems to be amplified, remove motor from vehicle and set on concrete floor or any solid base. If motor operates normally, then noise is normal due to natural noises being amplified by the motor mounting.

TM 9-741
146-147

MEDIUM ARMORED CAR T17E1

146. SERVICE OPERATIONS.

a. The steering gear must be removed from the hull to make adjustments. Due to the size and weight of this unit, the using arms should not attempt to remove or service it.

b. The only services that should be rendered by the using arms are normal maintenance operations including:

(1) Tighten steering gear-to-hull bolts.

(2) Tighten the pump-to-steering gear valve line connections.

(3) Tighten the steering gear valve to power cylinder line connections.

(4) Tighten steering gear body bolts.

c. **Tests on Steering Gear Motor Circuit.** When the steering gear motor does not operate and no internal motor trouble (smoke or odor) is indicated, the following tests will locate a defective steering gear motor switch or circuit breaker.

(1) To determine whether the steering gear motor switch is defective, perform the following tests.

(a) Connect one voltmeter lead to the input side of the switch and the other voltmeter lead to a ground. A reading on the voltmeter (21.6 to 26.4) indicates that power is being delivered to the input side of the switch. If the reading on the voltmeter is less than 21.6, a loose connection or open circuit is indicated.

(b) With the switch in the "ON" position, and with power being delivered to the switch, connect one voltmeter lead to the motor side of the switch and the other voltmeter lead to a ground. A reading above 21.6 on the voltmeter indicates a serviceable switch. A reading less than 21.6 indicates a defective switch.

(2) To determine a defective circuit breaker when it has been determined that the motor switch is serviceable, perform the tests in (1) above on the circuit breaker terminals.

(a) A voltmeter reading less than 21.6 on the input side of the circuit breaker indicates a loose or corroded connection at the circuit breaker or a broken line between the circuit breaker and the motor switch.

(b) A voltmeter reading less than 21.6 on the motor side of the circuit breaker indicates a defective circuit breaker.

147. STEERING GEAR MOTOR.

a. **Description.** The electric motor, which drives the hydraulic pump for the steering gear, is a compound wound machine having both a shunt

STEERING GEAR

and a series field. This type motor gives better control and performance than other types of motors. The motor is attached to a bracket which is welded to the forward end of the front axle tunnel by four cap screws. The commutator end of the motor is down, and the hydraulic pump mounts to the top end of the motor.

b. Steering Gear Motor Removal.

(1) Remove ammunition box from rack in front of assistant driver's seat.

(2) Remove the power input cable connector from the motor. NOTE: This connector has one pin on the motor side. There is a guide key in one connector half and a keyway in the other for proper location. The coupling halves are held together by a knurled threaded collar.

(3) Remove the three hydraulic pump to motor screws. NOTE: Do not disconnect or loosen any of the hydraulic lines.

(4) Remove the four motor base-to-mounting bracket cap screws and lock washers. NOTE: This motor weighs approximately eighty pounds; therefore it must be properly supported before removing the last two screws.

(5) Lower motor-to-hull floor and work it to the right toward the assistant driver's foot space, and back over the front axle tunnel.

(6) Any service on this unit must be handled by ordnance personnel.

c. Steering Gear Motor Installation.

(1) Place the motor in the hull; work it up past the assistant driver's seat into the foot space. Slide it to the left and block it up in position for entering the retaining bolts.

(2) Line up the bolt holes and install the four bolts which attach the motor to bracket.

(3) Install the three bolts which attach the hydraulic pump to motor.

(4) Attach the power input cable connector to the motor. NOTE: Make sure the cable connector is in correct position before tightening the collar.

(5) Check all hydraulic line connections to see that they are tight.

(6) Test motor to see that it works properly by trying the switch a time or two.

(7) Install ammunition box in rack in front of assistant driver's seat.

148. STEERING CONNECTING ROD (fig. 115).

a. Description.

(1) The steering connecting rod, which connects the steering gear pitman arm to the steering third arm, is made of two tubular sections.

TM 9-741

MEDIUM ARMORED CAR T17E1

The steering third arm end is smaller in diameter than the pitman arm end; it is threaded and screws inside the pitman arm end. A large lock nut is used to lock the two sections in position.

(2) The connections to the pitman arm and steering arm are made by ball and socket joints. The balls are attached to the pitman arm and steering third arm and mount in spring-loaded ball seats in the steering connecting rod. These joints are lubricated by regular fittings mounted in the connecting rod in line with the ball joints.

(3) A spring-loaded seal at each ball joint retains the lubricant and prevents dirt from entering. This seal consists of spring, spring seats, two spacers, inner seal retainer, outer seal retainer, and felt seal (fig. 115).

b. **Steering Connecting Rod (Removal)** (fig. 115).

(1) Remove the cotter pins.

(2) Remove the end plugs.

(3) Turn the steering wheel each way to push the end ball seats away from the ball. Pull the connecting rod off the ball at each end.

(4) Remove the ball seats, springs, spring bumpers, and dust seal assemblies.

(5) Clean all parts carefully and inspect them for wear. Replace all damaged parts.

c. **Assembly** (fig. 115).

(1) Install the dust seal spring outer retainer and spring seat assembly, inner retainer, felt seal and the two halves of the spacer on the pitman arm and third arm ball.

(2) Install spring bumpers, flat side first, the special springs, two ball seats at each end, and start the end plugs in the threaded ends of the rod.

(3) Assemble rod to pitman arm and third arm.

(4) Turn the adjusting plugs up tight and back them off from ½ to 1 turn. Install cotter pins.

TM 9-741

STEERING GEAR

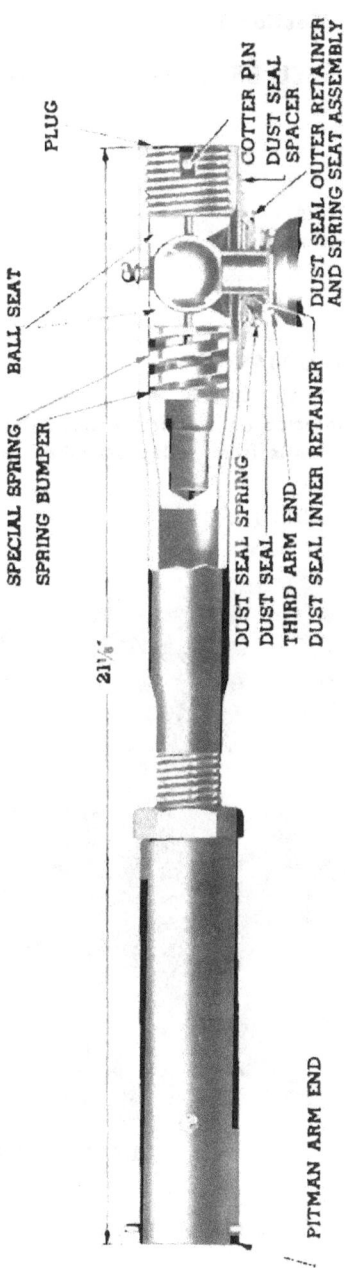

Figure 115 — Steering Connecting Rod

TM 9-741
149

MEDIUM ARMORED CAR T17E1

Section XXVIII

WHEELS, TIRES, WHEEL BEARINGS, AND TIRE PUMP

	Paragraph
Description	149
Trouble shooting	150
Maintenance and adjustments	151
Wheel and tire replacement	152
Wheel bearings	153
Tire pump	154

149. DESCRIPTION. (fig. 116).

a. **Wheels.** The wheels have 10.00cw. divided rims front and rear. A bead lock assembly is inserted on the rim between the beads of the casing to prevent the side walls from collapsing when the tire is run

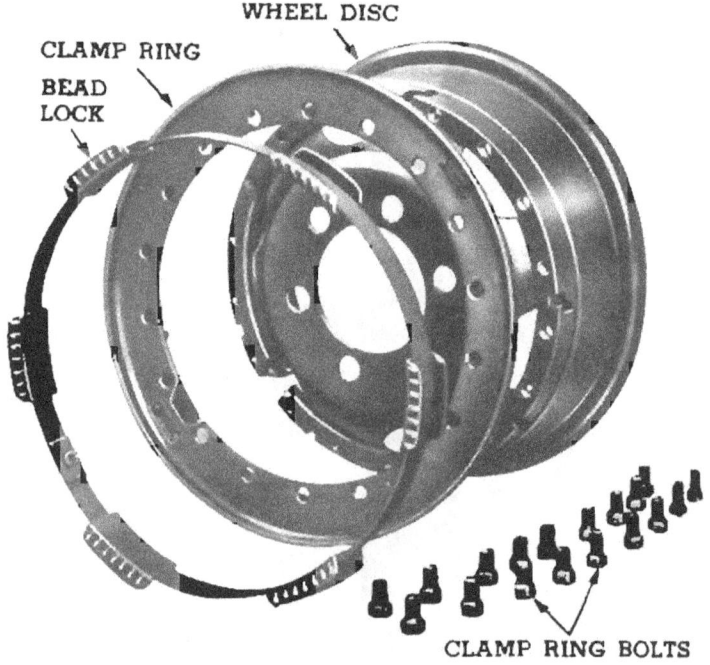

Figure 116 — Wheel Assembly

238

WHEELS, TIRES, WHEEL BEARINGS, AND TIRE PUMP

flat. A retainer is fastened to the bead lock assembly to prevent the valve stem from slipping into the tire when the tire is run flat.

b. Tires. Combat tires (14.00x20) are used front and rear. The inflation pressure is 70 pounds front and 80 pounds rear.

c. Wheel Bearings. The roller type bearings are interchangeable on all wheels. The hub and drum assemblies are interchangeable front and rear on the same side of the vehicle but not from one side to the other unless the proper wheel studs are used. Wheel studs with right-hand threads are used on the right side of the vehicle and left-hand threads on the left side.

150. TROUBLE SHOOTING.

a. Hard Steering.

Probable Cause	Probable Remedy
Tires under-inflated.	Inflate front tires to 70 pounds: rear, to 80 pounds.
Wheel bearings out of adjustment.	Readjust according to instructions in paragraph 153.
Wheel bearings scored or seized.	Report to ordnance personnel.
Lack of lubrication.	Lubricate steering gear, steering connectings rod, and the rod fittings.
Steering knuckle bearings improperly adjusted.	Report to ordnance personnel.
Improper steering adjustment.	Report to ordnance personnel.
Hydraulic system not operating properly.	Report to ordnance personnel.

b. Air Leakage.

Tube leaks.	Check for slow leaks in tube or at valve stem. Repair.
Valve cap missing.	Install new cap.
Valve core loose.	Remove cap and tighten core.
Valve core damaged.	Replace core.

c. Uneven Tire Wear.

Irregular.	Interchange tires to compensate for wear.
Side wear.	Check for improper camber and under-inflation.

MEDIUM ARMORED CAR T17E1

Probable Cause	Probable Remedy
Excessive wear.	Check for improper toe-in (par. 123).

d. Front Wheel Shimmy.

Improper tire inflation.	Inflate to 70 pounds front, 80 pounds rear.
Wheels loose on hubs.	Tighten wheel hub nuts (par. 152).
Improper toe-in.	Adjust (par. 123).
Loose front wheel bearings.	Adjust (par. 153).
Steering knuckle support trunnion bearing loose.	Report to ordnance personnel.

e. Wandering.

Axle shifted.	Report to ordnance personnel.
Loose front wheel bearings.	Adjust (par. 153).
Bent tie-rod.	Replace (par. 122).

151. MAINTENANCE AND ADJUSTMENTS.

a. **Tires and Tubes.** Tires should be repaired in accordance with conventional methods. Punctures and tears causing exposure of the cord or fabric should be vulcanized. Holes in bullet resisting and puncture sealing inner tubes should be repaired by cold patching. Hot patching or vulcanizing should not be attempted. Tires must be inflated equally and should not be operated under-inflated. Balanced tire pressures of the proper amount facilitate steering, improve riding conditions contribute toward safer driving and maximum tire mileage. Before pumping air into tubes, depress valve momentarily to let old air blow out any dirt in the valve. Keep caps on valves to prevent entrance of dirt and water. As oil and grease have a harmful effect on rubber, every attempt should be made to keep these substances from coming in contact with the tires.

b. **Wheels.** Check and tighten wheel stud nuts daily to make certain they do not loosen and "work" on the hub studs. Use the wrench provided in the tool kit for this purpose and do not use an extension on the handle or apply excessive force other than direct hand effort Successively tighten opposite nuts to prevent cocking wheel on stud Never use oil or grease on the wheel stud nuts.

152. WHEEL AND TIRE REPLACEMENT (fig. 117).

a. **Wheel Removal Procedure.**

(1) LOOSEN WHEEL NUTS. Loosen the ten wheel nuts with wrench provided in tool kit. NOTE: The wheel studs on the left side of vehicle

TM 9-741
152

WHEELS, TIRES, WHEEL BEARINGS, AND TIRE PUMP

RA PD 32516

Figure 117 — Removing Wheel and Tire Assembly

are left-hand threads, and the wheel studs on the right side of vehicle are right-hand threads. The bolts are stamped "L" for left and "R" for right.

(2) RAISE VEHICLE AND REMOVE WHEEL. With jack provided in tool kit, raise vehicle until the tire clears the ground. If a dolly similar to the one shown in figure 117 is available, place dolly under tire as shown. If a dolly is not available, a suitable sling can be made with a rope and chain hoist, provided that the rope from the hoist follows straight down the outside of the tire and loops around bottom of tire, up the inside, and forms a slip loop at the top and outside edge of tire. Remove the ten nuts and remove the wheel and tire. CAUTION: Be sure to float the wheel off the hub to prevent damaging the threads on hub bolts.

b. **Wheel Installation Procedure.**

(1) INSTALL WHEEL. Raise wheel and tire with dolly or chain hoist to a position that will permit the wheel to float on the hub bolts without damaging the bolts. Install the ten nuts and tighten them snugly.

(2) LOWER VEHICLE. Lower the jack and remove it. Tighten the ten wheel nuts securely. NOTE: Do not use an extension on the wrench handle that is provided in the tool kit.

241

TM 9-741
152

MEDIUM ARMORED CAR T17E1

RA PD 32505

Figure 118 — Removing Bead Lock

c. Tire Replacement. The wheel and tire assembly should be removed from the vehicle before attempting to change the tire.

(1) TIRE AND TUBE REMOVAL PROCEDURE.

(a) *Remove Clamp Ring.* Completely deflate the tire by removing the valve core. CAUTION: The tire must be completely deflated before loosening the nuts on the clamp ring to prevent possible injury if the clamp ring should be forced off due to the air pressure in the tire. Remove the eighteen nuts that attach the clamp ring to the wheel and pry off the clamp ring.

(b) *Remove Tire and Tube* (fig. 118). Press tire off of rim. Install the valve core in the tube and inflate the tube slightly to spread the tire beads. Distort the flexible band of the bead lock assembly at a point directly opposite the valve stem by pulling the bead lock toward the center of the tire. Then rotate the assembly to remove the bead lock retainer over and off the valve stem. Deflate the tube again and remove the tube and flap.

(2) TIRE AND TUBE INSTALLATION PROCEDURE.

(a) *Install Tube and Bead Lock* (fig. 118). Insert the tube in the casing with the valve stem at the red dot on the side wall of the casing.

242

WHEELS, TIRES, WHEEL BEARINGS, AND TIRE PUMP

Inflate the tube slightly and insert the flap. Start the bead lock over the valve stem and rotate the bead lock so that the retainer lugs will not interfere with or distort the valve stem while inserting the bead locks in the casing. Use tire tools and make sure that flap remains in the proper position.

(b) *Install Tire and Tube Assembly.* Deflate the tube by removing the valve core and press the tire and tube on the rim with the valve stem toward the outside of the wheel. Install the clamp ring, attaching the nuts on opposite sides. Turn down the nuts about one turn at a time opposite each other so as to equalize the pressure around the clamp ring. **NOTE:** It may be necessary to press the ring down to get the first two nuts started. Tighten all nuts securely, inflate tire to recommended pressure, and install the valve cap.

153. WHEEL BEARINGS.

a. Description. The wheel bearings are all interchangeable; therefore, the adjustment procedure is the same for all wheels.

b. Adjustment.

(1) Jack up vehicle at the wheel to adjust and place a stand jack under the axle.

(2) Remove the drive flange (fig. 119) on the front axle (par. 120) or remove the rear shaft (par. 126).

(3) Raise the lug of the wheel hub nut lock from the notch in the outer wheel hub adjusting nut. Remove the outer adjusting nut using the wrench provided with the vehicle, and remove the lock (fig. 120).

(4) Using the wheel bearing adjusting nut wrench, tighten the inner adjusting nut "wrench tight;" then back it off $\frac{1}{8}$ turn or 45 degrees minimum (fig. 120).

(5) Install adjusting nut lock and check the alinement of one of the three short lugs with the nearest notch in the nut. If further adjustment is necessary to secure alinement of a notch with a short lug, the nut should be backed off rather than tightened.

(6) Rotate hub and drum assembly to see that the bearings are seated properly and that the assembly turns freely.

(7) Punch the alined short lock lug into its notch in the adjusting nut. Install the outer adjusting nut and pull down tight to prevent any loosening of the inner adjusting nut. Punch one of the eight bent lugs on the adjusting nut lock securely into a corresponding notch in the outer adjusting nut. Recheck assembly for free rotation.

TM 9-741
153
MEDIUM ARMORED CAR T17E1

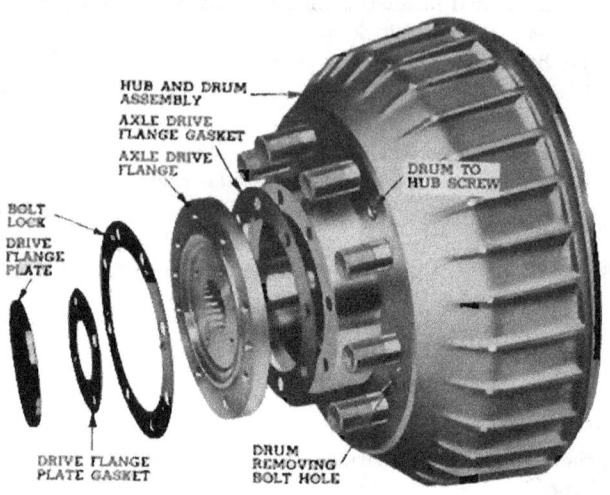

Figure 119 — Front Axle Drive Flange

Figure 120 — Adjusting Wheel Bearing

TM 9-741
153-154

WHEELS, TIRES, WHEEL BEARINGS, AND TIRE PUMP

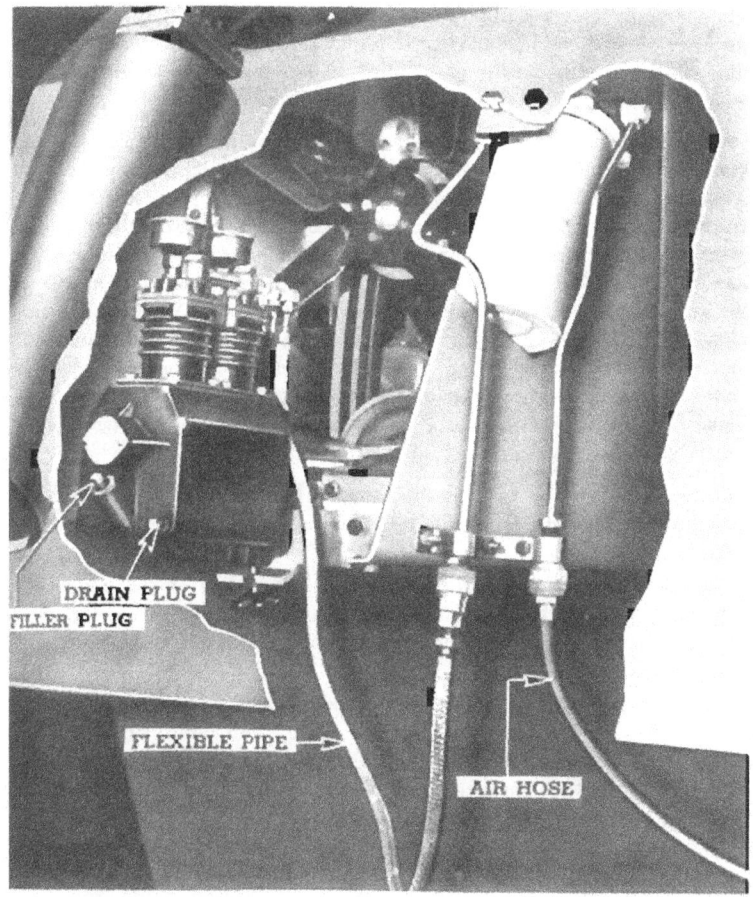

Figure 121 — Tire Pump Mounting

(8) Install axle shaft or drive flange and a new terneplate gasket.

(9) Using a new axle shaft or drive flange bolt lock, install axle shaft or drive flange bolts and tighten securely (100 to 120 pound-ft with torque wrench). Lock bolts.

154. TIRE PUMP (fig. 121).

a. **Description.** A power air pump is supplied with each vehicle for tire inflation. The pump and hose are carried in a metal box which is stowed in one of the side luggage boxes. The power unit consists of a

245

MEDIUM ARMORED CAR T17E1

two-cylinder engine-driven pump with the necessary attaching clamps and drive shaft, a flexible metal pipe for connecting the pump to the air filter which is permanently attached to the vehicle on the hull upper plate, and 22 feet of airhose for inflating the tires.

b. **Maintenance.** The oil in the crankcase should be changed at least twice a year, in the fall and spring. To drain the oil, remove the drain plug (fig. 121). When oil is drained, install drain plug. Remove the filler plug and fill with 1¼ pints of OIL, engine, seasonal grade. Install filler plug. Remove the air cleaner body caps from the pump at regular intervals and inspect the felts. If the upper felt has a crust of dirt or sand, check the lower felt and wash them in **SOLVENT**, dry-cleaning and dry them with air pressure. If crust of dirt will not wash off the felt, the top layer of felt may be peeled off, as the felts are laminated. When the pump is not in use, the dust covers should always be in place on the air filter inlet and outlet couplings as shown in figure 122. After using the pump, drain water from drain cock in filter.

c. **Power Tire Pump Installation** (fig. 121). Before installing the power pump, unscrew the filler plug, and see that the pump crankcase is full of oil.

(1) Attach the flexible metal pipe to the front end of the pump manifold by pressing the female coupling onto the male coupling. **NOTE:** To remove coupling, turn the knurled collar on the female coupling and pull down.

(2) Pass the pump unit up into the opening between the hull rear lower plate and the hull rear upper plate. Direct the pump propeller shaft through the guide plate and engage the drive pin extensions into the special drive cup on the engine crankshaft.

(3) Set pump on hull bracket as shown in figure 121. Engage the cover clamp eye bolt in the slot in the lower edge of the tire pump mounting bracket. Screw the cover clamp knob in tight against the bracket.

(4) Push the flexible metal pipe coupling onto the air filter inlet pipe. Push the air delivery hose coupling onto the air filter outlet pipe.

(5) Attach the air delivery hose chuck to the valve stem of the tire to be inflated and start and run the engine at about 600 to 800 revolutions per minute until the desired air pressure in the tire is obtained.

d. **Power Tire Pump Removal.**

(1) Disconnect the air hose chuck from the valve stem. Disconnect the air hose from the filter by turning the knurled knob and pulling down on the coupling.

TM 9-741
154

WHEELS, TIRES, WHEEL BEARINGS, AND TIRE PUMP

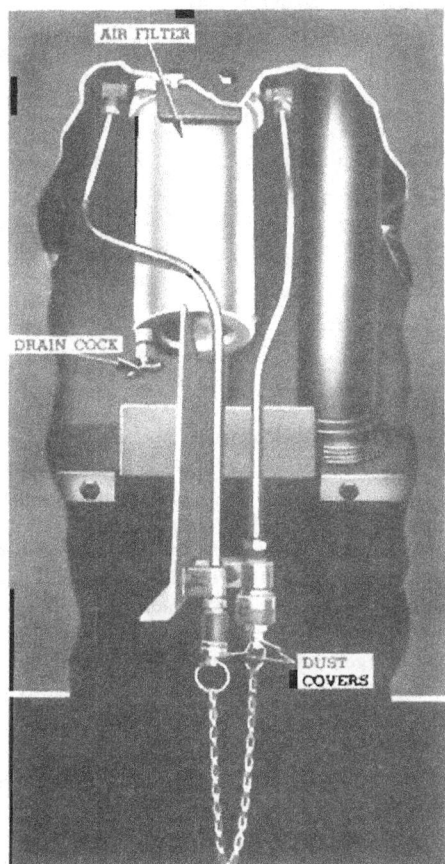

Figure 122 — Tire Pump Air Filter

(2) Disconnect the flexible metal pipe from the air filter by turning the knurled knob and pulling the connection down. Open drain cock on filter to drain the water.

(3) Unscrew eye bolt clamp and remove from bracket. Disconnect the flexible metal pipe from the manifold by turning the knurled knob and pulling down on the connection.

TM 9-741
155-156

MEDIUM ARMORED CAR T17E1

Section XXIX

BATTERIES AND STARTING SYSTEM

                                                                  Paragraph

Description 155
Battery inspection and maintenance...................... 156
Starting motor maintenance 157
Trouble shooting 158
Battery replacement 159
Starting motor replacement 160
Solenoid switch replacement 161

155. DESCRIPTION.

a. General. The battery and starting system consists of four batteries, two starting motors, two starting motor switches, cables, and the necessary wiring. The batteries supply the energy; the switches complete the circuit, allowing the battery energy to flow to the starting motor. The starting motors then deliver mechanical energy which does the actual work of cranking the engine. The batteries, however, perform other functions. The proper operation of the battery and starting system depends on the amount of current in the batteries and on the condition of the wiring and terminals. Therefore, it is of major importance to keep the batteries fully charged and the wiring terminals free of corrosion and tight. All of the wiring and cables are shielded to prevent interference with the radio operation; the shielding of the wiring is explained in paragraph 190.

b. Batteries. The batteries are 25-plate, 6-volt batteries and have a rating of 200 ampere hours each. The batteries are mounted in the floor of the hull under the front edge of the turret bracket, two batteries on each side of the front axle drive shaft tunnel. Access for inspection and filling the batteries is provided through a removable plate located in the floor of the turret basket.

c. Starting System. Each engine has its own starting system which consists of the starting motor, solenoid switch, starter button, and the necessary cables and wiring. Both systems use the same set of batteries

156. BATTERY INSPECTION AND MAINTENANCE.

a. General. The batteries should be tested and filled with distilled water once a week to obtain maximum service from them. Operation of the vehicle in zero or subzero temperatures makes frequent checking of the specific gravity of the solution imperative. Readings which would be satisfactory for normal temperatures will prove entirely unsatisfac

BATTERIES AND STARTING SYSTEM

tory in low temperatures and may result in freezing of the battery electrolyte. The temperatures at which electrolyte will freeze are as follows:

Specific Gravity	Temperature
1.180	−6 degrees F
1.200	−17 degrees F
1.220	−31 degrees F
1.240	−51 degrees F

The batteries must be kept tight in their retainers to prevent physical damage, and the terminals must be kept tight and free of corrosion.

b. **Test Battery.** Remove the three vent caps from each battery and check the specific gravity of the electrolyte solution with a reliable battery hydrometer. A fully charged battery will show a reading of 1.275 to 1.300. A fully discharged battery will read approximately 1.150. Should the reading be between 1.200 and 1.240, the battery should be recharged and the cause of the partially discharged condition investigated. If a battery or a cell does not read over 1.250 after the battery has been recharged, the battery should be replaced with a new or reconditioned battery.

c. **Filling Batteries.** After testing the batteries they should be filled with distilled water to a level of ¼ inch above the plates. If the temperature is below freezing, the water should be added before operating the vehicle. If the water is added to the elctrolyte and the vehicle is allowed to stand under these conditions, the water will freeze and the battery will be damaged. When replacing the caps, make sure that they are tight in the openings.

d. **Servicing Battery Terminals.** To clean the battery terminals, remove the cables from the batteries, scrape the corrosion from the battery terminals and the inside of the battery cables, and wash them off with a strong ammonia and water solution. After replacing the cables on the battery terminals, tighten the clamp bolt nuts tight and apply a coating of grease on the cable clamp and battery terminal.

157. STARTING MOTOR MAINTENANCE.

a. **Lubrication.** Lubricate the hinged cap oiler with 8 to 10 drops of engine oil every 5,000 miles.

b. **Cleaning Commutator.** The cover band should be removed periodically and the brushes and commutators inspected for dirt and roughness. If the commutator is dirty, it may be cleaned with paper, flint, No. 00; then blow out the dust with compressed air.

TM 9-741
157-160

MEDIUM ARMORED CAR T17E1

c. Testing Brush Springs. If weak springs are indicated, their tension may be checked with a spring scale. The tension necessary to lift the brush from the commutator is 24 to 28 ounces.

158. TROUBLE SHOOTING.

a. Discharged Battery.

Probable Cause	Probable Remedy
Loose or dirty terminals.	Clean and tighten terminals.
Generators not charging.	Refer to ordnance personnel.
Leak in wiring.	Check all wiring for short or open circuits.
Excessive use of starting motor due to hard starting.	Tune engine (par. 62).
Cells shorted.	Replace battery (par. 159).
Cells dry.	Fill and recharge battery (par. 156).

159. BATTERY REPLACEMENT.

a. Remove Batteries. Remove the two wing nuts from the triangular covers which cover the corner of the battery just inside of each entrance door. Lift out the two removable floor plates in the turret basket and remove all of the battery cables. NOTE: When removing negative terminal from right-hand set of batteries, wrap the end of cable in a rag to prevent the terminal from touching any metal which would cause a short circuit. Remove the four nuts that attach the retainer on the top of each pair of batteries and remove the retainers. Lift rear end of inside battery up and toward the rear until the front of the battery clears the opening in the basket; then lift the battery out. Push the outside battery over to the inside position and remove it in the same manner.

b. Install Batteries. Place one battery in the carrier through the hole in the basket and push it to the outside position; place other battery in the inside position. Install the battery retainer and tighten the four nuts securely. Install all of the battery cables; tighten the clamps securely and spread a coat of grease on the terminals. Install the floor plates in the basket and the triangular cover with wing nuts.

160. STARTING MOTOR REPLACEMENT.

a. Remove Starting Motor (figs. 36 and 37). Remove the transmission inspection plate and remove the lower bolt that attaches the starting motor to the flywheel housing. Remove the twenty-five bolts that attach the compartment cover to the hull and remove the cover.

BATTERIES AND STARTING SYSTEM

Remove the bolts that attach the coil to the coil mounting bracket and move the coil out of the way. Disconnect the cable and two wires from solenoid switch. Remove the top bolt that attaches the starting motor to the clutch housing and pull the starting motor back until it clears the opening in the flywheel housing. Tie a small rope around the drive end and thread the starting motor up and out of the opening which was made by removing the coil.

b. Install Starting Motor (figs. 36 and 37). Lower starting motor through opening made by removing the coil. Place motor in position in flywheel housing and install upper bolt and lock washer. Connect battery cable and the two wires to the solenoid switch. Place the coil in position on the mounting bracket and install the mounting bolts and lock washers. Install the lower attaching bolt at the bottom of the starting motor and install transmission inspection plate. Press the starting button to check the operation, and if satisfactory, install the compartment cover and tighten the attaching bolts securely.

161. SOLENOID SWITCH REPLACEMENT (fig. 36).

a. Remove Solenoid Switch. Remove the twenty-five bolts that attach the compartment cover to the hull and remove the cover. Disconnect the three wires from the solenoid switch. Remove the two nuts that attach the copper strap and remove the strap. Remove the pin from the shift lever linkage. Remove the four cap screws that attach the solenoid switch to the starting motor and remove the switch.

b. Install Solenoid Switch. Place switch in position on starting motor and install the four mounting cap screws. Install the copper strap and tighten the two nuts securely. Connect the three wires to the terminals: Install the pin to secure the shift lever linkage to the shift lever and secure with cotter pin. Install the compartment cover.

TM 9-741
162-163

MEDIUM ARMORED CAR T17E1

Section XXX

GENERATORS AND CONTROLS

	Paragraph
Description	162
Inspection and testing	163
Trouble shooting	164
Generator replacement	165
Generator regulator unit replacement	166

162. DESCRIPTION (figs. 35, 37, and 38).

a. Each engine is equipped with a generator which is driven by two fan belts. The generators are located between the engines. Each generator has a regulator unit in the circuit which controls the maximum voltage of the generator and keeps it from exceeding a predetermined value fixed by the setting of the regulators. The actual charging rate to the battery varies, depending on the state of charge in the battery. The current regulator controls the maximum amperage output of the generators and prevents them from exceeding 50 amperes, which is the setting of the current regulator, thereby preventing damage to the generator due to overload. The cut-out relay prevents the battery from discharging through the generator when the engine is not running. The function of the generator system may be summed up as follows: It converts a small amount of mechanical energy from the engine into electrical energy which is carried through the wiring to the batteries where it is stored for future use. Each generator and its system operates independently of the other generator. All of wiring in the generator circuit is shielded to prevent interference with the radio operation. The shielding of the wires is explained in paragraph 190.

163. INSPECTION AND TESTING.

a. **Inspection.** The cover band should be removed from the generator and the commutator inspected at regular intervals. If the commutator is dirty, it may be cleaned with PAPER, flint, No. 00, then blow out the dust with compressed air. If the commutator is rough, out-of-round, or has high mica, the generator should be replaced with a new or reconditioned generator. The spring tension checked with a spring scale should read 24 to 28 ounces to lift the brushes off the commutator.

b. **Testing.** Whenever the ammeter indicates that the generator is not operating properly, it will be necessary to make several tests to determine whether the trouble is in the generator, regulator unit, or battery. Because of the shielding on the wiring and terminals, it will

TM 9-741
163-164

GENERATORS AND CONTROLS

be necessary to make a jumper lead about 18 inches long with terminals that will fit the shield terminals on the generator and regulator filter box. The jumper lead should be made of insulated wire but it does not need to be shielded.

(1) LOW OR NO CHARGING RATE TEST.

(a) Check the circuit for loose connections, corroded battery terminals, and loose or corroded ground straps. The high resistance resulting from these conditions will prevent normal charge from reaching the battery. If the entire charging circuit is in good condition, and the battery is not fully charged, then either the regulator or generator is at fault.

(b) With a jumper wire, connect the field and armature terminals of the generator, increase the speed of the engine momentarily and check the output. If the output increases, the regulator is at fault and should be replaced. If the output does not increase, a further check is necessary.

(c) Remove the armature wire from the generator and connect the field and armature terminals with a jumper wire; increase the engine speed momentarily and flash the armature terminal with a screwdriver. If no spark occurs, the trouble is in the generator, and it should be replaced.

(2) UNSTEADY OR LOW OUTPUT TEST.

(a) Check drive belt tension as instructed in paragraph 86.

(b) Check brush spring tension with spring scale which should read 24 to 28 ounces to lift brush from commutator.

(c) Check commutator for roughness, grease, dirt, high mica, out-of-round and burned bars. If any of these conditions exist, the generator should be replaced.

164. TROUBLE SHOOTING.

a. No Output.

Probable Cause	Probable Remedy
Burned commutator bars.	Replace generator (par. 165).
Dirty commutator.	Clean commutator with paper, flint No. 00, and blow out dust.
Loose fan belts.	Adjust fan belts (par. 86).
Faulty ammeter.	Replace ammeter (par. 180).
Faulty filter.	Replace filter.
Defective generator.	Replace (par. 165).

MEDIUM ARMORED CAR T17E1

Probable Cause	Probable Remedy
b. Unsteady or Low Output.	
Fan belts loose.	Adjust fan belts (par. 86).
Dirty commutator.	Clean commutator with paper, flint, No. 00, and blow out dust.
Defective generator.	Replace (par. 165).
c. Excessive Output.	
Defective generator.	Replace generator (par. 165), and generator regulator (par. 166).
d. Noisy Generator.	
Loose mounting.	Tighten mounting bolts.
Worn, dry or dirty bearings.	Replace generator (par. 165).
Improperly seated brushes.	Replace generator (par. 165).
e. Low or No Charging Rate, or Excessive Charging Rate.	
Generator regulator adjustment incorrect.	Replace regulator (par. 166).

165. GENERATOR REPLACEMENT.

a. Removal Procedure.

(1) REMOVE COMPARTMENT COVER. Remove the twenty-five bolts that attach the compartment cover to the hull and remove the cover.

(2) REMOVE FAN SHROUD. Remove the four bolts at the top that attach the shroud to the brackets; then remove the three bolts on each side of the shroud. Crawl under opening at rear of hull and remove the diagonal brace in the opening nearest to the center partition. Remove four bolts in left compartment, or remove the seven bolts in right compartment that attach the shroud at the bottom and remove the shroud.

(3) REMOVE FAN BLADES. Remove the four bolts that attach the fan blades to the pulley and remove the blades.

(4) REMOVE GENERATOR. Remove the bolt in the slotted brace at each end of the generator, push the generator toward the engine, and remove the fan belts. Disconnect the two terminals on top of generator and remove the wires. Remove the four bolts on the bottom of the generator that attach the generator to the saddle. Tie a rope around the pulley and slide generator to the rear and down through opening in rear of hull.

b. Installation Procedure.

(1) INSTALL GENERATOR. Lift generator through opening at bottom and rear of hull and into position on mounting saddle and install the four bolts and lock washer. CAUTION: Be sure to use the same length bolt

TM 9-741
165

GENERATORS AND CONTROLS

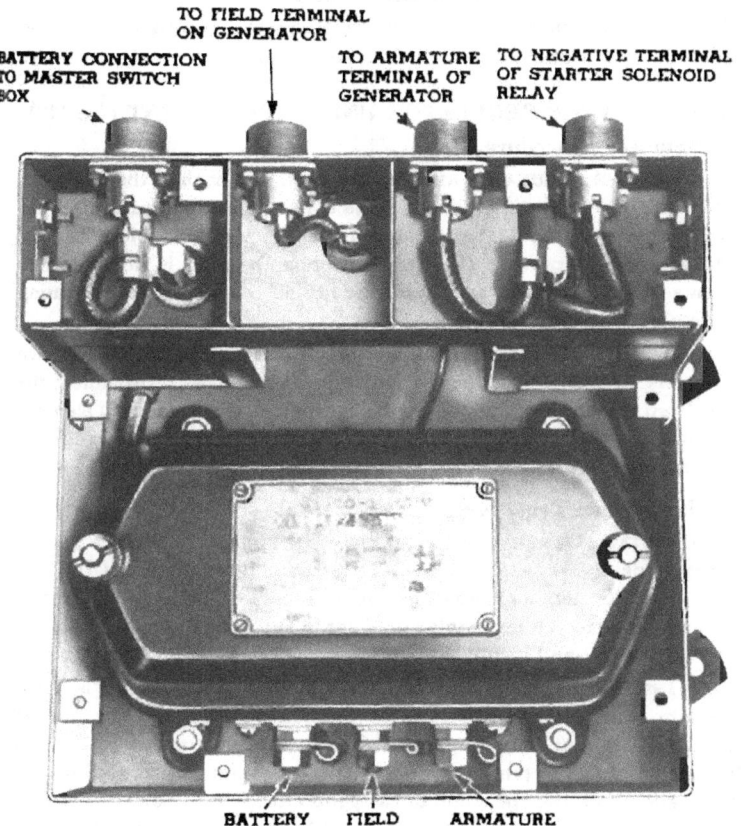

Figure 123 — Generator Regulator Connections

that was originally furnished and an extra thick lock washer. Install bolts in the slotted brace at each end of generator. Connect the two wire connections at top of generator. Install the fan belt over the pulleys.

(2) INSTALL FAN BLADES. Place the fan blades in position on the pulley and install the four bolts and lock washers. Tighten the bolts securely. Adjust fan belt (par. 86).

MEDIUM ARMORED CAR T17E1

(3) INSTALL FAN SHROUD. Place the fan shroud in position and install all the bolts and lock washers. Tighten the bolts securely. Install the diagonal brace and tighten the bolts securely.

(4) INSTALL COMPARTMENT COVER. Place the compartment cover in position and install the twenty-five bolts. Tighten the bolts securely.

166. GENERATOR REGULATOR UNIT REPLACEMENT (fig. 123).

a. Removal Procedure.

(1) REMOVE REGULATOR AND FILTER BOX. Tag each of the four wires so that they can be installed in the same connectors; then disconnect the four terminals on top of box, and remove the wires. Insulate battery lead with tape to prevent short circuit while the lead is disconnected. Remove the two bolts on each side that attach the box to the mounting brackets and lift out the box.

(2) REMOVE THE GENERATOR REGULATOR UNIT (fig. 123). Remove the screws that attach the cover and remove the cover. Disconnect the wires from the three terminals at the bottom of the regulator. Remove the four bolts that attach the generator regulator to the box and lift out the generator regulator.

b. Installation Procedure.

(1) INSTALL GENERATOR REGULATOR UNIT (fig. 123). Place generator regulator in box with terminals toward the bottom. Install the four mounting bolts and lock washers; tighten the bolts securely. Connect the three wires to the terminals and tighten the nuts securely. Place the cover on the box and install the screws.

(2) INSTALL GENERATOR REGULATOR AND FILTER BOX. Place the generator regulator and filter box in position on the mounting bracket and install the four mounting bolts; tighten the bolts securely. Connect the four wires to the same terminals they were removed from. Start the engine and check the operation of the generator. If satisfactory, place the engine compartment cover in position and install the twenty-five bolts. Tighten the bolts securely.

TM 9-741
167

Section XXXI

LIGHTING SYSTEM

	Paragraph
Description	167
Trouble Shooting	168
Head lamp replacement	169
Head lamp sealed beam assembly replacement	170
Head lamp aiming	171
Blackout marker lamp	172
Auxiliary blackout drive lamp sealed beam assembly replacement	173
Tail and stop lamp assembly replacement	174
Tail and stop lamp unit replacement	175
Instrument panel bulb replacement	176
Dome lamp bulb replacement	177

167. DESCRIPTION.

a. General. The lighting system consists of two head lights, two blackout marker lights that are attached to the top of the head lights, a blackout tail and stop light, a blackout tail, service tail, and stop light, and auxiliary blackout drive light and blackout marker lamp, four instrument panel lights, two dome lights in the driver's compartment and two dome lights in the turret. A portable inspection lamp is included in the tool kit. An automatic circuit breaker is connected in the lighting circuit which breaks the circuit in the event the load exceeds 20 amperes. In addition to the light switches and automatic circuit breaker switches, a master switch is incorporated which is located to the rear and above the right entrance door (fig. 18).

b. Head Lamps. The head lamps are of the sealed beam type, the reflector, bulb and lens are sealed as a single unit. The lamps are mounted on special brackets at the front of the hull and are protected by steel straps extending over the top of each assembly. The special mounting brackets incorporate a release button, so that the head lamps can be removed quickly and stored inside the vehicle. Plugs are carried in the storage brackets to plug the openings in the brackets when the lamps are removed. The head lamps operate on 24 volts and have a candlepower of 50 watts.

c. Blackout Marker Lamps. The blackout marker lamps are fastened to the top of the head lamps. The bulb is a 24-volt, 3-candle-

MEDIUM ARMORED CAR T17E1

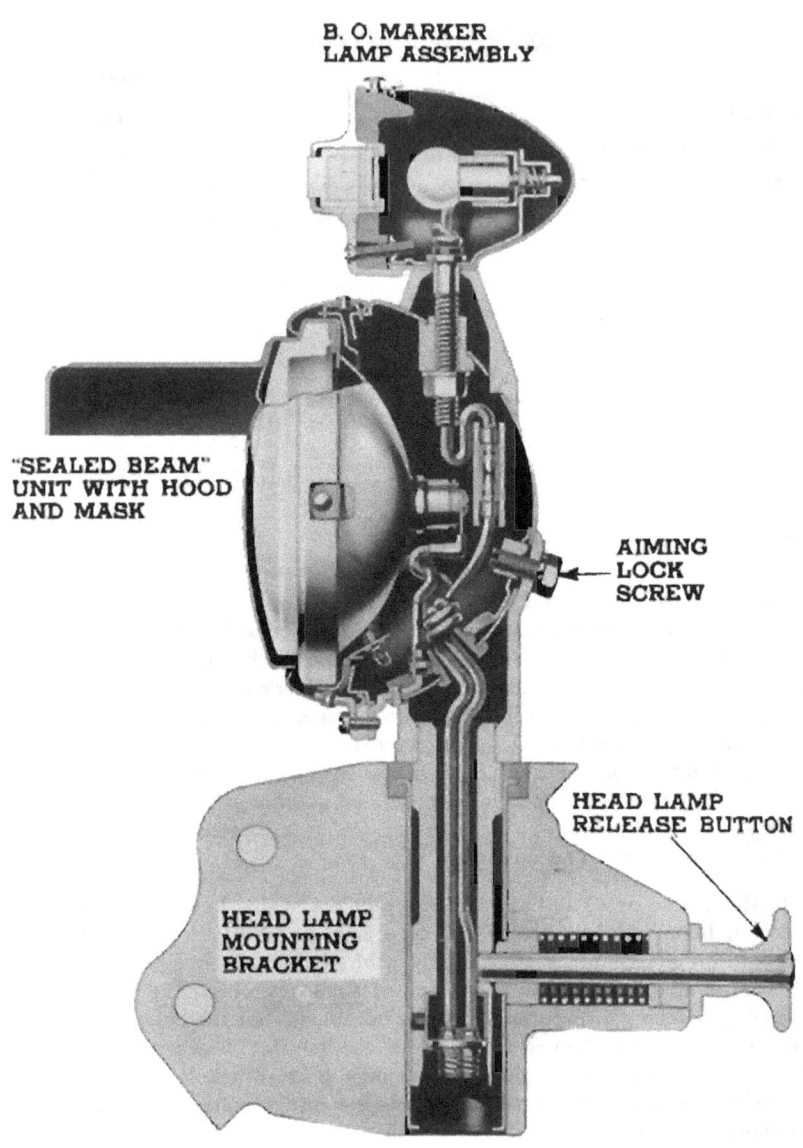

Figure 124 — Blackout Driving Lamp

TM 9-741
167

LIGHTING SYSTEM

power, single contact bayonet base bulb. The bulbs in the marker lamps can be replaced.

d. **Auxiliary Blackout Drive and Marker Lamp** (fig. 124). The auxiliary blackout drive and marker lamp is carried in a storage bracket on the left side of the hull. This lamp is used in place of the left head lamp for driving in a blackout. The diffused beam pattern does not have a hot spot concentrated on the road. The beam pattern casts illumination from 30 to 100 feet in front of the vehicle. The blackout drive lamp is equipped with a "sealed" unit which has a 6-volt, 10-candlepower bulb. A resistence in series with the lamp permits its use on a 24-volt system. The marker lamp is the same as the marker lamp mounted on the head lamps. The drive light incorporates a hood and mask which diffuses the beam and prevents detection of the beam from overhead.

e. **Blackout Tail and Stop Lamp.** A combination blackout tail and stop lamp is mounted on the rear of the vehicle on the right side, the blackout tail and stop lamps each consist of a housing, a 24-volt, 3-candlepower bulb soldered to the housing, a filter, and lens. When the lamp bulb in either lamp burns out, the housing unit must be replaced.

f. **Blackout Tail and Service Tail and Stop Lamp.** The blackout tail and service tail and stop lamp is mounted on the rear of the vehicle on the left-hand side. It is constructed in the same manner as the blackout tail and stop lamp, the difference being in the service tail and stop lamp unit which contains a double filament bulb, 6-candlepower for the taillight and 32 candlepower for the stop light and a lens made of ruby glass.

g. **Instrument Panel Lights.** The instrument panel lights are all controlled by the "PANEL LIGHTS" switch, and the brilliance of the lights from bright to out can be varied by turning the light switch knob. The bulbs are 24-volt, 3-candlepower single contact, bayonet base bulbs, having a double filament in series, giving a life of approximately 1,000 hours.

h. **Dome Lamps.** Two dome lamps are provided in the driver's compartment, located over the heads of the driver and assistant driver. These lamps are controlled by toggle switches integral with the lamps. Two dome lamps are provided in the turret which are controlled by a switch located on the turret control box and toggle switches that are integral with each lamp so that one or both lights can be used. The dome lamp bulbs are 24-volt, 3-candlepower, single contact, bayonet base bulbs.

i. **Inspection Lamp.** The inspection lamp, which is carried in the tool kit, is portable and is provided with 15 feet of cord. It can be connected

TM 9-741
167-171

MEDIUM ARMORED CAR T17E1

in either of the windshield wiper outlets on the instrument panel or at the inspection lamp socket on the turret control switch box. The bulb is a 24-volt, 15-candlepower, double contact, bayonet base bulb.

168. TROUBLE SHOOTING.

a. Lights Burn Dim.

Probable Remedy	Probable Remedy
Loose connections.	Clean and tighten connections.
Burned switch contacts.	Replace switch (par. 180).
Corroded battery terminals.	Clean and tighten battery terminals (par. 156).
Weak battery.	Charge or replace battery and check generator charging rate.

b. Lights Do Not Burn.

Bulb burned out.	Replace bulb or sealed beam unit.
Circuit breaker switch open.	Close switch.
Open circuit.	Install wire or repair broken wire.

169. HEAD LAMP REPLACEMENT.

a. Remove Head Lamp. Pull out the release button under the instrument panel on the forward bulkhead to release the left head lamp and lift the lamp out of the socket. Pull out the release button just ahead of the assistant driver on the forward bulkhead and lift out the right head lamp.

b. Install Head Lamp. Place head lamp in position and be sure that release plunger fits into recess in head lamp.

170. HEAD LAMP SEALED BEAM ASSEMBLY REPLACEMENT.

a. Remove Sealed Beam Unit. Remove the single screw that retains the lens rim and remove the rim. Lift the unit from the shell and remove the screw from the terminal at the bulb base.

b. Install Sealed Beam Unit. Connect wire to terminal at base of bulb. Place unit in position and install rim. Tighten screw securely.

171. HEAD LAMP AIMING (fig. 125).

a. Preliminary Procedure. Secure a light colored piece of cloth about 5 feet high and 8 feet long, stretch the cloth over a wooden frame, and draw a black vertical line down the middle; then draw two more vertical lines, 18 inches on each side of the center line (fig. 125). Measure 24 inches up from the bottom and drive a nail on each side of the screen; let nail stick out about an inch. Put about 15 more nails on each

LIGHTING SYSTEM

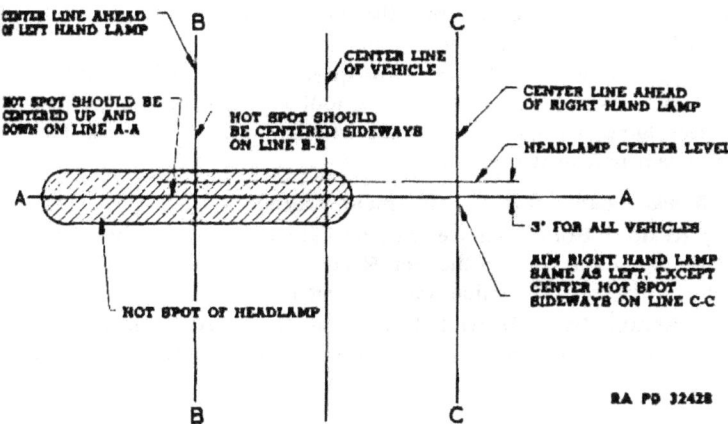

Figure 125 — Headlamp Aiming Diagram

side spaced 1 inch apart; this will make the screen universal so that it can be used on other vehicles. Secure a piece of black tape about 2 feet longer than the screen is wide and put a weight on each end of the tape to keep it stretched when it is hung on the nails on each side of the screen. Set up a stall in a dark corner so that the screen can be placed 25 feet in front of the vehicle to be checked; the floor should be level and the screen should be set so that the horizontal tape line is level and the vertical center line is in line with the center of the vehicle.

b. Aiming Procedure.

(1) CHECK AIMING OF LIGHT BEAM. Set car and screen in position as outlind in previous paragraph. Measure distance from the floor to the center of the head lamp and set the horizontal tape on the screen 3 inches less than this measurement from the floor. Turn on the head lights, cover one lamp, and check the location of the beam on the screen. The center line of the hot spot should be centered on the intersection of the vertical and horizontal lines (fig. 125). Check aiming of the other head lamp in the same manner.

(2) ADJUST HEAD LIGHTS. Loosen the aiming lock screw (fig. 124) on the bracket and move the head of the lamp body in its bracket until the beam is aimed as described in the previous paragraphs; then tighten the three cap screws.

172. BLACKOUT MARKER LAMP.

a. Replacement Procedure.

(1) REMOVE MARKER LAMP. Remove sealed beam unit from head lamp as instructed in paragraph 170. Pull out marker lamp wiring

connection. Remove the nut from the post that holds marker lamp on the head lamp and lift off the marker lamp.

(2) INSTALL MARKER LAMP. Place spacer in position on head lamp and place marker lamp in position. Install nut, line up marker lamp with head lamp, and tighten nut. Push wire in connector and install the sealed beam unit in the head lamp.

b. Marker Lamp Bulb Replacement Procedure.

(1) REMOVE BULB. Remove the single screw that retains the rim to the lamp body and remove the rim. Reach inside and turn bulb to left to release bayonet connection and pull out bulb.

(2) INSTALL BULB. Insert bulb in socket and turn to right to lock bayonet connection. Place rim in position and install the rim retaining screw.

173. AUXILIARY BLACKOUT DRIVE LAMP SEALED BEAM ASSEMBLY REPLACEMENT. (fig. 124).

a. Remove Sealed Beam Unit. Remove the rim retaining screw at the bottom of the rim and remove the rim. Disconnect the wire from the connection in the center of the sealed beam unit.

b. Install Sealed Beam Unit. Connect the wire to the connector on the back of the unit. Place unit in position in shell and install rim and retaining screw.

174. TAIL AND STOP LAMP ASSEMBLY REPLACEMENT.

a. Remove Lamp Assembly. Remove the bolts that attach the mounting bracket to the hull. Push the wire connections in and turn to the left to release the bayonet connections and pull out wires. Remove the nuts that retain the lamp to the bracket and remove the lamp.

b. Install Lamp Assembly. Place the bracket on the lamp and tighten the two nuts. Push wires in sockets and turn to right to lock the bayonet connections. Place assembly in position on hull and install the two mounting bolts. Tighten bolts securely.

175. TAIL AND STOP LAMP UNIT REPLACEMENT.

a. Remove Lamp Unit Assembly. Release bayonet connection at rear of lamp. Remove the two screws that retain the rim to the lamp body and remove the rim. Pull out the inner housing and bulb assembly.

b. Install Lamp Unit Assembly. Put bulb assembly in place in the lamp body and install the rim and two screws. Connect bayonet connections at back of lamp.

LIGHTING SYSTEM

176. INSTRUMENT PANEL BULB REPLACEMENT.

a. **Remove Bulbs.** Remove the snap covers over the bulbs in the instrument panel. Reach in the opening, push in, and turn the bulb to the left to release the bayonet connection and pull out the bulb.

b. **Install Bulb.** Push bulb in socket and turn it to the right to lock the bayonet connection. Push snap cover in place in the instrument panel.

177. DOME LAMP BULB REPLACEMENT.

a. **Remove Bulb.** Remove the two screws that retain the rim to the lamp body. Push in and turn bulb to left to release bayonet connection and pull out bulb.

b. **Install Bulb.** Push bulb in socket and turn to right to lock bayonet connections. Place rim in position and install the two screws.

TM 9-741
178-180

MEDIUM ARMORED CAR T17E1

Section XXXII

INSTRUMENTS AND GAGES

Paragraph

Instrument panel 178
Instrument panel removal 179
Removal of instrument panel units 180
Instrument panel installation 181
Fuel gage 182
Engine heat indicator 183
Engine oil pressure gage 184

178. INSTRUMENT PANEL.

a. Most of the instruments and gages are grouped on the instrument panel (fig. 17).

b. The instruments which pertain to each power plant are duplicated (one set for each power plant). A brief explanation of the use of these instruments and gages will be found in paragraph 6.

179. INSTRUMENT PANEL REMOVAL.

a. In order to service any of the units mounted on the instrument panel, it is necessary to remove the instrument panel. Remove the two screws and lift off plate from top of wheel. Unscrew the steering wheel nut and pull the steering wheel off with puller KM-J 1618 M 6. Disconnect the seven wiring harness conduit couplings from the right end of the instrument panel, the two from the left end of the instrument panel, and one from the lower left side back of panel. Disconnect stop light wire at stop light switch by loosening shield clamp bolt and disconnecting sleeve connectors. Remove the bolts from the three instrument panel brackets, disconnect the speedometer cable, and remove the instrument panel.

180. REMOVAL OF INSTRUMENT PANEL UNITS.

a. Remove the four screws which retain the speedometer set cable hole cover, the screws which retain the two instrument panel back shields, and remove the shields.

b. The three ammeters, two oil gages, two temperature indicators, speedometer and fuel gage are held to the instrument panel by U-clamps. Disconnect the wires from the unit, remove the nuts which attach the U-clamps, and remove the instrument from the face side of the instrument panel. The U-clamps with one resistor are used on the oil pressure

264

INSTRUMENTS AND GAGES

gages and fuel gages while two resistors are used on the temperature gages. NOTE: To remove the speedometer, it is necessary to pull off the reset button and remove the reset button retaining nut.

c. To remove the ignition switches, disconnect the ignition wires from switch, turn the switch to "ON" position, release the switch button by entering wire in hole on face of switch, turn switch to left, and remove switch button. Remove the retaining nut on the face side of panel, with a spanner wrench, and remove switch from the back side of instrument panel.

d. The blackout driving light switch can be removed by disconnecting the wires, loosening the button retaining screw, unscrewing the button and retaining nut. Remove the switch from the back side of panel.

e. The starter buttons are removed by disconecting the wires, removing the retaining nut on the face of the panel, and removing the buttons from the back side of panel.

f. To remove the lighting switch, it is necessary to disconnect the wires. Note the markings on the switch body and mark the wires removed so that they can be installed correctly. Loosen the button retaining screw and unscrew the button. Loosen the cap screw which retains the blackout safety catch body, press down on the safety button, and remove the safety catch assembly. Remove the nut which retains the switch and remove the switch.

g. To remove lighting and fuel pump circuit breakers from right end of panel disconnect wires, remove retaining bolts and take circuit breaker from back of panel.

h. To replace the instrument lamp bulbs, remove the cover plates from face of panel, and install new bulb.

i. Any of the units removed can be installed by reversing the removal procedure.

j. Install instrument panel back shield and speedometer reset cable plate. NOTE: Place instrument panel ground strap under right top cover screw.

181. INSTRUMENT PANEL INSTALLATION.

a. Place the instrument panel in position, connect the speedometer cable, and install the panel bracket bolts, placing the ground strap under the right bracket top bolt. Connect the ten wiring harness connections and tighten the conduit couplings. Connect the stop light wiring and shield. Install the steering wheel and tighten the nut securely. Install plate and secure with two screws.

TM 9-741
182

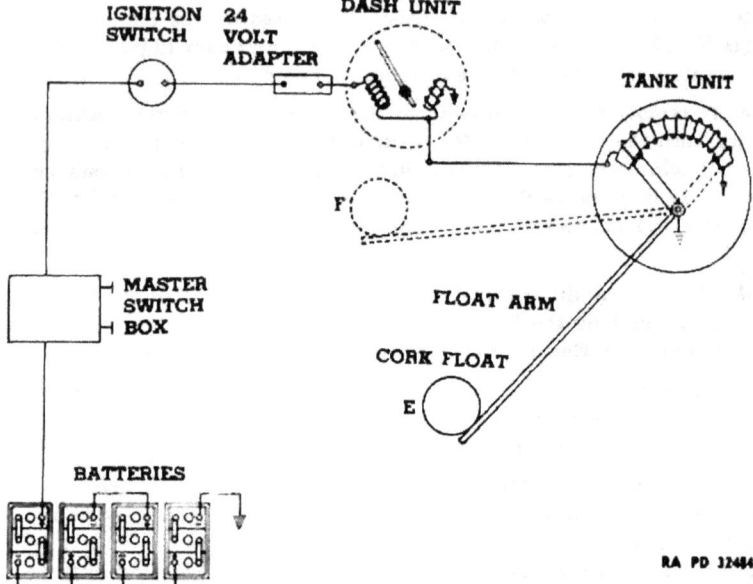

Figure 126 — Fuel Gage Circuit

182. FUEL GAGE.

a. Description.

(1) The fuel gage is composed of two units: the dash or indicating unit which is mounted on the instrument panel and the tank unit located in the fuel tank at the rear of the hull. Gages are not used for the jettison tanks as they feed to the main tank by gravity. The dash unit has a scale graduated in fractions between "EMPTY" and "FULL." The graduation figures, letters, and pointer are luminous. The current for the gage passes through the master switch and ignition switch; therefore, the master switch and one ignition switch must be "ON" for the gage to register.

(2) Figure 126 shows the fuel gage circuit. It is important that all connections be kept tight. If the external electrical circuits are good, the only trouble would probably be in the dash or tank units.

b. Testing Dash and Tank Units.

(1) The following is a procedure for checking the fuel gage units, using KM-KMO 204 fuel gage tester (fig. 127). A similar tool can be made from a new unit by providing leads from the unit for attachment to the wires. CAUTION: Care must be used when checking units to avoid short circuiting and burning the units.

INSTRUMENTS AND GAGES

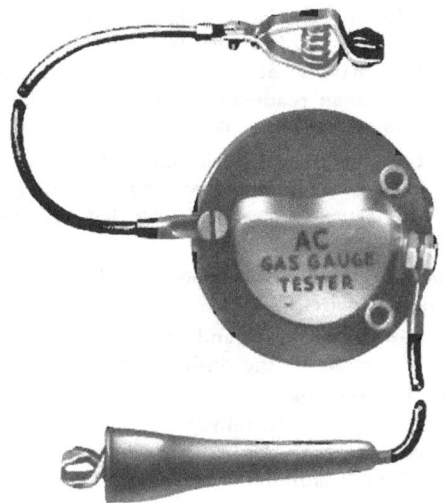

RA PD 32485
Figure 127 — Fuel Gage Tester

(2) Turn ignition switches "OFF." Disconnect tank unit wire from dash unit.

(3) Attach the red wire of the tester to this terminal and black wire to a good ground.

(4) Turn ignition switch "ON," move tester arm up and down. Dash unit should register "FULL" and "EMPTY" if it is serviceable. If so, turn ignition switch "OFF" and reconnect tank wire.

(5) If dash unit does not register at all on above test, before replacing it make certain that it is getting current from the ignition switches. This can be checked by connecting a 6-volt test lamp from the positive terminal (adapter terminal) to ground. Replace dash unit if lamp lights.

(6) If dash unit is serviceable, check the wiring between dash and tank units.

(7) Disconnect the tank unit wire at the tail lamp and fuel tank terminal box.

(8) Attach the red wire of the tester to the wire running to the dash unit and the black wire to ground.

(9) Turn ignition switch "ON," move tester arm up and down observing dash unit indication which should move from "EMPTY" to "FULL" if the wiring is serviceable.

MEDIUM ARMORED CAR T17E1

(10) If on this test the dash unit reads "EMPTY" at all times or the reading is noticeably lower than that during the check at the dash unit, look for shorts or leaks in wiring between dash unit and terminal box. If dash unit reads above "FULL" at all times or if it reads higher at "EMPTY" and "FULL" than readings obtained when checking at the dash, look for poor connections or break in the wiring.

(11) If dash unit and wiring are serviceable, remove tank unit. Clean away dirt that may have accumulated around tank unit terminal, as some types of dirt may cause an electrical leak that will cause an error in reading.

(12) Connect tank unit to the wire leading to dash; ground the tank unit with a short piece of wire.

(13) Turn ignition switch "ON" and move the float arm up and down. If this unit is serviceable, the dash unit will give corresponding "EMPTY" and "FULL" readings.

(14) If tank unit is serviceable, reinstall in the tank. If not, replace with a new unit. Check new tank unit as above before installing in tank. NOTE: Always check tank units for freedom of movement of the float arm by raising it to various positions and observing that it will fall to "EMPTY" position in every instance.

c. Unit Replacement.

(1) The replacement of the dash unit is covered in paragraph 180 b.

(2) To replace the tank unit, remove the twenty-five bolts which attach the left engine compartment cover and remove the cover.

(3) Shut off the jettison fuel tanks and drain the fuel to a point below the top of main fuel tank.

(4) Remove the fifteen cap screws which retain the inner half of the left engine fan shroud and remove shroud.

(5) Remove the two cover bolts and remove cover that is over gage unit.

(6) Disconnect the two wires from the fuel gage and remove the five clutch head screws.

(7) Remove the gage from top of tank.

(8) Reverse the above procedure for installation.

183. ENGINE HEAT INDICATOR.

a. Description.

(1) The temperature of each engine cooling system is recorded on an indicator on the instrument panel. The left indicating unit records the left engine cooling solution temperature while the right unit indicates the temperature of the right engine cooling solution.

TM 9-741
183

INSTRUMENTS AND GAGES

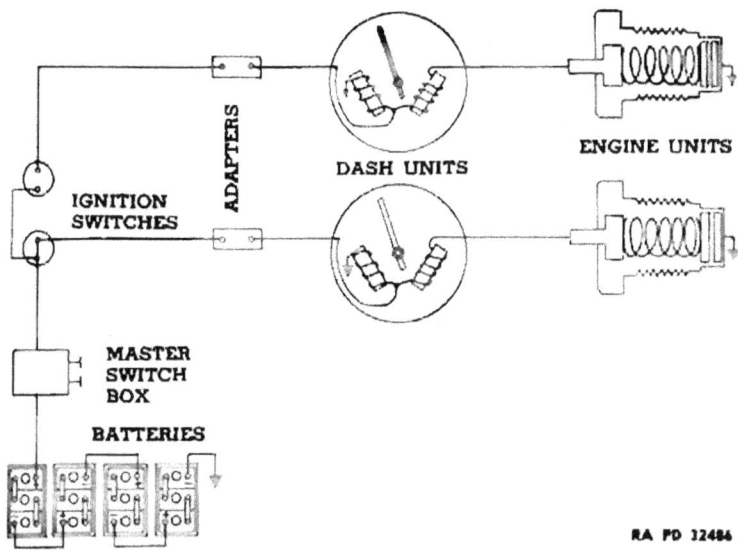

Figure 128 — Engine Temperature Indicator Circuits

(2) Each engine indicating system consists of two units: the dash unit and the thermo or operating unit connected to each other and the electrical system by wires (fig. 128).

(3) The faces of the dash units are graduated in degrees Fahrenheit. The graduation marker "240" on the dial is orange luminous paint while the others are white luminous paint.

(4) The engine thermo units are mounted in the right side of the thermostat housing. They have no moving parts and require no service.

b. Maintenance.

(1) If either of the units of the two gages fails to operate, it should be replaced.

(2) The replacement of the dash unit was covered in paragraph 180 b.

(3) To replace a thermo unit, drain the cooling system to a point below the unit. CAUTION: Turn ignition switches "OFF" before making or changing connections.

(4) Disconnect the wire from the unit by unscrewing the coupling and pulling the wire out.

(5) Unscrew the coupling assembly and remove it from the thermostat housing.

269

MEDIUM ARMORED CAR T17E1

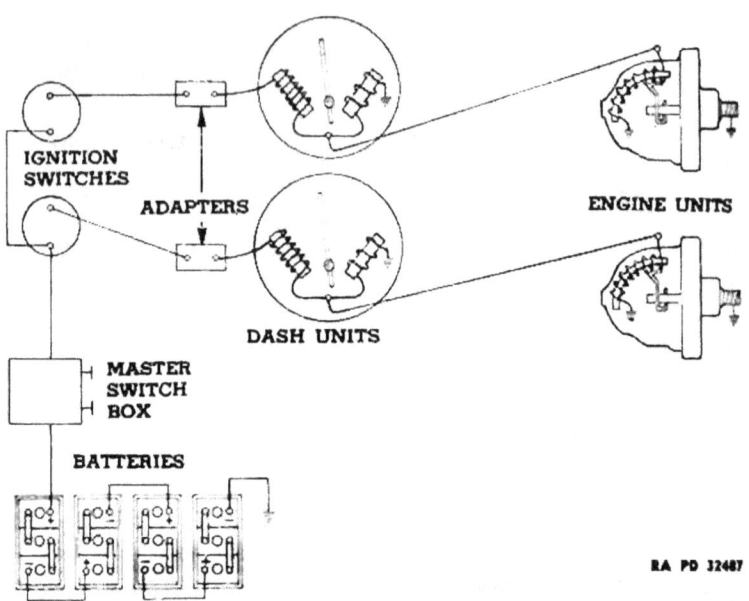

Figure 129 — Engine Oil Gage Circuits

(6) Reverse the above procedure to install the gage. CAUTION: Do not use sealing compound when installing the thermo unit as it will increase the resistance and cause incorrect reading.

184. ENGINE OIL PRESSURE GAGE.

a. Description.

(1) An electric oil pressure gage is used to register the oil pressure of each engine lubricating system. The left gage on the instrument panel is for the left engine; the right gage, for the right engine.

(2) Each engine pressure gage consists of two units, the dash unit and the engine unit which is mounted on the left side of the engine and is connected in the oil passage.

(3) The dial has a scale graduated in pounds per square inch. The graduations, figures, and pointer are identified with white luminous paint. Figure 129 shows the oil gage circuits. The master switch and ignition switch are in the circuit and must be "ON" to complete a circuit.

b. Service Operations.

(1) The replacement of dash units was covered in paragraph 180 b.

INSTRUMENTS AND GAGES

(2) To remove either engine unit, disconnect the lead wire by unscrewing the connection and pulling the wire from terminal.

(3) Remove the engine unit by unscrewing it from the engine block. NOTE: Do not start the engine with this unit removed.

(4) Reverse the above procedure for installation. NOTE: Sealing compound must not be used when installing the engine unit as it will increase the resistance in the circuit and cause inaccurate reading.

MEDIUM ARMORED CAR T17E1

Section XXXIII

ELECTRICAL ACCESSORIES

	Paragraph
Siren	185
Siren replacement	186
Siren switch replacement	187
Electric windshield wipers	188
Wiper motor replacement	189

185. SIREN (fig. 2).

a. Description. The warning siren is located on the top of the vehicle just ahead of the left front corner of the turret. A foot switch to operate the siren is located at the left of the steering gear within reach of the driver's left foot. The siren runs at 7,000 revolutions per minute.

186. SIREN REPLACEMENT.

a. Remove Siren. Disconnect the wire leads on the siren. Remove the four mounting screws and remove the siren.

b. Install Siren. Place siren in position and install the four mounting screws. Connect the wire leads to the siren.

187. SIREN SWITCH REPLACEMENT.

a. Remove Switch. Remove the bolts that attach the foot switch shield to the floor. Remove the two wires from the terminals. Remove the two bolts that attach the switch to the shield and remove the switch.

b. Install Switch. Place switch in shield and install the two attaching bolts. Connect the two wires to the terminals. Mount switch shield to floor and tighten bolts securely.

188. ELECTRIC WINDSHIELD WIPERS.

a. Description. Two auxiliary windshields with electric windshield wipers attached are carried in the hull for use in noncombatant areas. The windshields are installed in the front vision door openings by assembling the lower edge of the frame under clips at the bottom of the opening and fastening the top with the special slide clips provided. Outlets for the electric wipers are located at the top right side of the instrument panel. These outlets are always in circuit when the master switch is "ON," as they are directly connected to the battery terminal of the main light switch. A single wire with metal shielding leads from each wiper to the outlets. The shielding acts both as a ground connection for the

ELECTRICAL ACCESSORIES

motor and also as a shielding to prevent radio interference. The electric wiper motors are equipped with integral switches and dual wiper blades, one inside and the other outside.

b. **Servicing Wiper Motors.** The only service that can be performed on the wiper motors is to clean the commutator. To clean the commutator, remove the two Phillips screws that attach the cover and remove the cover. Clean the commutator with a piece of PAPER, flint, No. 00.

189. **WIPER MOTOR REPLACEMENT.**

a. **Remove Motor.** Remove the outside blade. Remove the two bolts that attach the bracket frame and lift out motor and bracket. Remove the inner wiper blade and remove the bracket from the motor.

b. **Install Motor.** Place motor in place on bracket and install the mounting screws. Install inner blade and attach motor and bracket to frame. Install outer blade.

MEDIUM ARMORED CAR T17E1
Section XXXIV
RADIO SUPPRESSION

	Paragraph
Description	190
Radio suppression maintenance	191

190. DESCRIPTION.

a. General. Radio suppression equipment is installed on this vehicle to control the radiation of radio interference by the electrical equipment of the vehicle and also to control noise in the vehicle equipped with a radio set. The suppression system eliminates or controls to a minimum the radio interference emanating from the vehicle over the range of 0.5 to 30 megacycles. No attempt has been made to suppress interference that may exist outside of this range. The radio suppression system must be maintained in perfect condition because vehicles that radiate interference can be easily located by the use of short wave listening devices. Radio suppression is accomplished by the use of resistor suppressors, filters and condensers, and shielding of the wires. The suppression system consists of fourteen suppressors, eight filters, eight condensers, and complete shielding of electric wiring, which is grounded with plated bolts, nuts and shakeproof lock washers. The filters in the generator regulator box are shown in figure 130.

b. Suppressors. Resistor suppressors are installed on each high tension lead to the spark plugs and to the leads from the coils to the center of the distributor caps. The function of the resistor suppressor is to change the radiation characteristics of the high tension system to inhibit the more disturbing components of the system.

c. Filters.

(1) IGNITION COIL FILTERS (fig. 131). Two filters, one for each coil, are enclosed in the ignition coil filter box which is located on the right side of the center partition in the engine compartment. A wiring harness, containing two wires, leads from the instrument panel ignition switch wiring harness. These wires are connected to the top of each filter. The terminals on the bottom of the filters extend through the box, and single wires lead from these terminals to the ignition coils.

(2) GENERATOR FILTERS (fig. 132). Two filters of the type shown in figure 132 are enclosed in each of the filter and regulator boxes in each engine compartment. The top terminal of the left-hand filter is connected to the circuit breaker in the main switch box and the lower terminal is connected to the battery or left terminal of the regulator unit. The top terminal of the right-hand filter is connected to the armature terminal on the generator, and the bottom terminal is connected to armature or

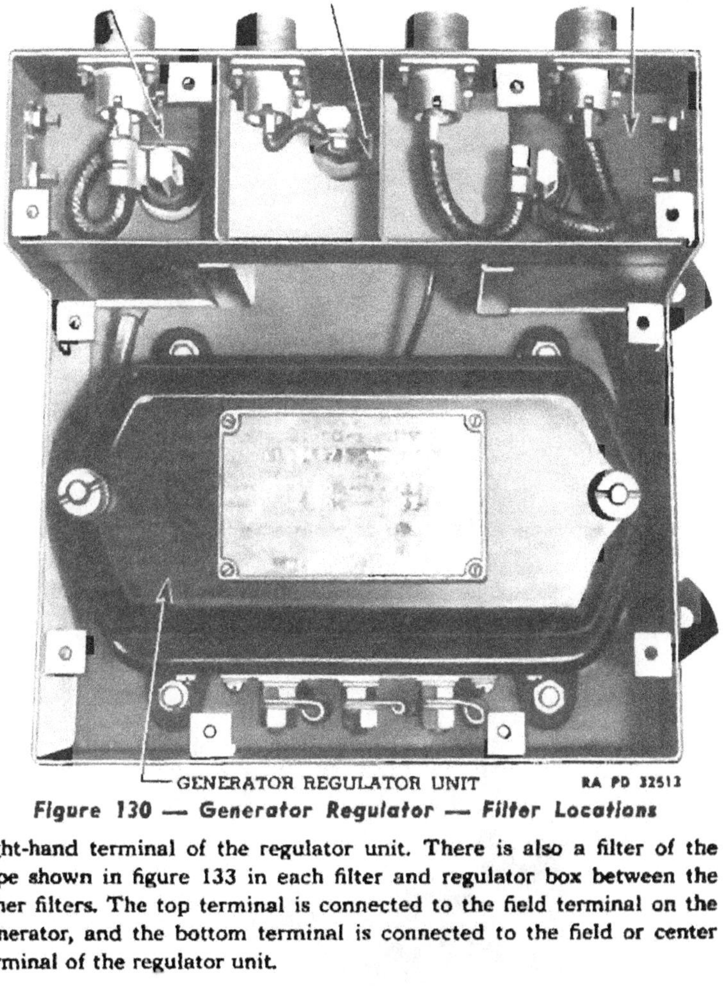

TM 9-741
190

RADIO SUPPRESSION

BATTERY FILTER   FIELD FILTER   ARMATURE FILTER

GENERATOR REGULATOR UNIT   RA PD 32512

Figure 130 — Generator Regulator — Filter Locations

right-hand terminal of the regulator unit. There is also a filter of the type shown in figure 133 in each filter and regulator box between the other filters. The top terminal is connected to the field terminal on the generator, and the bottom terminal is connected to the field or center terminal of the regulator unit.

d. **Condensers.**

(1) TURRET MOTOR CONDENSERS. There are four condensers used in the turret motor. They are located on the four brush holders and are connected between the brush holders and the rear bearing retainer.

(2) STEERING GEAR MOTOR CONDENSER. The steering gear motor condenser is located in the steering gear motor and is connected between the positive (+) brush holder lug and the rear frame of the steering gear motor.

(3) FUEL PUMP CONDENSER. The fuel pump condenser is located on

MEDIUM ARMORED CAR T17E1

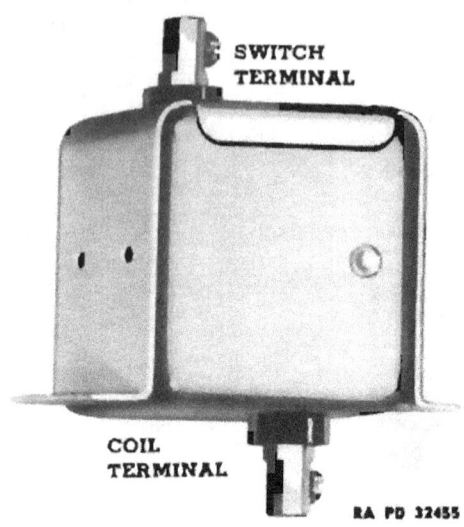

Figure 131 — Ignition Coil Filter

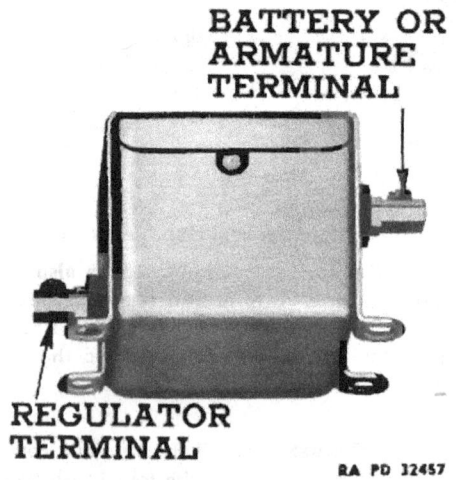

Figure 132 — Generator Filter — Battery and Armature Circuits

the fuel pump mounting flange and is connected between the positive (+) and negative (−) terminals on the pump.

(4) WINDSHIELD WIPER MOTOR CONDENSERS. A condenser is placed in each of the windshield wiper motors and is connected between the positive (+) and negative (−) terminals of the motor.

RADIO SUPPRESSION

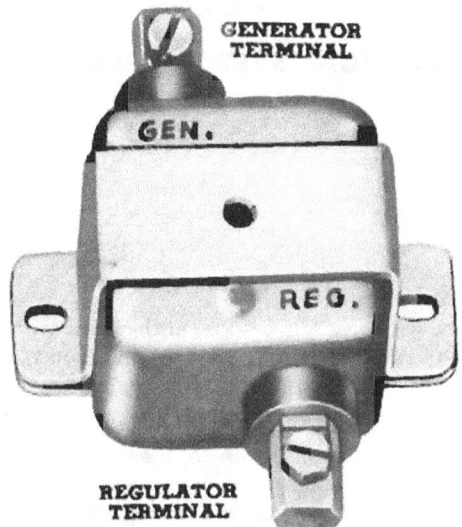

Figure 133 — Generator Filter — Field Circuit

e. **Wire Shielding.** The entire wiring system is shielded by covering the wires with flexible conduit or solid metal tubing. The conduit and tubing are grounded by plated metal clips attached with plated screws, shakeproof lock washers, and nuts.

191. RADIO SUPPRESSION MAINTENANCE.

a. **General.** The success of the radio suppression system depends entirely on tight connections. All clips, attaching bolts, nuts, and shakeproof lock washers must be kept clean and tight. A faulty filter or condenser will cause a noise in whichever circuit is affected. A loose connection will have practically the same effect as if there were no suppression being used. Therefore, it is imperative that the terminals be kept clean and tight at all times to eliminate radio interference.

b. **Checking the Suppression System.**

(1) Check all ground straps and wire shielding clips to make sure they are in good condition and that each is properly attached to the parts being grounded.

(2) Check all plated bolts and nuts to make sure they are tight and that shakeproof lock washers are in place.

TM 9-741
192

MEDIUM ARMORED CAR T17E1
Section XXXV
ELECTRICAL SYSTEM WIRING

	Paragraph
Description	192
Turret slip ring assembly	193
Trouble shooting	194
Wiring connection tables	195

192. DESCRIPTION.

a. General. The wiring system consists of the wires, wiring harnesses, various switches, circuit breakers, filters, switch boxes, and terminal boxes, all of which are connected to the various electrical units of the vehicle.

b. Wiring. The wiring is made up in a series of wiring harnesses and covered with flexible conduit. Two main power lines lead from the batteries to the master switch box, and wiring harnesses lead to the various operating units, or to terminal boxes from which supplementary harnesses lead to the units. The wiring in the various harnesses can be traced by the various colors of the insulation.

c. Switches. The master switch is located to the rear and above the right entrance door inside the hull (fig. 134). When the two handles on the left side of the switch are in the "OFF" position, none of the electrical units will operate. The upper switch handle controls the 12-volt circuit to the turret slip ring terminal box and radio and telephone terminal box. To turn the switches "OFF", pull the handles and turn in either direction until the handle stops. To turn the switches "ON", turn the handle until it is horizontal and the switch drops into position. The other mechanical switches are explained in paragraph 6.

d. Circuit Breakers. The wiring circuits and electrical units are protected against overloads by individual overload circuit breakers that open the circuit affected in the event of a short circuit or heavy overload.

(1) LIGHTING SWITCH CIRCUIT BREAKER (fig. 135). The lighting switch circuit breaker is located on the right instrument panel support. A push button extends through the support within reach of the driver. If the load exceds 20 amperes, the push button "throws out." The unit can be reset manually by pushing the button in. This unit protects all circuits controlled by the main light switch.

(2) GENERATOR CUT-OUT RELAY CIRCUIT BREAKER (fig. 134). Two generator cut-out relay circuit breakers are located in the master switch

TM 9-741
192

ELECTRICAL SYSTEM WIRING

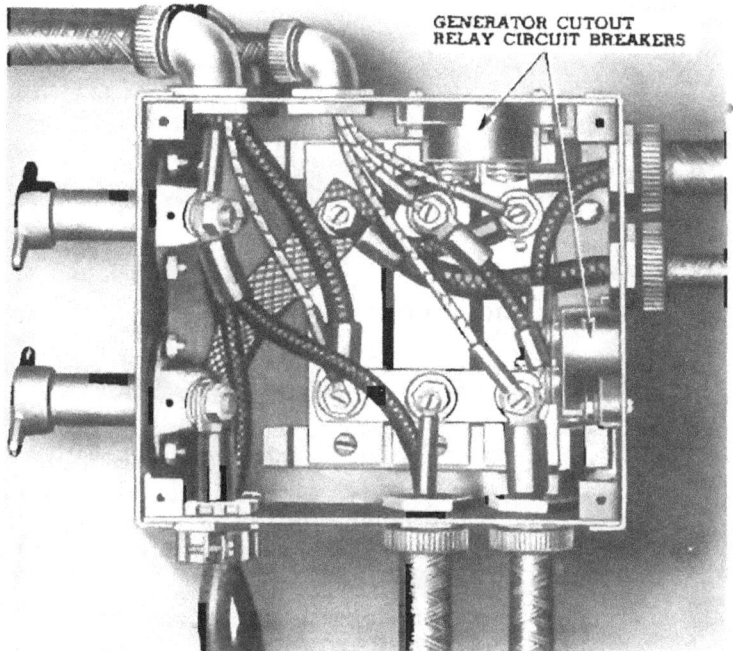

Figure 134 — Master Switch Box

box, one for each generator. If the load exceeds 70 amperes, the unit clicks on and off. The units reset themselves automatically after the overload condition is corrected.

(3) FUEL PUMP CIRCUIT BREAKER. The fuel pump circuit breaker is located on the right instrument panel support. A push button extends through the support within reach of the driver (fig. 135). If the load exceeds 2 amperes, the circuit breaker "throws out." The unit can be reset manually by pushing the button in.

(4) STEERING GEAR MOTOR CIRCUIT BREAKER (fig. 144). The steering gear motor circuit breaker is located in the steering gear motor switch box. If the load exceeds 120 amperes, the circuit breaker "throws out." The unit can be reset manually by pushing in the button on the front of the box.

(5) SIREN CIRCUIT BREAKER (fig. 144). The siren circuit breaker is located in the steering gear motor switch box. If the load exceeds 15 amperes, the circuit breaker "throws out." The unit can be reset manually by pushing in the button.

MEDIUM ARMORED CAR T17E1

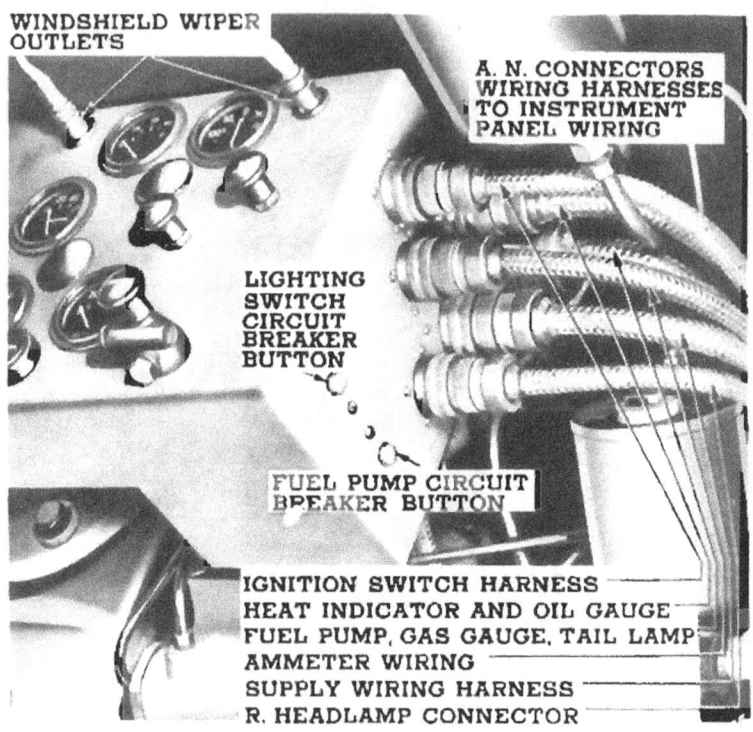

Figure 135 — Instrument Panel Connections

(6) TURRET LIGHT SWITCH AND CIRCUIT BREAKER (fig. 150). The turret light switch and circuit breaker are a part of the turret light switch on the turret control switch box. If the circuit load exceeds 6 amperes, the switch toggle lever is thrown to the "OFF" position. The toggle lever can be reset manually by pushing the lever over.

(7) TURRET TRAVERSE CONTROL MOTOR RELAY SWITCH (fig. 149). The turret traverse control motor relay switch is located in the turret motor relay switch box. If the load exceeds 100 amperes, the overload control will open and the motor will stop. To restart the motor, lift the start and reset switch button on the turret control switch box. The overload control device is automatic and will close as soon as the overload condition is corrected.

TM 9-741
193

ELECTRICAL SYSTEM WIRING

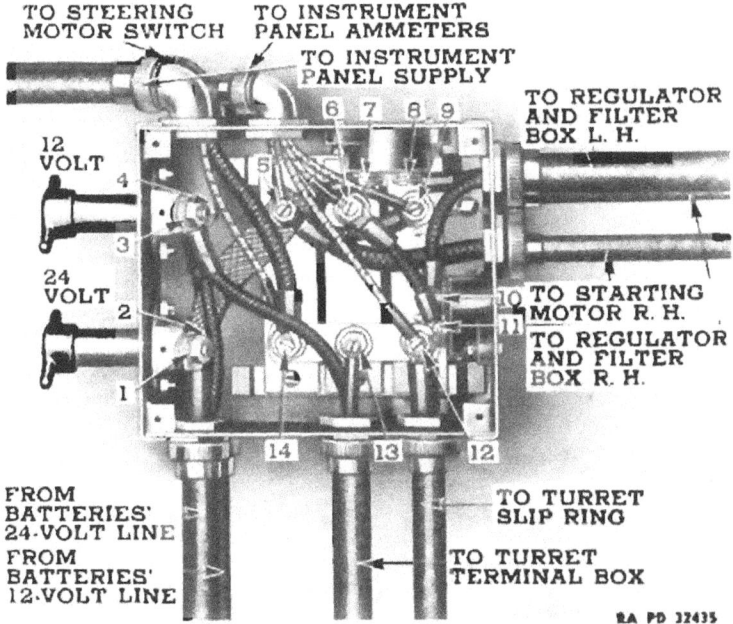

Figure 136 — Master Switch Box Connections

193. TURRET SLIP RING ASSEMBLY.

a. **Description.** The turret slip ring assembly provides a means of transmitting current from the vehicle hull to the revolving turret for the operation of the turret traversing system, lights, gun firing solenoids, and radio and telephone apparatus. The assembly is mounted in the center of the turret basket floor and fastened to the floor with four machine screws. The outer section containing the brushes and leads to the turret basket mechanism revolves with the basket. The inner section, containing the armature and leads from the master switch box and turret slip ring terminal box, remains stationary. A forked yoke welded to the front axle tunnel engages a boss on the lower flange of the assembly and prevents the inner section from turning. A ground strap from this boss to the yoke affords a ground connection for the assembly. A directional arrow is mounted on top of the assembly and fastened to the armature. This arrow points toward the front of the vehicle and does not rotate with the turret, allowing the occupants of the fighting compartment to orient their position in relation to the vehicle direction.

TM 9-741
194-195

MEDIUM ARMORED CAR T17E1

194. TROUBLE SHOOTING.

a. Lights Will Not Burn.

Probable Cause	Probable Remedy
Bulbs burned out.	Replace burned out bulbs.
Switches not turned on.	Turn on switches.
Circuit breakers off.	Reset circuit breakers and check for trouble.
Corroded terminals.	Clean and tighten terminals.
Loose connections.	Tighten connections.
Disconnected wires.	Connect wires.
Faulty switches.	Replace switches.
Short circuit.	Eliminate short.

b. Wiring System Circuit Breaker Will Not Stay Closed.

Short circuit in wiring.	Check and correct.
Overload in circuit caused from binding.	Free up parts that are binding.
Faulty circuit breaker.	Replace circuit breaker.

195. WIRING CONNECTION TABLES.

a. Master Switch Box (fig. 136). The terminals in the switch box shown in the illustration are numbered for explanatory purposes only in connection with the description in the following table:

Terminal No.	Connects to	Wire Color	Wire Size No.
1	Batteries (main 24-volt line)	Black cable	1
2	Terminal 5 in Master Switch box	Copper strap	
3	Terminal 5 in turret slip ring terminal box (12-volt lead)	Black cable	6
4	Batteries (main 12-volt line)	Black cable	6
5	Terminal 2 in master switch box	Copper Strap	2
	Starting motor solenoid switch, right-hand	Black cable	1
	Instrument panel ammeter wiring harness receptacle, terminal A	Red	14
6	Terminal 11 in master switch box	Black with red tracer	6

282

ELECTRICAL SYSTEM WIRING

Terminal No.	Connects to	Wire Color	Wire Size No.
	Instrument panel ammeter wiring harness receptacle, terminal B	Natural with red tracer	14
7	Terminal 9 in master switch box	Black with red tracer	6
8	Filter in left-hand generator regulator and filter box. Connects to B terminal on generator regulator	Black cable	6
9	Instrument panel ammeter wiring harness, terminal C	Natural with red cross tracer	14
	Terminal 7 on circuit breaker for left-hand generator charging circuit	Black with red tracer	6
10	Filter in right-hand generator regulator and filter box. Connects to B terminal on generator regulator	Black cable	6
11	Terminal 6 in master switch box	Black with red tracer	6
12	Instrument panel ammeter wiring harness receptacle, terminal D	Natural with black tracer	14
	Turret slip ring positive (+) terminal	Black cable	2
13	Positive (+) terminal in turret slip ring terminal box (24-volt lead)	Red cable	6
14	Steering motor switch terminal 1	Red cable	4
	Instrument panel supply wiring harness receptacle, terminal A	Natural with red and black cross tracer	10

b. **Main Light Switch** (fig. 137). The main light switch is located on the lower right corner of the instrument panel. Figure 137 illustrates the terminal connections on the main light switch. The markings as illustrated are stamped on the side of the switch body.

MEDIUM ARMORED CAR T17E1

c. **Main Light Switch Connection Table** (fig. 137).

Terminal Marking	Connects to	Wire Color	Wire Size No.
BAT	Lighting switch circuit breaker	Natural	14
	Right-hand windshield wiper socket	Natural	14
	Left-hand windshield wiper socket	Natural	14
SW	Stop lamp switch	Natural	14
SS	Stop lamp switch	Natural	14
S	Instrument panel tail lamp wiring harness assembly, terminal D	Natural	14
BS	Instrument panel tail lamp wiring harness assembly, terminal E	Natural with green cross tracer	14
HT	Right-hand lamp connector, terminal B	Natural with black and red cross tracer	14
	Instrument panel tail lamp wiring harness assembly, terminal F	Natural with black and red cross tracer	14
	Left-hand lamp connector, terminal B	Natural with red and green cross tracer	14
BHT	Blackout driving light switch	Natural with red tracer	14
	Right-hand lamp connector, terminal A	Natural with green tracer	14
	Instrument panel tail lamp wiring harness assembly, terminal C	Natural with green tracer	14
	Left-hand lamp connector, terminal C	Natural	14

d. **Instrument Panel Supply Assembly and Connections** (fig. 138). The wiring to and from the instrument panel is through a series of "AN" (Army, Navy) connectors. The connectors consist of receptacles to which are attached leads to each of the instruments and plugs to which are connected the shielding harnesses and wires. The receptacles

ELECTRICAL SYSTEM WIRING

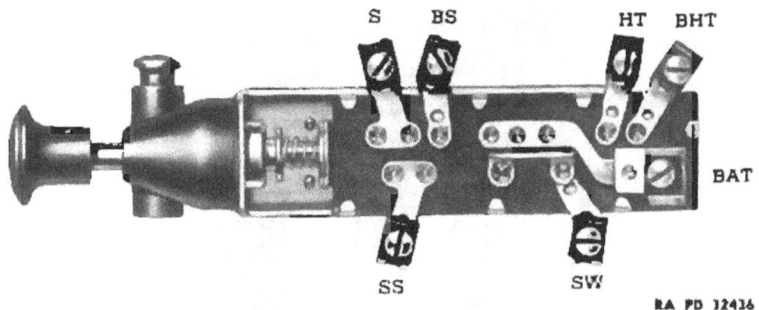

Figure 137 — Main Light Switch Terminals

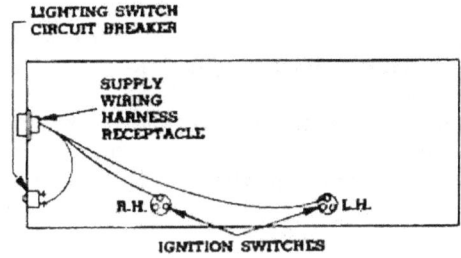

Figure 138 — Instrument Panel Supply Wiring Harness Connections

and plugs are marked for each connection and halves are keyed so they will assemble correctly.

Terminal Marking	Connects to	Wire Color	Wire Size No.
A	Lighting switch circuit breaker	Natural with red and black cross tracer	10
	Right-hand ignition switch, upper terminal	Natural with red tracer	12
	Left-hand ignition switch, upper terminal	Natural with red tracer	12
B	No connection		

TM 9-741
195

MEDIUM ARMORED CAR T17E1

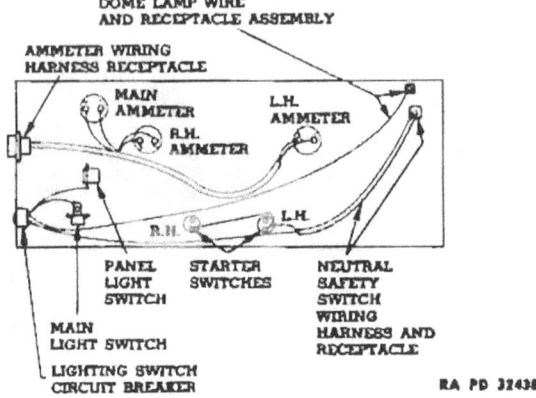

Figure 139 — Ammeter Wiring Harness Connections

e. **Instrument Panel Ammeter Wiring Assembly** (fig. 139).

Terminal Marking	Connects to	Wire Color	Wire Size No.
A	Main ammeter negative (−) terminal	Red	14
B	Right-hand ammeter postive (+) terminal	Natural with red tracer	14
C	Left-hand ammeter positive (+) terminal	Natural with red cross tracer	14
D	Main ammeter positive (+) terminal	Black	14
	Right-hand ammeter negative (−) terminal	Natural with black tracer	14
	Left-hand ammeter negative (−) terminal	Natural with black cross tracer	14

f. **Instrument Panel Dome Lamp Wiring Connector** (fig. 139). One terminal of dome lamp connector is connected to the upper terminal on the light switch circuit breaker.

g. **Instrument Panel Neutral Safety Switch Wiring Connector** (fig. 139).

A	Left-hand starting switch	Natural with black tracer	14
	Right-hand starting switch	Natural with black tracer	14
B	Lighting switch circuit breaker	Natural with black and red cross tracer	14

ELECTRICAL SYSTEM WIRING

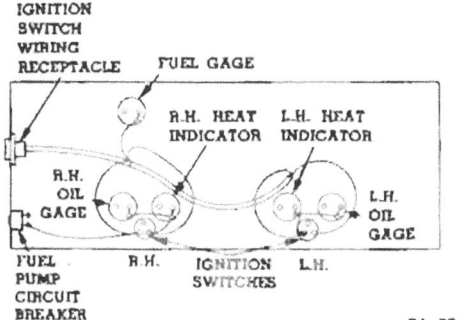

Figure 140 — Ignition Switch Wiring Connections

h. Instrument Panel Ignition Switch Wiring Assembly (fig. 140).

Terminal Marking	Connects to	Wire Color	Wire Size No.
A	Left-hand ignition switch, left terminal	Natural with black tracer	14
	Left-hand heat indicator adapter	Natural with black cross tracer	14
	Left-hand oil gage adapter	Natural	14
B	Right-hand ignition switch, right terminal	Natural with black cross tracer	14
	Right-hand heat indicator adapter	Natural with black tracer	14
	Right-hand oil gage adapter	Natural	14
C	No connection		

NOTE: This harness also contains wire connections as follows:

	Left-hand ignition switch, right terminal to right-hand ignition switch, left terminal	Natural with red tracer	14
	Right-hand ignition switch, left terminal to fuel pump circuit breaker	Natural with red tracer	14
	Ignition switch conecting wire to fuel gage adapter	Natural with black tracer	14

MEDIUM ARMORED CAR T17E1

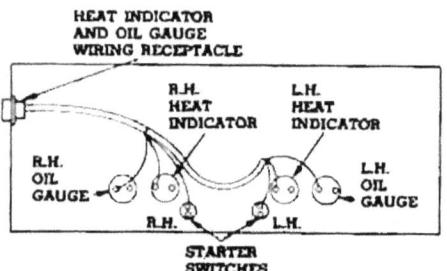

Figure 141 — Heat Indicator and Oil Gage Connections

l. Instrument Panel Heat Indicator and Oil Gage Wiring Assembly (fig. 141).

Terminal Marking	Connects to	Wire Color	Wire Size No.
A	Right-hand oil gage, engine terminal	Natural with red cross tracer	14
B	Right-hand heat indicator, engine terminal	Natural with black cross tracer	14
C	Left-hand heat indicator, engine terminal	Natural with black tracer	14
D	Left-hand oil gage, engine terminal	Natural with red tracer	14
E	Left-hand starter switch button	Natural with black and red cross tracer	14
F	Right-hand starter switch button	Natural	14

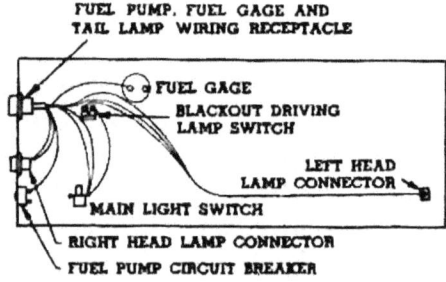

Figure 142 — Fuel Pump, Fuel Gage and Tail Lamp Connections

TM 9-741
195

ELECTRICAL SYSTEM WIRING

j. **Instrument Panel Fuel Pump, Fuel Gage and Lamp Wiring Harness Assembly (fig. 142).**

Terminal Marking	Connects to	Wire Color	Wire Size No.
A	Fuel gage tank terminal	Natural with black tracer	14
B	Fuel pump circuit breaker	Natural with red cross tracer	14
C	Left head lamp connector, terminal E	Natural (opt.)	14
	Right head lamp connector, terminal A	Natural with green tracer	14
	Main light switch, terminal BHT	Natural with green tracer	14
	Blackout driving light switch	Natural with red tracer	14
D	Main light switch, terminal S	Natural with green cross tracer	14
E	Main light switch, terminal BS	Natural	14
F	Left head lamp connector, terminal B	Natural with red and green cross tracer	14
	Right head lamp connector, terminal B	Natural with black and red cross tracer	14
	Main light switch, terminal HT	Natural with black and red cross tracer	14

NOTE: There is also a wire in the harness leading from left head lamp connector as follows:

| A | Blackout driving light switch | Natural with red tracer | 14 |

k. **Instrument Panel, Stop Light, and Windshield Wiper Wiring (fig. 143).**

Panel light switch	Panel lights	Natural with black cross tracer	14
Stop lamp	Main light switch terminal HT	Natural with black and red cross tracer	14
Windshield wiper socket	Main light switch terminal BAT	Natural	14

289

TM 9-741
195

MEDIUM ARMORED CAR T17E1

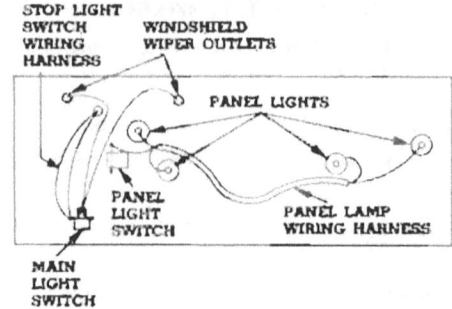

Figure 143 — Panel Lights, Stop Light and Windshield Wiper Connections

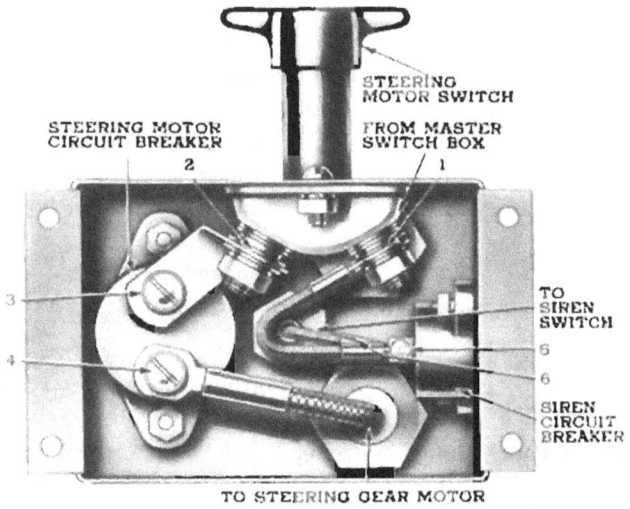

Figure 144 — Steering Motor Switch Box

ELECTRICAL SYSTEM WIRING

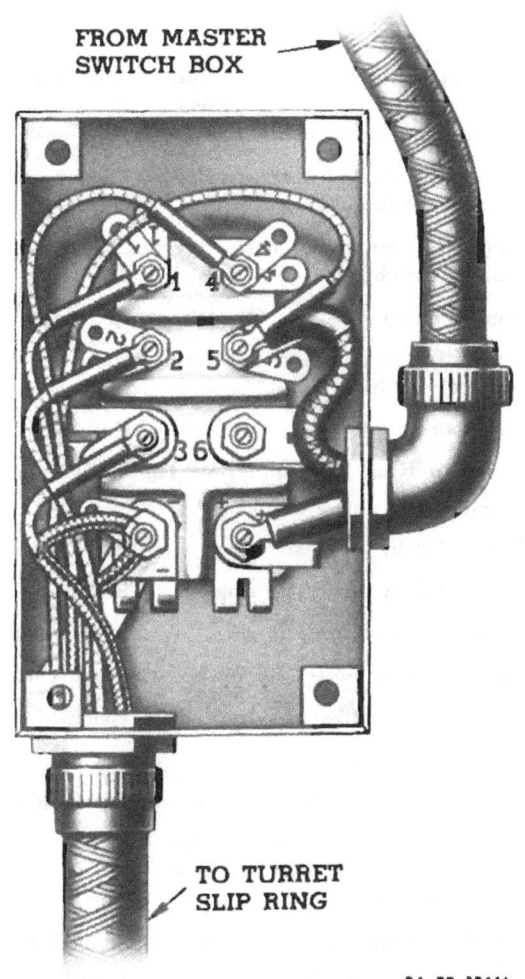

Figure 145 — Turret Slip Ring Terminal Box

TM 9-741
195

MEDIUM ARMORED CAR T17E1

l. Steering Motor Switch Box Connection Table (fig. 144). The terminals in the illustration are numbered for explanatory purposes only.

Terminal No.	Connects to	Wire Color	Wire Size No.
1	Terminal 14 in master switch box	Red cable	4
	Terminal 5 on siren circuit breaker in steering motor switch box	Natural	14
2	Jumper to terminal 3 on steering motor circuit breaker	Copper strap	
3	Jumper to terminal 2 on steering motor switch	Copper strap	
4	Steering motor	Red cable	4
5	Jumper to terminal 1 on steering motor switch	Natural	14
6	Siren foot switch	Natural	14

m. Turret Slip Ring Terminal Box Table (fig. 145).

Terminal No.	Connects to	Wire Color	Wire Size No.
1	Terminal 1 in turret slip ring	Orange	18
2	Terminal 2 in turret slip ring	Blue	18
3	Terminal 3 in turret slip ring	Shielded cable	18
4	Terminal 4 in turret slip ring	Shielded cable	18
5	Terminal 3 in master switch box	Black cable	6
	Terminal 5 in turret slip ring	Black	12
6	No connection in this box		
+	Terminal 13 in master switch box	Red cable	6
−	Ground in box (shielding on wires attached to terminals 3 and 4, grounded to this terminal)	Strap	

n. Turret Slip Ring Brush Connection Table (fig. 146).

Terminal No.	Connects to	Wire Color	Wire Size No.
1	Terminal 1 in radio and phone terminal box	Orange	18
2	Terminal 2 in radio and phone terminal box	Blue	18
3	Terminal 3 in radio and phone terminal box	Shielded cable	18

ELECTRICAL SYSTEM WIRING

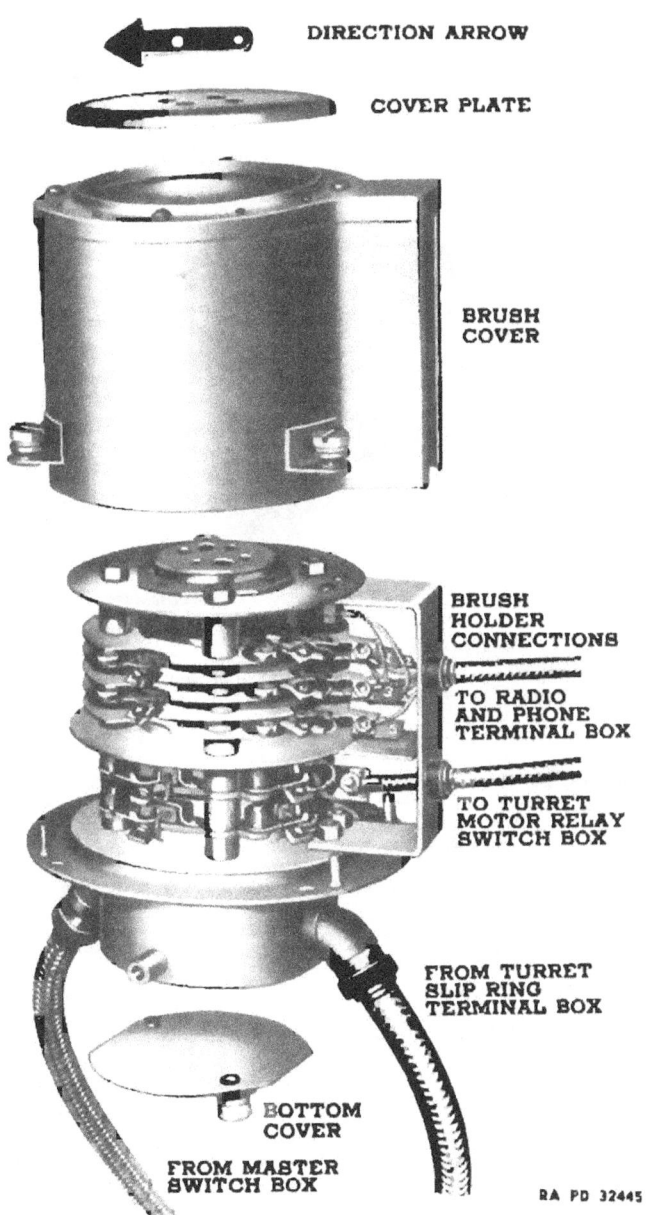

Figure 146 — Turret Slip Ring Brush Connections

TM 9-741
195

MEDIUM ARMORED CAR T17E1

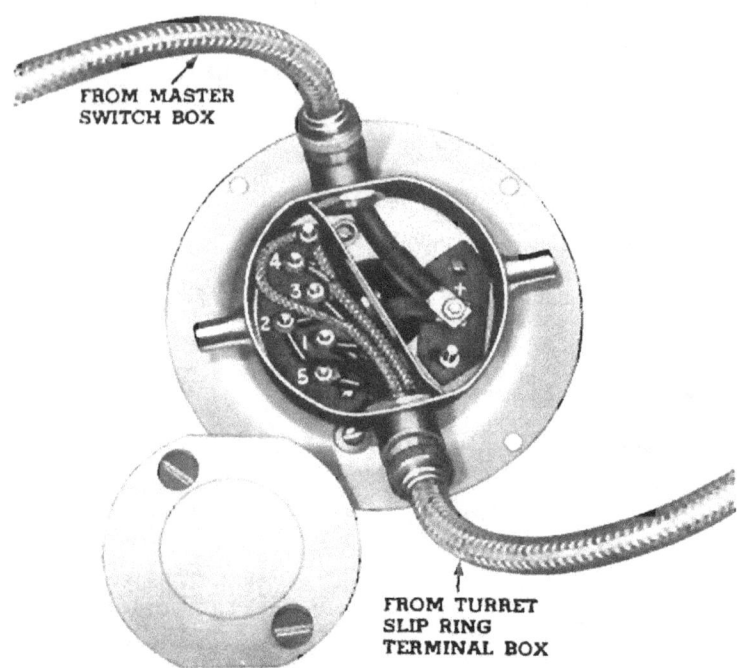

Figure 147 — Turret Slip Ring Armature Connections

Terminal No.	Connects to	Wire Color	Wire Size No.
4	Terminal 4 in radio and phone terminal box	Shielded cable	18
5	Terminal 5 in radio and phone terminal box	Black	12
+	Terminal L1 in turret motor relay switch box	Black cable	2
−	Ground		

o. **Turret Slip Ring Armature Connection Table** (fig. 147).

1	Terminal 1 in turret slip ring terminal box	Orange	18

294

TM 9-741
195

ELECTRICAL SYSTEM WIRING

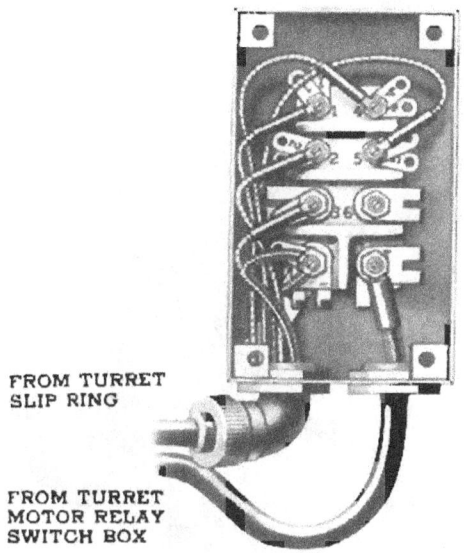

Figure 148 — Radio and Phone Terminal Box

Terminal No.	Connects to	Wire Color	Wire Size No.
2	Terminal 2 in turret slip ring terminal box	Blue	18
3	Terminal 3 in turret slip ring terminal box	Shielded cable	18
4	Terminal 4 in turret slip ring terminal box	Shielded cable	18
5	Terminal 5 in turret slip ring terminal box	Black	12
+	Terminal 12 in master switch box	Black cable	2

p. Radio and Phone Terminal Box and Connection Table (fig. 148). The terminal box is located on the wall of the turret basket. Leads are provided for 12- and 24-volt circuits for radio and phone installations. Connector leads are supplied to the turret slip ring terminal box for radio and phone outlets.

TM 9-741
195

MEDIUM ARMORED CAR T17E1

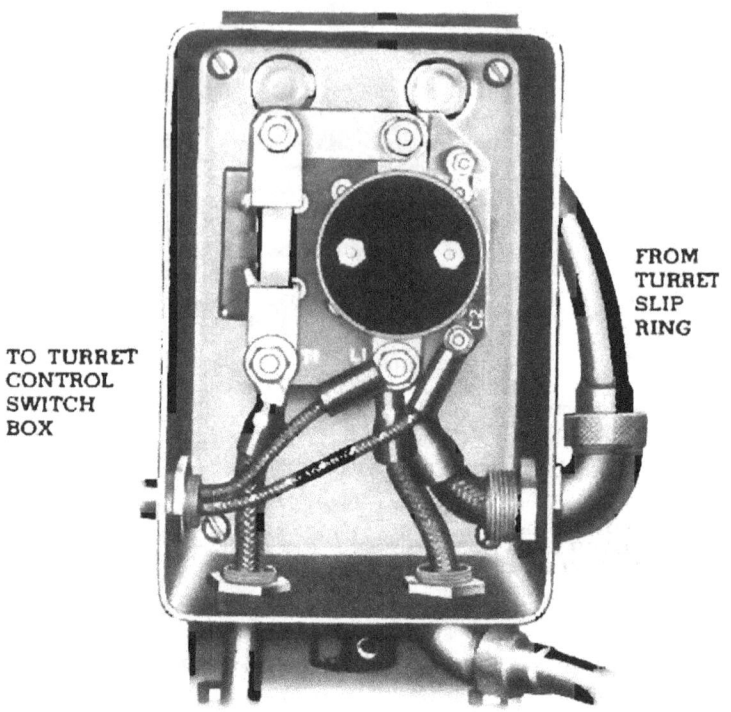

FROM TURRET SLIP RING

TO TURRET CONTROL SWITCH BOX

TO TURRET MOTOR

RA PD 32448

TO RADIO AND PHONE TERMINAL BOX

Figure 149 — Turret Motor Relay Switch Box

Terminal No.	Connects to	Wire Color	Wire Size No.
1	Terminal 1 in turret slip ring	Orange	18
2	Terminal 2 in turret slip ring	Blue	18
3	Terminal 3 in turret slip ring	Shielded cable	18
4	Terminal 4 in turret slip ring	Shielded cable	18
5	Terminal 5 in turret slip ring	Black	12
6	No connection in this box		
+	Terminal L1 in turret motor relay switch box	Black cable	6
−	Ground		

296

ELECTRICAL SYSTEM WIRING

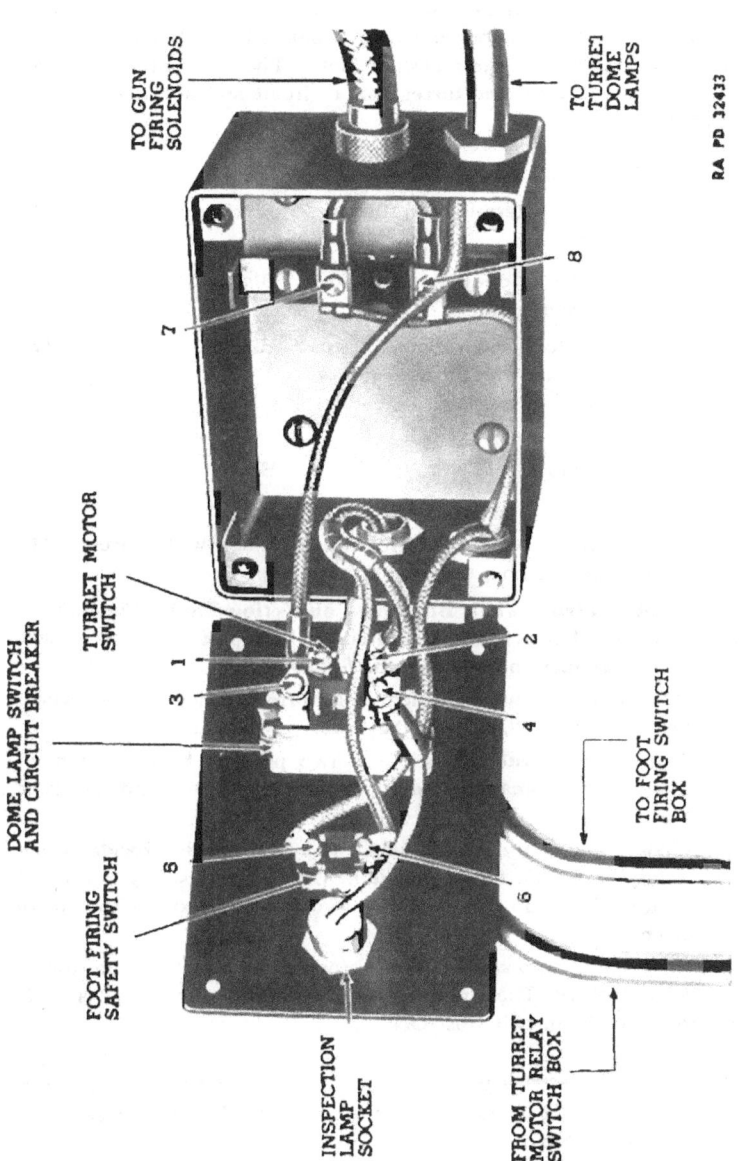

Figure 150 — Turret Control Switch Box

TM 9-741

MEDIUM ARMORED CAR T17E1

q. Turret Motor Relay Switch Box and Connection Table (fig. 149). The turret motor relay switch box is located at the left side of the turret traversing system electric motor. The box contains an overload circuit breaker for the turret motor circuit and a solenoid switch for this circuit.

Terminal Marking	Connects to	Wire Color	Wire Size No.
L1	Positive (+) terminal in turret slip ring	Black cable	2
	Positive (+) terminal in radio and phone terminal box	Black cable	6
L1	Turret motor switch, dome lamp switch, and foot firing safety switch in turret control switch box	Black	12
T1	Turret motor	Blue	4
C1	Ground		
C2	Turret motor switch in turret control switch box, terminal 1	Black with tracer	14

r. Turret Control Switch Box and Connection Table (fig. 150). The turret control switch box is located at the top of the turret basket in front of the gunner and contains the following:

(1) Turret motor switch, which is a spring loaded toggle switch. The upper position is "ON" and the lower position is "OFF."

(2) Dome lamp switch, which is a two position toggle switch incorporating a circuit breaker. The upper position is "ON" and the lower position is "OFF."

(3) Foot firing safety switch, which is a two position toggle switch with a safety cover. To turn the switch "ON," lift the safety cover and lift the toggle lever up. To turn the switch "OFF," strike the safety cover down.

(4) Inspection lamp socket, which provides an outlet for the trouble lamp in the tool kit. The circuit to the inspection lamp socket is "ON" when the dome lamp switch is "ON."

Terminal Marking	Connects to	Wire Color	Wire Size No.
1	Terminal C2 in turret motor relay switch box	Black with tracer	14
2	Terminal L1 in turret motor relay switch box	Black	12

TM 9-741
195

ELECTRICAL SYSTEM WIRING

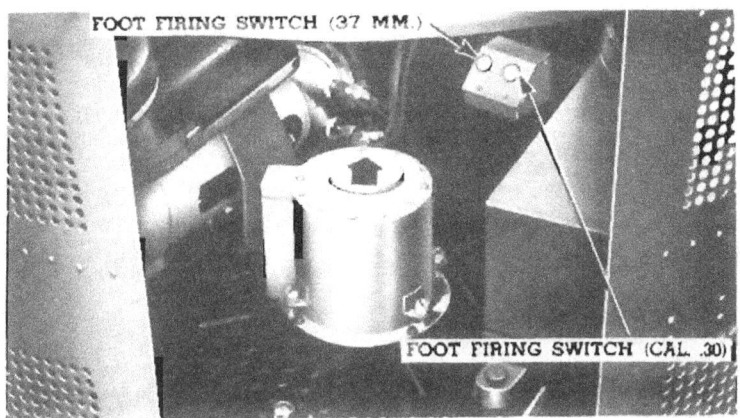

Figure 151 — Foot Firing Switches

Terminal No.	Connects to	Wire Color	Wire Size No.
3	Dome lamps	Black	14
4	Terminal L1 in turret motor relay switch box	Black	12
	Inspection lamp socket	Black	14
5	37-mm and cal. .30 foot firing switch	Black	14
6	Terminal L1 in turret motor relay switch box	Black	12
7	37-mm foot firing switch	Red	14
	37-mm gun firing solenoid	Red	14
8	Cal. .30 foot firing switch	Yellow	14
	Cal. .30 gun firing solenoid	Yellow	14

s. **Foot Firing Switch Box** (fig. 151). The foot firing switch box is located on the floor of the turret basket near the gunners right foot. Two foot switches operate the gun firing solenoids: the left switch operates the 37-mm gun; and the right switch operates the cal. .30 machine gun.

t. **Engine Terminal Box and Connection Table** (fig. 152). The engine terminal box is located on the left side of the center partition

299

MEDIUM ARMORED CAR T17E1

in the engine compartment. A wiring harness containing six wires leads to the box, and six harnesses, carrying one wire each, lead to the various electrical units. The terminals in the box are not numbered but for explanatory purposes, the illustration (fig. 152) is numbered and explained in the following table:

Terminal No.	Connects to	Wire Color	Wire Size No.
1	Heat indicator dash unit, left-hand	Natural with black tracer	14
	Heat indicator engine unit, left-hand	Natural with black tracer	14
2	Oil gage dash unit, left-hand	Natural with red tracer	14
	Oil gage engine unit, left-hand	Natural with red tracer	14
3	Starting motor push button, left-hand	Natural with black and red cross tracer	14
	Starter switch relay, left-hand	Natural with black and red cross tracer	14
4	Heat indicator dash unit, right-hand	Natural with black cross tracer	14
	Heat indicator, engine unit, right-hand	Natural with black cross tracer	14
5	Oil gage, dash unit, right-hand	Natural with red cross tracer	14
	Oil gage, engine unit, right-hand	Natural with red cross tracer	14
6	Starting motor push button, right hand	Natural	14
	Starter switch relay, right-hand	Natural	14

u. **Tail Lamp and Fuel Tank Terminal Box and Connection Table** (fig. 153). The tail lamp and fuel tank terminal box are located at the left side of the hull in the engine compartment. A wiring harness containing six wires leads to the box from the instrument panel. Two harnesses lead from the box: one contains two wires to the fuel tank; the other contains four wires to the left-hand tail lamp terminal box.

ELECTRICAL SYSTEM WIRING

The terminals are not marked in the box but for explanatory purposes, they are marked in the illustration (fig. 153).

Terminal No.	Connects to	Wire Color	Wire Size No.
1	HT terminal on main light switch	Natural with black cross tracer	14
	ST terminal in tail lamp terminal box, left-hand	Natural with black cross tracer	14
2	S terminal on main light switch	Natural	14
	SS terminal in tail lamp terminal box, left-hand	Natural	14
3	BHT terminal on main light switch	Natural with green tracer	14
	BOT terminal in tail lamp terminal box, left-hand	Natural with green tracer	14
4	BS terminal on main light switch	Natural with green cross tracer	14
	BOS terminal in tail lamp terminal box, left-hand	Natural with green cross tracer	14
5	Fuel gage dash unit	Natural with black tracer	14
	Fuel gage tank unit	Natural with black tracer	14
6	Fuel pump circuit breaker on instrument panel	Natural with black and red cross tracer	14
	Fuel pump	Natural with black and red cross tracer	14

v. **Tail Lamp Terminal Box Left-hand and Connection Table** (fig. 154). The left-hand tail lamp terminal box is located on the left side of the hull, back of the engine compartment. A wiring harness containing four wires leads to the box from the tail lamp and fuel tank terminal box. A wiring harness containing two wires leads to the tail lamp terminal box, right-hand, and an outlet at the back of the box leads direct to the left-hand rear lamp. The terminals in the box are not marked but for explanatory purposes, they are marked in the illustration (fig. 154).

MEDIUM ARMORED CAR T17E1

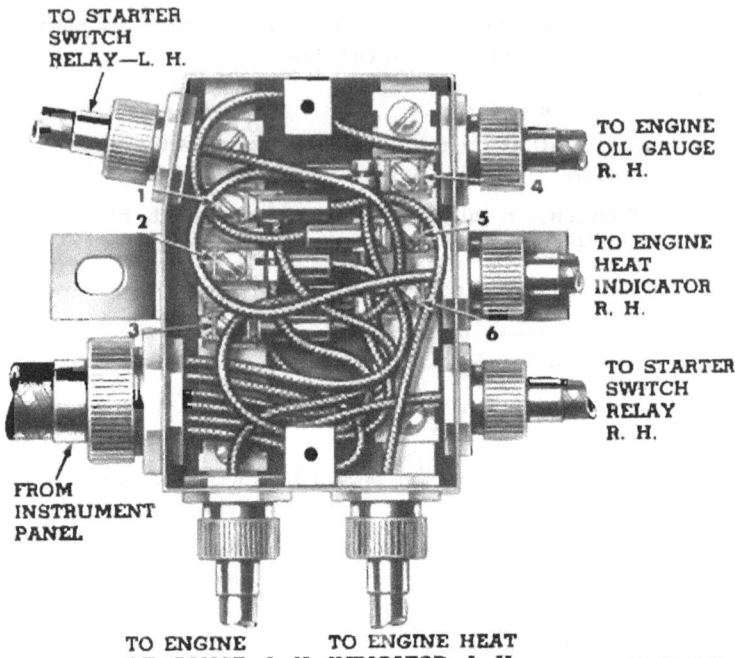

Figure 152 — Engine Terminal Box Connections

Terminal Marking	Connects to	Wire Color	Wire Size No.
ST	Terminal 1 in tail lamp and fuel tank terminal box	Natural with black cross tracer	14
	Service tail light in left-hand rear lamp	Natural with black cross tracer	14
SS	Terminal 2 in tail lamp and fuel tank terminal box	Natural	14
	Service stop light in left-hand rear lamp	Natural	14
BOT	Terminal 3 in tail lamp and fuel tank terminal box	Natural with green tracer	14
	BOT terminal in tail lamp terminal box, right-hand	Natural with green tracer	14
	Blackout tail lamp in left-hand rear lamp	Natural with green tracer	14

TM 9-741
195

ELECTRICAL SYSTEM WIRING

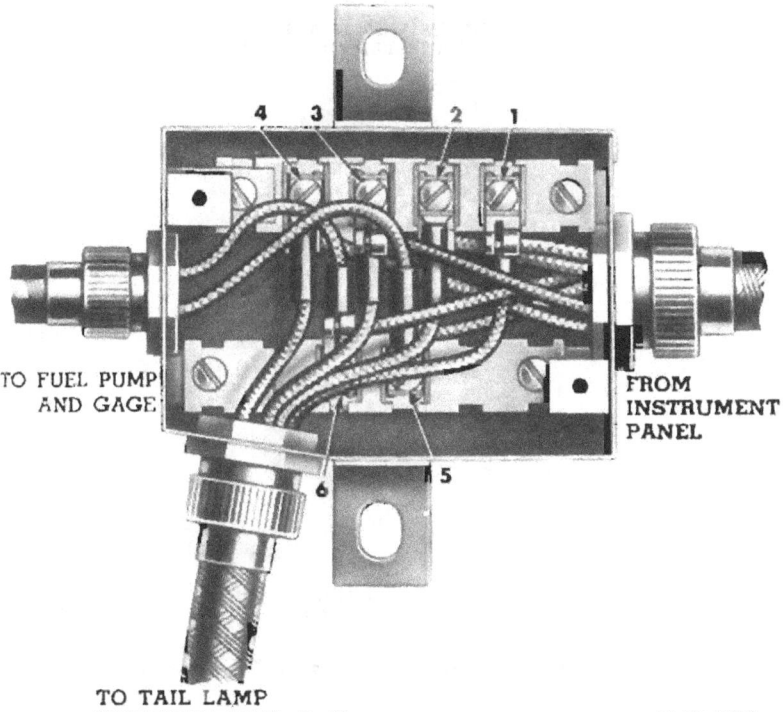

Figure 153 — Tail Lamp and Fuel Tank Terminal Box

Terminal Marking	Connects to	Wire Color	Wire Size No.
BOS	Terminal 4 in tail lamp and fuel tank terminal box	Natural with green cross tracer	14
	BOS terminal in tail lamp terminal box, right-hand	Natural with green cross tracer	14

w. **Tail Lamp Terminal Box Right-Hand and Connection Table** (fig. 155). The right-hand tail lamp terminal box is located at the right side of the hull, back of the engine compartment. A wiring harness containing two wires leads to the box from the tail lamp terminal box, left-hand. An outlet at the back of the box provides connections for the right-hand rear lamp. The terminals in the box are not marked but for explanatory purposes, they are marked in illustration (fig. 155).

TM 9-741
195

MEDIUM ARMORED CAR T17E1

Figure 154 — Left-Hand Tail Lamp Terminal Box

Terminal Marking	Connects to	Wire Color	Wire Size No.
BOT	BOT terminal in left-hand tail lamp terminal box	Natural with green tracer	14
	Blackout tail lamp in right-hand rear lamp	Natural with green tracer	14
BOS	BOS terminal in left-hand tail lamp terminal box	Natural with green cross tracer	14
	Blackout stop light in right-hand rear lamp	Natural with green cross tracer	14

x. **Fuel Pump and Fuel Gage Terminal Box and Connection Table** (fig. 156). The fuel pump and fuel gage terminal box is located on top of the main fuel tank and shield the terminals at the fuel pump and fuel gage. A wiring harness containing two wires leads from the tail lamp and fuel tank terminal box. The terminals in the box are not numbered but for explanatory purposes, they are numbered in the illustration (fig. 156).

TM 9-741
195

ELECTRICAL SYSTEM WIRING

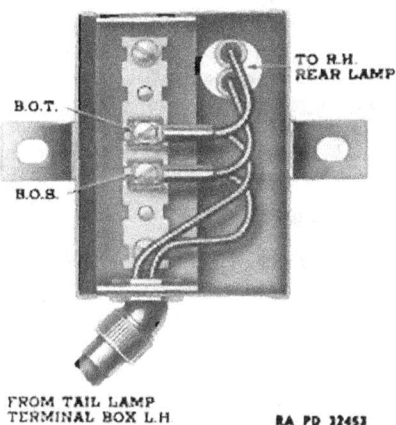

Figure 155 — Right-Hand Tail Lamp Terminal Box

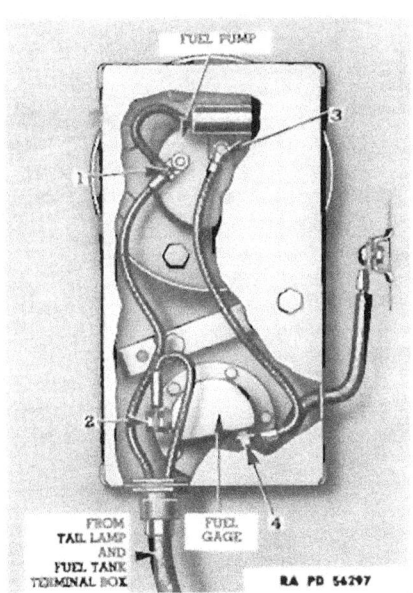

Figure 156 — Fuel Pump and Fuel Gage Terminal Box

MEDIUM ARMORED CAR T17E1

Terminal No.	Connects to	Wire Color	Wire Size No.
1	Terminal 6 in tail lamp and fuel tank terminal box	Natural with black and red cross tracer	14
	Condenser pigtail		
2	Terminal 5 in tail lamp and fuel tank terminal box	Natural with black tracer	14
3	Condenser, ground		
	Negative (—) on fuel gage, (terminal 4)	Natural	14
4	Negative (—) terminal on fuel pump	Natural	14
	Ground Terminal	Natural	14

Section XXXVI

TURRET AND TRAVERSING SYSTEM

	Paragraph
Description	196
Operation	197
Trouble shooting	198
Turret electric motor replacement	199
Control valve assembly replacement	200
Traversing gear assembly replacement	201
Hydraulic motor replacement	202
Hydraulic pump replacement	203

196. DESCRIPTION.

a. **Turret and Basket** (figs. 157 and 158). The turret is a one-piece casting which rotates on a ball bearing race and is protected against direct hits and lead splash. The turret ball ring assembly consists of an outer race and ring gear, inner race, one hundred and twenty 1¼-inch balls and twelve ball retainers. The outer race and ring gear are bolted to the hull while the turret is bolted to the inner race. The turret is retained in place by the turret ring. Lubrication of the ball ring is done by means of three lubrication fittings spaced around the inner circumference of the inner race. The turret basket is attached to the turret ball inner race by a ring of bolts around its top circumference. The traversing mechanism and gun elevating mechanism are contained within the turret and turret basket. Electric connections for the turret basket are made through a collector ring to permit 360 degrees rotation of the turret.

b. **Turret Traversing Systems** (fig. 157). The turret traversing systems which rotate the turret consist of two separate systems: the manual control and the hydraulic system, both operating through a gear fastened to the hull. The two systems operate independently of each other and are controlled by a gear shift lever (1, fig. 157) which serves to control the drive desired. When the shift lever is up, the hydraulic system is engaged, and when the lever is down, the manual system is engaged. If desired, the turret can be locked in any position.

(1) HYDRAULIC TURRET TRAVERSING SYSTEM (fig. 157). The hydraulic system consists of an electric motor, which drives the hydraulic pressure pump, an oil reservoir, hydraulic motor, gear box, control valves, control handle, hydraulic tubing, hoses and fittings. Turning the control handle counterclockwise rotates the turret to the left; turning the control handle clockwise rotates the turret to the right. The speed at which the turret will traverse is from 0 to 3 revolutions per

TM 9-741
MEDIUM ARMORED CAR T17E1

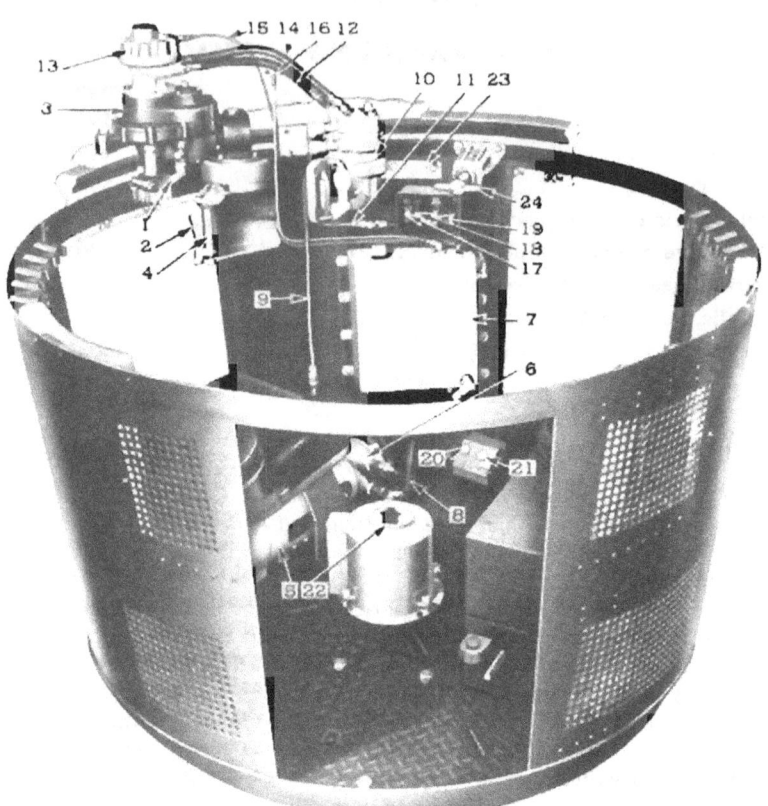

1. TRAVERSE SHIFT LEVER
2. HANDWHEEL BRAKE TRIGGER
3. TRAVERSING GEAR BOX
4. HANDWHEEL CRANK
5. ELECTRIC MOTOR
6. HYDRAULIC PUMP
7. OIL RESERVOIR
8. SUCTION HOSE
9. PRESSURE LINE
10. HYDRAULIC CONTROL VALVE
11. EXHAUST LINE
12. SHORT MOTOR HOSE
13. HYDRAULIC MOTOR
14. LONG MOTOR HOSE
15. MOTOR DRAIN LINE
16. RESERVOIR BREATHER
17. TURRET MOTOR SWITCH
18. DOME LAMP SWITCH
19. FOOT FIRING SAFETY SWITCH
20. FOOT FIRING SWITCH (37 MM.)
21. FOOT FIRING SWITCH (30 CAL.)
22. VEHICLE DIRECTIONAL INDICATOR
23. TURRET DIRECTION INDICATOR SCALE
24. TURRET LOCK

RA PD 32503

Figure 157 — Turret Traversing System

TM 9-741
196

TURRET AND TRAVERSING SYSTEM

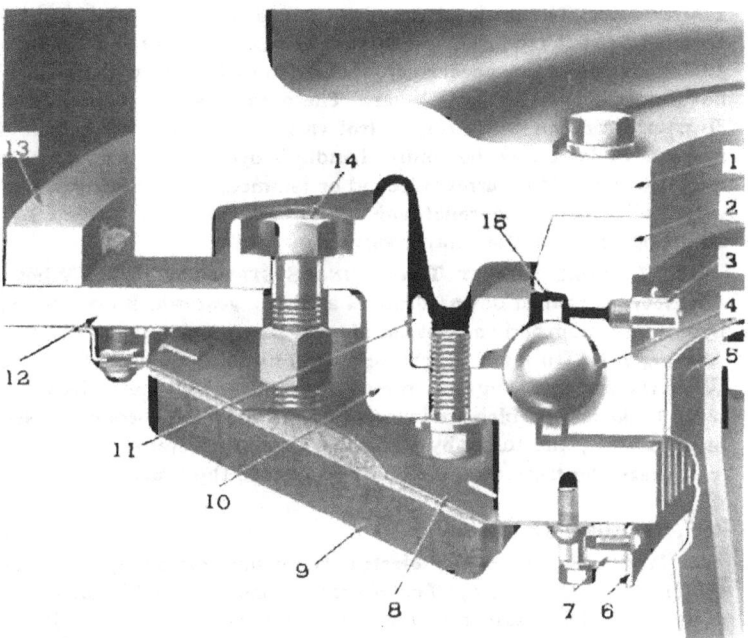

1. TURRET
2. BALL INNER RACE
3. LUBRICATION FITTING
4. BALL
5. TURRET BASKET
6. TURRET DIRECTION INDICATOR SCALE
7. INDICATOR SCALE MOUNTING BLOCK
8. BALL RING PAD RETAINER
9. BALL RING PAD
10. OUTER RACE AND RING GEAR
11. BALL RING
12. HULL
13. HULL FRONT ROOF PLATE TURRET RING GUARD
14. BALL OUTER RACE ATTACHING BOLT
15. BALL RETAINER

RA PD 32502

Figure 158 — Turret and Ball Ring Assembly

minute and is governed by the amount the control handle is turned. Springs have been built into the pistons in the control handle to aid in returning the handle to neutral. The handle is locked in the neutral position by a plunger which is released when the trigger is squeezed. The traversing system is operated by hydraulic pressure and incorporates an oil reservoir tank that is mounted on the basket below the

MEDIUM ARMORED CAR T17E1

control valve. The oil level gage is attached to the oil filler cap. Whe[n] the oil level is at the "FULL" mark, the reservoir contains 2 gallons c[f] oil. The oil should be checked frequently to make sure that the oil leve[l] never goes below the "LOW" mark. The pump maintains approximatel[y] 50 pounds pressure with the control valve in the neutral position. Th[e] pressure builds up as the control handle is opened and may go as hig[h] as 600 pounds if the turret is locked or jammed, at which time the relie[f] valve will open. The relief valve pressure can be regulated by a[n] adjusting screw on the control valve (par. 200).

(2) MANUAL TURRET TRAVERSING SYSTEM (fig. 157). When th[e] shift lever is pushed down, it moves a sliding gear which disengages th[e] hydraulic system and engages the manual traversing system. The manu[al] rotating of the turret is accomplished by turning the handwheel cran[k]. When the system is engaged in manual control, the turret is locked by [a] set of brake shoes which are incorporated in the handwheel crank assem[-] bly. To rotate the turret by using the handwheel crank, it is necessar[y] to squeeze the trigger on the crank to release the brakes.

197. OPERATION.

a. Hydraulic. Start the electric traversing motor by turning th[e] switch on (17, fig. 157). The operator should keep his hand on th[e] switch until he is sure that the motor is running normally, that is [a] constant speed. If the motor speed continues to increase, the switc[h] should be turned off and the trouble located, as the motor speed coul[d] increase to a point that would cause damage to the motor. After th[e] motor is running normally, push the shift lever (1, fig. 157) up t[o] engage the hydraulic traversing mechanism. Make certain that turr[et] lock (24, fig. 157) is disengaged. Squeeze trigger on control handle an[d] turn handle to right or left.

b. Manual. Push gear shift lever down, see that turret lock [is] released, squeeze trigger on handwheel crank, and turn crank to rig[ht] or left.

198. TROUBLE SHOOTING.

a. Faulty Operation of Turret Traversing System.

Probable Cause	Probable Remedy
Control handle does not operate properly.	Notify ordnance personnel.
Oil leakage.	Tighten screws, caps, or fittings. [If] leak still exists, refer to or[d]nance personnel.
Turret operates sluggish.	Notify ordnance personnel.

TURRET AND TRAVERSING SYSTEM

Probable Cause	Probable Remedy
Motor starts but turret will not rotate, chain coupling noisy.	Tighten Allen set screws at chain coupling, or refer to ordnance personnel.
Motor speed too fast or too slow.	Check hydraulic pressure and/or refer to ordnance personnel.
b. Inoperative Turret Motor.	
Switch and/or circuit breaker not "ON".	Turn switch to "ON" position and/or push in circuit breaker reset button.
Motor armature does not rotate, but motor hums and vibrates, or circuit breaker "throws out" continually.	Replace motor (par. 199).
Motor whines and vibrates from excessive speed.	Replace motor (par. 199).
Noisy motor.	Replace motor (par. 199).

199. TURRET ELECTRIC MOTOR REPLACEMENT.

a. Removal Procedure.

(1) REMOVE POWER LINE. Remove the four screws from the end plate at rear of motor and remove the end plate. Disconnect the main power input line from the insulated input terminal at the rear of motor.

(2) LOOSEN COUPLING. Rotate the motor until the set screws are visible through the holes in the right side of motor front bracket. Loosen the rear set screw.

(3) DISCONNECT MOTOR FROM PUMP. Remove the four pump to motor front bracket cap screws and lock washers. Slip the pump off the motor. NOTE: Do not disconnect the hydraulic hoses.

(4) REMOVE THE MOTOR. Remove the four motor-to-turret basket floor bolts and lock washers and slip the motor out from under the gunner's seat from rear of seat to front.

b. Installation Procedure.

(1) INSTALL MOTOR. Place motor in position under gunner's seat and install the four lock washers and bolts that attach the motor to the floor.

(2) CONNECT PUMP AND POWER LINE. Push pump and coupling in position on motor. NOTE: Be sure Woodruff keys are in place. Install the four lock washers and bolts that attach pump to motor. Tighten the Allen set screws in coupling. Connect power line at rear of motor and install the end cover.

TM 9-741
200-202

MEDIUM ARMORED CAR T17E1

200. CONTROL VALVE ASSEMBLY REPLACEMENT (fig. 157).

a. **Removal Procedure.**

(1) DISCONNECT HOSES. Remove the motor hoses from control valves, drain fluid from hoses, and plug the ends of the hoses. Disconnect the pressure and exhaust hoses from valve tubes, drain the hoses, and plug the ends.

(2) REMOVE CONTROL VALVE ASSEMBLY. Remove the four cap screws that attach the valve body to the basket and remove the control valve assembly.

b. **Installation Procedure.**

(1) INSTALL CONTROL VALVE BODY ASSEMBLY (fig. 157). Place valve body in position and install the four bolts and lock washers. Install the long motor hose in valve body port marked "CYL. 2," and the short motor hose in port marked "CYL. 1." Connect the exhaust hose to the upper port of the valve body marked "OUT." NOTE: Hoses should be crossed when installed. Install a pressure gage (MTM-M3-27) in the pressure port marked "IN," using a short nipple to connect the gage tee to the valve body. Assemble the pressure hose to pressure tube and the tube to tee of gage.

(2) ADJUST RELIEF VALVE. Lock basket to turret. Remove the relief valve acorn lock nut. Start the pump motor and turn the control handle either way as far at it will go. Using a screwdriver, turn the relief valve adjusting sleeve until a reading of 600 pounds on the gage is obtained. Install lock nut and tighten securely. Check pressure again, and if satisfactory remove the gage and replace the fitting. Check the oil level in the reservoir and fill to "FULL" mark. Check for oil leaks.

201. TRAVERSING GEAR ASSEMBLY REPLACEMENT.

a. **Remove Gear Assembly.** Disconnect the two hoses at top of hydraulic motor and plug the ends of the hoses. Disconnect drain tube at top of motor and plug the tube. Remove the locking wires from the two lower cap screws securing mechanism and pinion guard and remove screws. Remove the two upper cap screws and remove the mechanism. NOTE: Right screw is very close to body and may have to be removed with punch and hammer.

b. **Install Gear Assembly.** Place assembly in position on turret and install the four mounting bolts. Insert lock wire in lower bolts. Connect drain tube and the two hoses to the hydraulic motor. NOTE: Hoses should be crossed when installed. Start traversing motor and let run a couple of minutes. Check the oil level in the reservoir.

202. HYDRAULIC MOTOR REPLACEMENT.

a. **Remove Hydraulic Motor.** Disconnect the two hoses and drain line and plug the ends of hoses and drain line. Remove the four

TURRET AND TRAVERSING SYSTEM

cap screws that attach the pump to the mounting plate and lift off the motor.

b. Install Hydraulic Motor. Place motor in position on plate and install the four cap screws and lock washers. Connect the drain line. Connect the hoses. NOTE: Make sure that hoses are crossed when installed. Start traversing motor and let it run a couple of minutes; then check the oil level in the reservoir.

203. HYDRAULIC PUMP REPLACEMENT.

a. Removal Procedure.

(1) DRAIN OIL. Remove the drain plug from the bottom of the reservoir tank and drain the oil into a drain pan.

(2) REMOVE PUMP. Rotate the pump until the set screws are visible through the hole in the right side of motor front bracket and loosen the front set screw. Disconnect the hydraulic hoses from the pump and plug the ends. Remove the four pump-to-motor front bracket bolts and slip pump off of motor.

b. Installation Procedure.

(1) INSTALL PUMP. Slip pump in coupling; be sure that Woodruff key is in place in shaft. Install the four mounting bolts and lock washers. Connect the two hoses and tighten connections securely. Tighten Allen set screws.

(2) FILL RESERVOIR. Install drain plug in reservoir and fill tank with 2 gallons of OIL, hydraulic. Start traversing motor and let it run a minute or two, then check the oil level in tank and all connections for oil leaks.

TM 9-741
204-206

MEDIUM ARMORED CAR T17E1
Section XXXVII
HULL

	Paragraph
Description	204
Vision doors	205
Periscope mounting	206
Hull doors	207
Hull drain plugs	208

204. DESCRIPTION.

a. The hull of the vehicle is made of welded armor plate, ¾-inch thick at the driver and crew compartments, and ⅜-inch thick at the engine compartment.

b. The hull is divided into two sections by a bulkhead. In the forward part is a compartment for the driver and assistant driver and the basket that is attached to the turret. The rear section is the engine compartment.

c. The spring shackle brackets are attached directly to the hull. A tunnel is provided lengthwise for propeller shaft clearance. A tunnel is also provided for each of the two axles. This results in a materially lower floor and center of gravity.

d. Brackets are attached to each side of the hull to support the two jettison fuel tanks and two luggage boxes. Other clips and brackets on the outside of the hull provide a means of attaching certain tools and equipment.

205. VISION DOORS.

a. Two vision doors which are attached to the outside of the front sloping armor can be raised or lowered from the inside by pulling down on the hand levers attached to the doors after releasing the locking catch (fig. 159). When pulled to the full open position, a spring-loaded catch engages the lever to lock doors in position. To release the door for closing, pull out on the thumb screw of lever locking catch; lower door and door locking catch will snap in place. CAUTION: These doors are extremely heavy and the operator should take a firm hold on the extreme end of handle to control the door movement.

206. PERISCOPE MOUNTING.

a. Three periscopes are mounted in the driver's compartment, two for the driver and one for the assistant driver. Figure 160 shows the method of mounting the seals and locking thumb screws.

TM 9-741
206-208

HULL

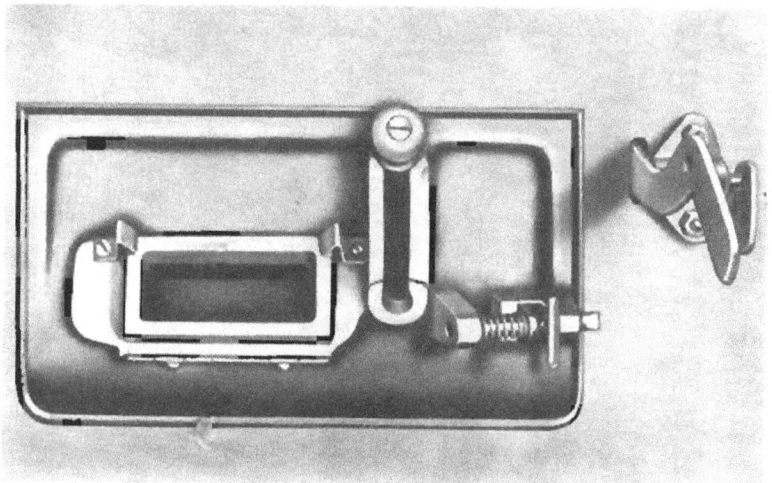

Figure 159 — Front Direct Vision Door

RA PD 32480

b. Each periscope is mounted in a ball-shaped retainer that permits tilting. This ball joint is held by two retaining plates secured together by eight screws. The retaining plates revolve in the castings welded to the top of the hull.

c. To install periscopes, place in retainer, and slide up into position. Lock thumb screw in place. The cover for the periscope opening is spring-loaded and will raise automatically. Periscope may be lowered until cover closes without removing from holder.

207. HULL DOORS.

a. Figure 161 shows one of the entrance doors. There is one located on each side of the hull. The locking bolt is held in the off position with a spring-loaded detent ball. To remove detent ball and spring, remove bolt on lower side of latch support.

208. HULL DRAIN PLUGS.

a. Six spring-loaded drain plugs, three in the engine compartment, and three in the crew compartment, are provided in the floor to drain water from the hull (fig. 162).

b. The plugs are located as follows:

(1) CREW COMPARTMENT. These drain plugs are operated by pressing down on button on end of valve stem with the foot.

MEDIUM ARMORED CAR T17E1

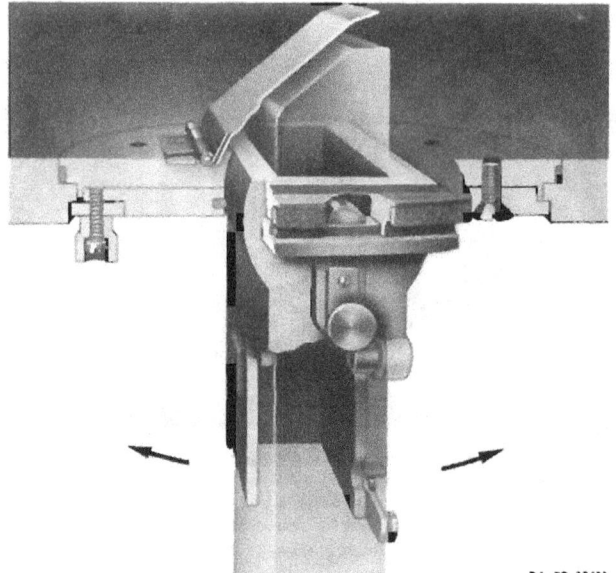

Figure 160 — Periscope Mounting

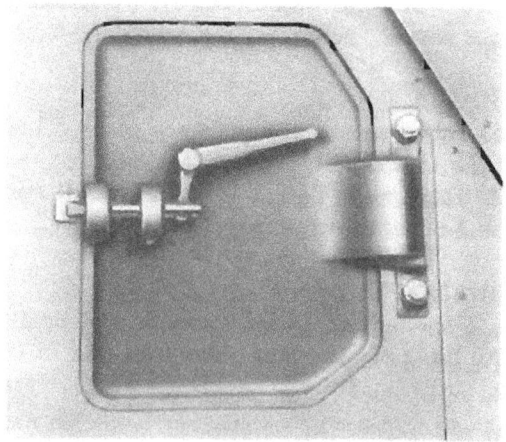

Figure 161 — Hull Door (Inside View)

TM 9-741
HULL

Figure 162 — Hull Drain Plug Locations

(a) Right forward corner near assistant driver's right foot.

(b) At rear of assistant driver's seat.

(c) At rear of driver's seat.

(2) ENGINE COMPARTMENT.

(a) One plug each side of division wall near center bulkhead, operated by pulling levers set in forward side of bulkhead.

(b) Right rear corner between fuel tank and side bulkhead, operated by lever from outside of engine compartment.

TM 9-741
209

MEDIUM ARMORED CAR T17E1

Section XXXVIII

GYROSTABILIZER

	Paragraph
Control of gun	209
Starting the equipment	210
Adjusting the stiffness rheostat	211
Adjusting the recoil rheostat	212
Checking stabilizer operation	213
Trouble shooting	214
Charging with oil	215
Removing air from system	216
Maintenance	217
Oil level	218
Lubrication	219
Cleanliness	220
Trunnion friction and gun balance	221
Gyro control gear box adjustment	222
Replacing gyro control	223
Replacing oil pump and/or motor	224
Replacing control box	225
Replacing piston and cylinder	226
Replacing recoil switch	227
Replacing gear box and mounting bracket	228
Replacing wiring and shielded conduit	229
Replacing clutch parts	230
Replacing flexible shaft	231
Replacing flexible shaft gear box	232
Replacing control link	233
Repairing and replacing damaged oil lines	234

209. CONTROL OF GUN.

CAUTION: The stabilizer should be used only when the vehicle is in motion. To avoid drain on the batteries when the vehicle is standing still, the gun should be controlled with the hand elevating gears in mesh. This action automatically disconnects the stabilizer by means of the disengaging switch.

a. To get best results from the stabilizer, the vehicle must be run at a constant speed. Complete cooperation and understanding between driver and gunner are essential.

GYROSTABILIZER

b. When the stabilizer is in operation and the hand elevating gears are out of mesh, the gun is elevated or depressed in the usual manner by turning the handwheel. This action changes the angular relation between the gun and the gyro control, and the gun automatically takes up the desired position as the gyro control immediately returns to its normal vertical position.

c. When the vehicle is moving and oscillating or pitching normally, and the stabilizer is operating normally, it will keep the gun very near its aimed angular position within its free range of angular movement, which is determined by its mounting in the vehicle. Therefore, after the gun is aimed, the stabilizer must be allowed to control the aimed position of the gun, the gunner turning the handwheel only when necessary to reaim the gun when the target moves, when the vehicle changes direction, or when the elevation of the vehicle (other than that caused by normal pitching) changes. CAUTION: The handwheel should not be turned after the gun has reached its maximum limits of travel in the elevated or depressed plane. Continued turning of the handwheel with the gun against either stop will only displace the gyro control from its vertical position and cause an excessive overload on the oil pump and battery.

210. STARTING THE EQUIPMENT.

a. Be sure the oil reservoir shows at least two-thirds full of OIL, hydraulic.

b. Set the stiffness rheostat at zero (fig. 163).

c. Take the hand elevating gears out of mesh by pulling the knurled lock plunger knob out and moving the worm gear assembly down and out of mesh with the sector (fig. 167). When the worm gear is completely out of mesh, the lock plunger will engage the upper of the two lock holes in the worm gear housing.

d. Turn the handwheel until the gyro control is approximately vertical to avoid bumping occupants with the gun or damaging the gyrostabilizer mechanism when the gyro control starts and turns vertical.

e. Start the oil pump by turning the turret switch to the "ON" position.

f. Start the gyro-control motor by throwing toggle switch to the "ON" position. Closing this switch starts the motors and lights the pilot light on the control box. The pilot light indicates that the gyro control circuit is energized. Allow at least 1 minute for the gyro motors to come up to speed.

CAUTION: In cold weather the oil should be permitted to warm up to obtain the best performance. In subzero weather allow 1½ minutes

TM 9-741
210-212

MEDIUM ARMORED CAR T17E1

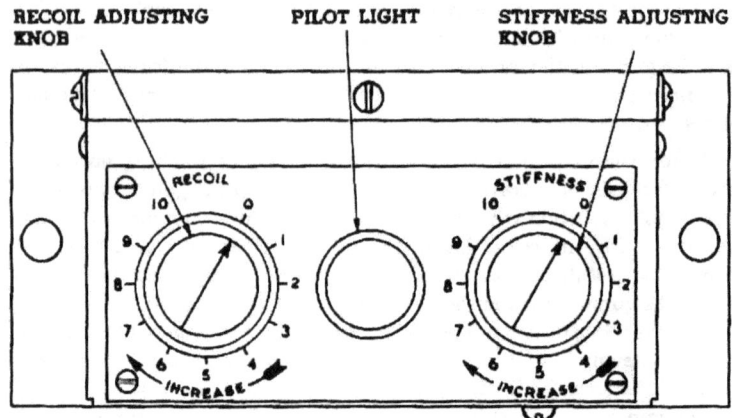

Figure 163 — Gyrostabilizer Control Box

running time for each degree of temperature below 0 F; for example, a total running time of 30 minutes at minus 20 F.

211. ADJUSTING THE STIFFNESS RHEOSTAT.

a. The stiffness rheostat, located in the control box (fig. 163), regulates the effectiveness or sensitivity of the stabilizer. Turn the control knob of the stiffness rheostat slowly to the right or clockwise. A vigorous vibration of the gun indicates too stiff an adjustment. Failure of the gun to remain in its aimed or set position indicates insufficient stiffness. If the gun starts to vibrate as the stiffness rheostat knob is adjusted, turn the knob in the opposite direction until the vibration is eliminated. It may be necessary for the operator to change the stiffness adjustment from time to time as operating conditions vary.

212. ADJUSTING THE RECOIL RHEOSTAT.

a. The recoil rheostat, located in the control box (fig.163), provides an adjustment for maintaining the position of the gun during recoil. This adjustment must be made by trial and error while the gun is being fired. Set the recoil rheostat knob on No. 5 position. If the breech of the gun drops during recoil, turn the knob to the right or clockwise. If the breech rises, turn the knob to the left. If faulty operation is obtained from the stabilizer during recoil at all positions of the knob, check for looseness in the mounting and for correct adjustment of recoil switch.

GYROSTABILIZER

213. CHECKING STABILIZER OPERATION.

a. The stabilizer should be checked for effectiveness with the vehicle in motion and before the vehicle is actually used in combat.

(1) Start the equipment.

(2) Choose a suitable location for a trial run of the vehicle. The terrain should be average rough, with no great slopes and sufficient acreage to permit adequate cruising time.

(3) Operate the vehicle over the average rough terrain at a normal speed. CAUTION: Do not attempt to check the stabilizer operation while the vehicle is starting or stopping or making sharp turns. The driver should be instructed to operate the vehicle in a straight line and at a constant speed.

(4) Aim the gun in the usual manner, using the horizon or some fixed object as the target.

(5) If necessary, adjust the stiffness rheostat.

(6) If possible, fire the gun and check the recoil adjustment.

214. TROUBLE SHOOTING.

a. If the gun does not fluctuate to any appreciable extent above or below the horizon, the stabilizer is operating satisfactorily. If the gun does not operate satisfactorily, proceed as follows:

(1) Check the oil level in the oil reservoir (fig. 179) which should be at least two-thirds full at all times.

(2) Check for air in the oil system.

(3) Check the vehicle voltage which must be maintained between 24 and 28 volts.

(4) Check for excessive friction in the gun mounting and unbalance in the gun.

(5) Check the external electrical connections for looseness or broken wires. Resolder loose connections and replace defective wiring.

(6) Check for lost motion between the gyro control and the handwheel. If necessary, adjust gear box.

(7) Check for loose mounting between gyro control, mounting bracket, and gun. Tighten all mounting bolts and screws.

(8) Check for excess looseness in the cylinder pivot pins and tighten if necessary.

(9) Check lubrication and, if necessary, add grease to the fittings on the stabilizer equipment and the turret assembly.

(10) If there is excessive oil leakage around the piston rod replace piston and cylinder.

MEDIUM ARMORED CAR T17E1

(11) Check the location and operation of the disengaging and recoil switches.

(12) Check clutch and replace parts, if necessary. NOTE: One of the most common causes of unsatisfactory operation is the condition called "hunting." This may be a vigorous vibration of the gun or a slow movement up and down. Many of the above-mentioned conditions will cause "hunting," the most common ones being air in the oil system, looseness in cylinder pivots, loose gyro control mounting, lost motion in gear box worm or worm bracket, excess friction in trunnion bearings, and high voltage.

b. If the trouble has not been corrected after the above checks, replace the various components in the following order, until the source of trouble is found: gyro control, oil pump, piston and cylinder, recoil switch, disengaging switch, control box, and pump motor.

215. CHARGING WITH OIL.

a. When charging the system with oil, it is very important for proper operation of the stabilizer that all air trapped in the system be removed.

b. The following procedure, therefore, must be adhered to:

(1) Use OIL, hydraulic.

(2) Heat oil to 150 to 200 F, if possible.

(3) Oil may be poured directly into the reservoir or pumped in with pump type oil can by removing the filler plug, or it may be added under a small amount of pressure. To get this pressure proceed as follows: Provide a filler can with 3-foot feed line, shut-off valve at reservoir connection end, and $\frac{3}{8}$-inch union below shut-off valve. Remove oil supply line from reservoir. Connect filler can feed line to oil reservoir line.

(4) Make certain that the turret switch is in the "OFF" position.

(5) Loosen the oil return line, remove small hexagon plugs, and loosen two bleeder valves on cylinder (fig. 164).

(6) Add oil to system until a flow, free of bubbles, is obtained from the return line. Tighten this connection permanently.

(7) After a solid flow of oil is obtained from bleeder valves, tighten finger tight.

(8) Loosen top bleeder valve. Push breech slowly to the uppermost position and tighten bleeder valve after a solid flow of oil is obtained.

(9) Loosen lower bleeder valve. Push breech slowly to lowest position and tighten bleeder valve after a solid flow of oil is obtained.

(10) Repeat steps (8) and (9).

GYROSTABILIZER

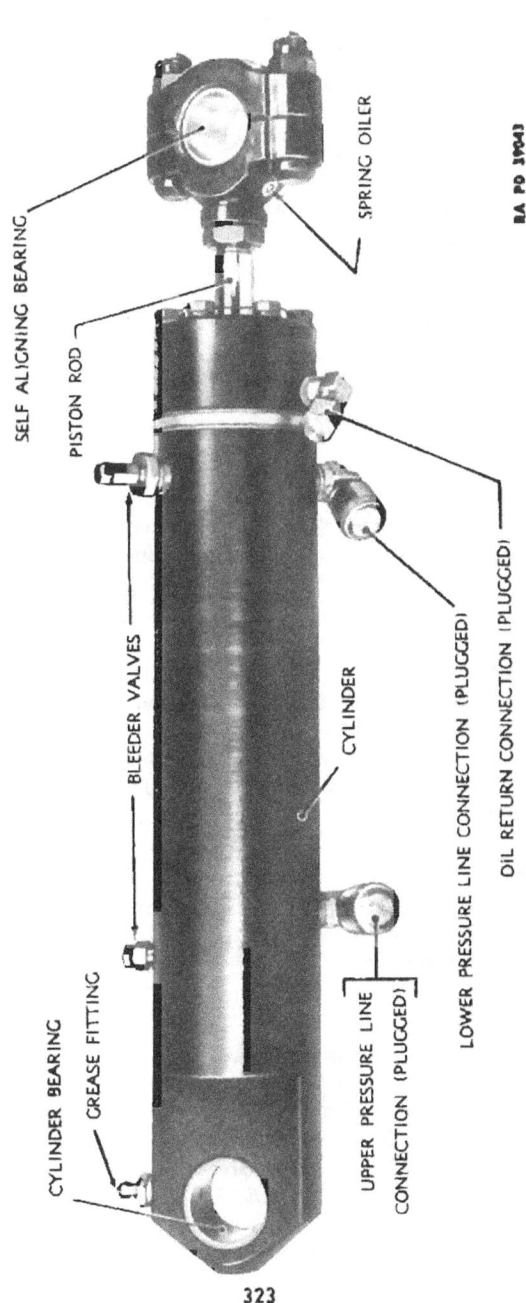

Figure 164 — Piston and Cylinder Assembly

TM 9-741
215-219

MEDIUM ARMORED CAR T17E1

(11) Remove pressure supply line and connections, if used, and reconnect oil reservoir supply line.

(12) Work gun up and down slowly until no more signs of air appear in oil reservoir.

(13) Run the pump approximately 10 minutes.

(14) Loosen both bleeder valves with pump running.

(15) After a solid flow of oil is obtained, tighten both valves permanently and stop motor.

(16) Recheck and fill oil reservoir approximately two-thirds full.

216. REMOVING AIR FROM SYSTEM.

a. To determine if the system is free of air:

(1) Lock gun in fixed position.

(2) Turn turret switch 'ON." If oil level drops, there is air in the system.

b. To remove air trapped in the system, turn the turret switch to the "OFF" position, disengage hand elevating mechanism, and work gun slowly up and down from 5 to 10 minutes. Then repeat the check in a above. If air still remains trapped in system, repeat purging procedure, paragraph 215 b step (8) through (10).

217. MAINTENANCE.

a. The incorrect adjustment or defective operation of any part of the stabilizer will cause unsatisfactory operation of the equipment as a whole. It is essential, therefore, that the operating instructions be carefully read before attempting to install, operate, adjust, or repair any part of the stabilizer.

218. OIL LEVEL.

a. The oil reservoir (fig. 179) should be approximately two-thirds full of OIL, hydraulic, at all times. If the oil level drops from day to day, check for leaks in the system.

219. LUBRICATION.

a. One Zerk type grease fitting and two spring oilers are provided on those parts of the equipment that require periodical greasing and oiling. One spring oiler is located on the piston rod end (figs. 164, 165, and 166) and the other on the gyro control gear box (fig. 170). The grease fitting is located on the cylinder. Grease and oil approximately every 50 hours of operation, using GREASE, OD. No. 0, and OIL, engine, seasonal grade.

TM 9-741
219

GYROSTABILIZER

Figure 165 — Stabilizer Installed

325

MEDIUM ARMORED CAR T17E1

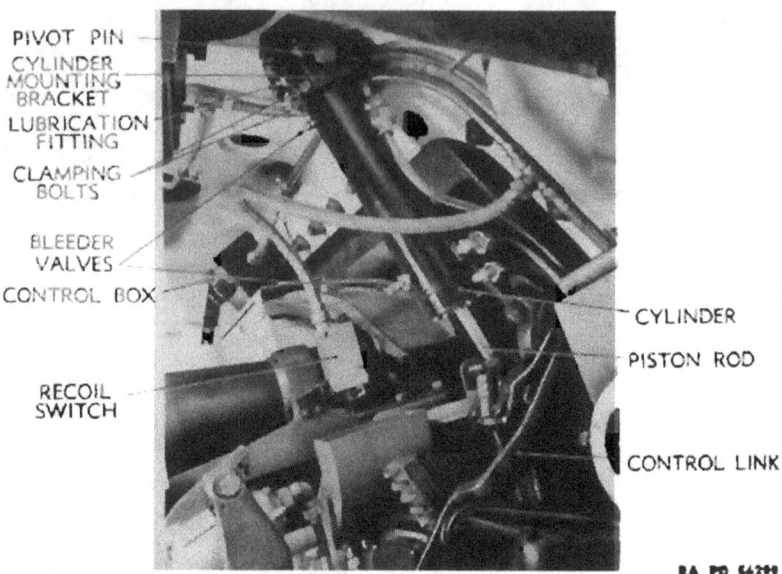

Figure 166 — Piston and Cylinder Installed

220. CLEANLINESS.

a. All parts of the stabilizer must be kept clean for satisfactory operation. Moving parts which are not protected from dust should be wiped daily. The filler hole cap must be kept on the oil reservoir to prevent the entrance of dust or other foreign material.

221. TRUNNION FRICTION AND GUN BALANCE.

a. The performance of the stabilizer will be impaired if the amount of friction in the trunnion bearings or the unbalance of the gun exceeds a predetermined value.

b. Inspection, to determine that this limit is not exceeded, may be made as follows:

(1) Disconnect the piston rod end (fig. 167) from the mounting bracket by removing pivot pin. It will be necessary to have machine gun out while removing this pin.

(2) Install machine gun and empty shell rack and fill machine gun ammunition tray with ammunition.

(3) Load the 37-mm gun or add weight equivalent to shell in breech.

(4) Disengage manual elevating mechanism.

TM 9-741
GYROSTABILIZER

- A — PISTON AND CYLINDER ASSEMBLY
- B — PISTON ROD END MOUNTING BRACKET
- C — CONTROL LINK
- D — GEAR BOX
- E — GYRO CONTROL UNIT
- F — MOUNTING BRACKET
- G — FLEXIBLE SHAFT
- H — DISENGAGING SWITCH
- J — SWITCH PLUNGER
- K — FLEXIBLE SHAFT GEAR BOX

RA PD 23308A

Figure 167 — Location of Gyro Control, Gear Box, Piston and Cylinder, and Mounting Brackets

TM 9-741
221-222

MEDIUM ARMORED CAR T17E1

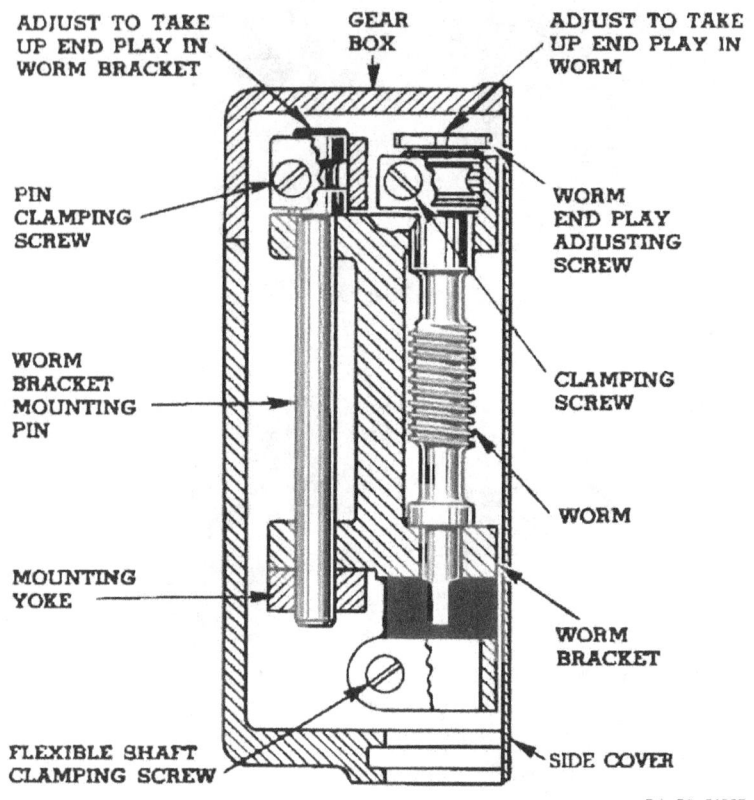

Figure 168 — Gear Box — Worm and Worm Yoke Pin Adjustment

(5) Attach a spring scale to the end of the muzzle of the gun and determine that the pull required to move the gun upward or downward does not exceed 2 pounds and that the difference in pull required to move gun upward and to move it downward does not exceed 1 pound.

c. Adjust balance by adding or removing laminations from counterweight.

222. GYRO CONTROL GEAR BOX ADJUSTMENT (figs. 168 and 169).

a. **Worm Bracket End Play Adjustment.** Remove gear box cover (fig. 170). Loosen the clamping screw (fig. 168). Adjust the end play by

TM 9-741
222

GYROSTABILIZER

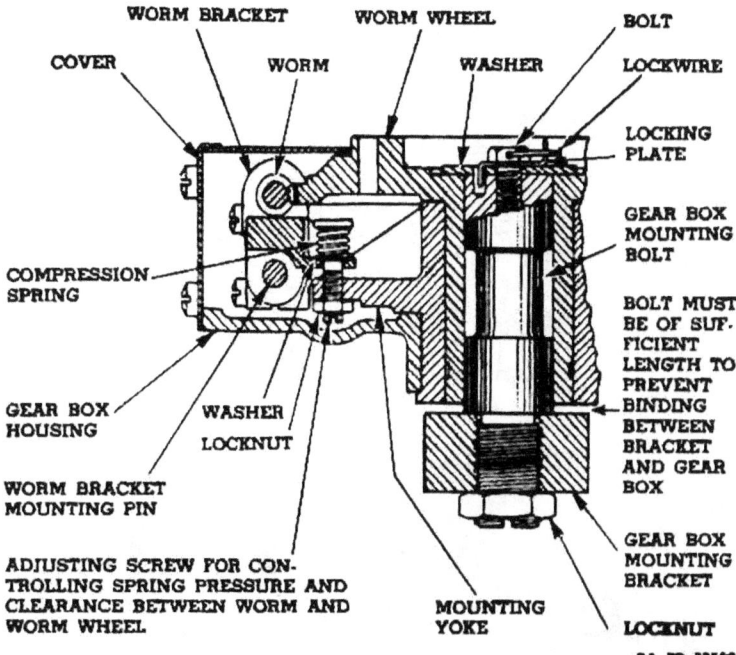

Figure 169 — Gear Box — Worm and Worm Wheel Adjustment

moving the worm bracket mounting pin up or down. The bracket must work freely and without appreciable end play. Tighten the clamping screw securely after making adjustments.

b. **Worm End Play Adjustment.** Remove the gear box cover. Loosen the clamping screw which locks the adjusting screw in the worm bracket (fig. 168). Adjust the screw to take out end play. Relock in position by tightening the clamping screw.

c. **Controlling the Mesh of the Worm and Worm Wheel** (fig. 169).

(1) The meshing of the worm wheel with the worm is controlled by the compression of a spring (fig. 169).

(2) To adjust this spring compression, first remove the gear box housing.

(3) Loosen the adjusting screw lock nut (fig. 169) and turn the adjusting screw to vary the spring compression.

(4) There must be at least $\frac{1}{32}$-inch movement between the worm and worm wheel, against the compression of the spring. The screw must

TM 9-741

MEDIUM ARMORED CAR T17E1

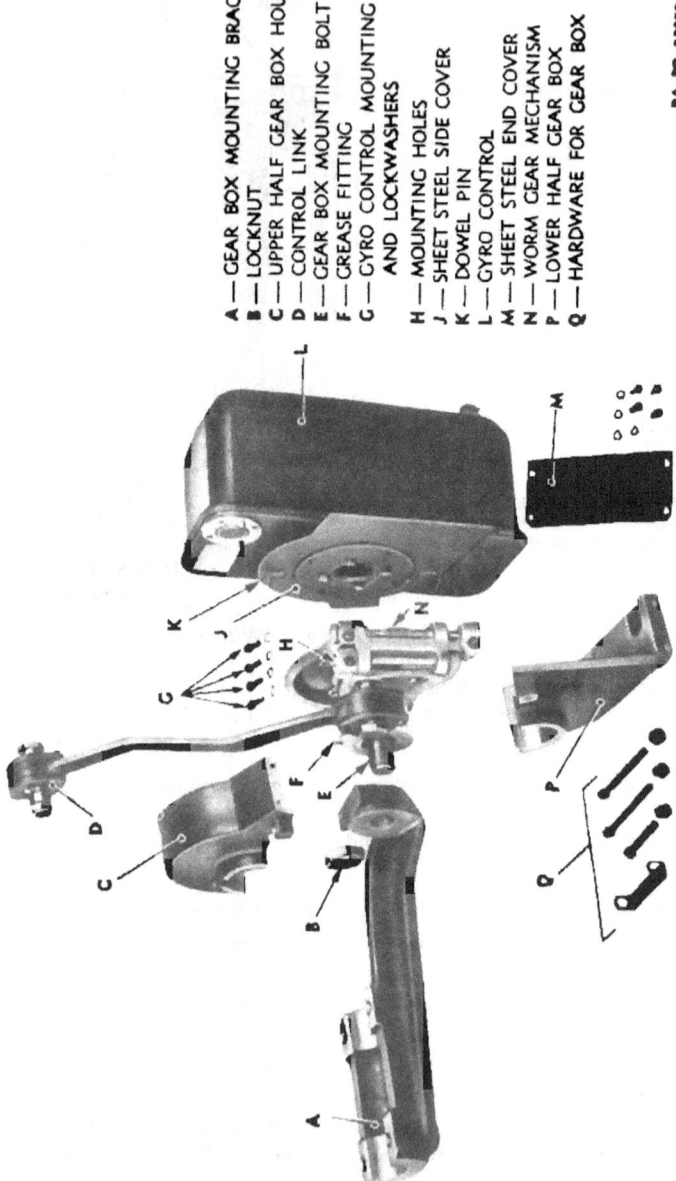

A — GEAR BOX MOUNTING BRACKET
B — LOCKNUT
C — UPPER HALF GEAR BOX HOUSING
D — CONTROL LINK
E — GEAR BOX MOUNTING BOLT
F — GREASE FITTING
G — GYRO CONTROL MOUNTING BOLTS AND LOCKWASHERS
H — MOUNTING HOLES
J — SHEET STEEL SIDE COVER
K — DOWEL PIN
L — GYRO CONTROL
M — SHEET STEEL END COVER
N — WORM GEAR MECHANISM
P — LOWER HALF GEAR BOX
Q — HARDWARE FOR GEAR BOX

Figure 170 — Gyro Control and Gear Box Parts

GYROSTABILIZER

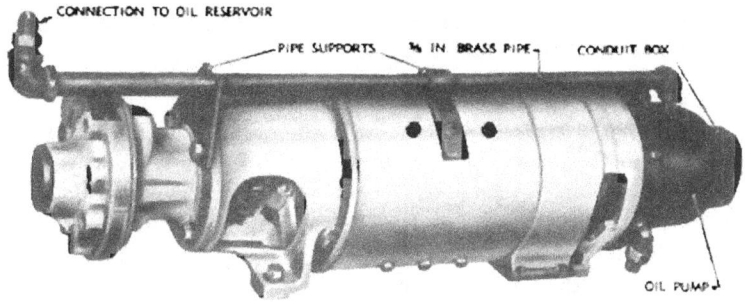

Figure 171 — Oil Pump and Motor Assembly

not be tightened to the extent that binding develops between the worm and worm wheel.

(5) Tighten the lock nut securely.

d. Clearance between Gyro Control Gear and Mounting Bracket. The clearance between the gyro control gear box and the gear box mounting bracket (fig. 169) is governed by the length of the gear box mounting bolt. If binding occurs between gear box and bracket, replace the mounting bolt.

223. REPLACING GYRO CONTROL.

CAUTION: The gyro control should be handled with extreme care at all times. Severe shock by dropping or jarring may injure the internal parts and cause erratic operation of the system. Under no conditions are the ordnance seals to be broken without proper authority.

a. In the event of improper operation remove the gyro control as follows:

(1) Turn stabilizer and turret switches to "OFF" position.

(2) Press release lever and remove multi-prong plug from receptacle on gyro control base.

(3) Remove gear box end cover (fig. 170).

(4) Remove gear box clamp and two halves of gear box.

(5) Remove the four gyro control mounting bolts on worm wheel and remove control unit.

b. To install replacement unit, reverse the removal procedure and check the operation of the complete stabilizer.

MEDIUM ARMORED CAR T17E1

224. REPLACING OIL PUMP AND/OR MOTOR.

a. The oil pump and motor should be removed from the vehicle as an assembly when either requires replacement.

b. This assembly should be removed as follows:

(1) Turn the turret and stabilizer switches to the "OFF" position.

(2) Disconnect oil lines from pump and cap them to avoid oil loss (fig. 171). A suitable container should be used when performing this operation to catch any oil that might drain from the fittings.

(3) Cap the oil connections on the oil pump.

(4) Remove the four motor mounting screws.

(5) Disconnect the turret traverse pump from the motor.

(6) Raise the oil pump end of the motor sufficiently to disconnect the wiring.

(7) Remove the terminal box cover and disconnect the three wires, green, yellow, and white, from the oil pump terminals.

(8) Disconnect the electrical lead to the motor.

(9) Lift the motor and oil pump assembly from the vehicle.

c. The oil pump may be removed from the motor by first removing the four Allen head screws and then tapping the base plate of the oil pump with a light hammer while pulling on the oil pump.

d. Installation of the oil pump and motor is the reverse of removal operations with the following exceptions:

(1) Be careful when replacing the pump on the motor that the two shafts mesh properly.

(2) Additional oil must be added to the system and air removed (par. 215).

(3) Check the complete stabilizer for operation (par. 213).

225. REPLACING CONTROL BOX.

a. Throw the turret and stabilizer switches to the "OFF" position. Remove top cover and disconnect wiring. Remove mounting screws and remove control box. To install replacement unit, reverse the removal procedure. Be sure to reconnect the wires according to the wiring diagram. Check the operation of the complete stabilizer.

226. REPLACING PISTON AND CYLINDER (figs. 164, 165, 166, and 172).

a. If parts become damaged, bearings or pivot pins worn, or leaks occur around the piston rod, replace the unit as follows:

GYROSTABILIZER

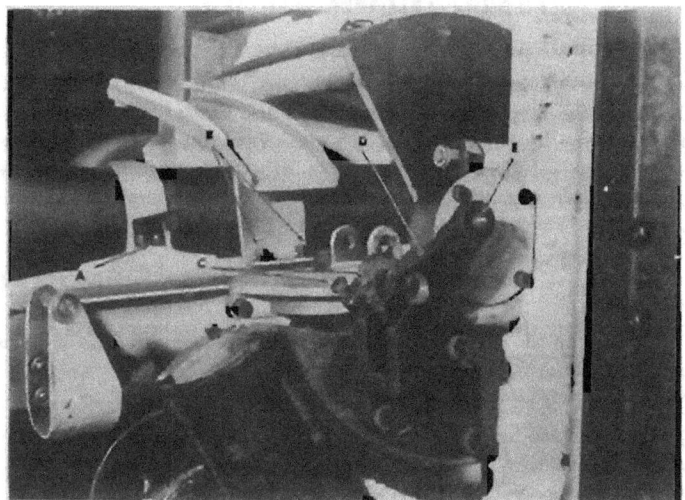

A — RECOIL SWITCH CONTACT BRACKET C — PIVOT PIN CLAMPING BOLTS
B — RECOIL SWITCH MOUNTING BRACKET BOLTS D — PISTON ROD END MOUNTING BRACKET
E — CONTROL LINK MOUNTING HOLE

RA PD 23317

Figure 172 — Piston Rod End Mounting Bracket

(1) Turn turret and stabilizer switches to "OFF" position.

(2) Disconnect and plug the three oil line connections. Make provisions to catch oil lost when disconnecting lines.

(3) Remove the four pivot pin clamp bolts.

(4) Spring piston rod end mounting bracket with screwdriver and slide out pivot pin.

(5) Spring cylinder mounting bracket, remove pivot pin, and take out piston and cylinder.

b. The installation of a new piston and cylinder is the reverse of the removal operations with the following exceptions:

(1) Recharge system with oil and remove air.

(2) Check the operation of the complete stabilizer.

227. REPLACING RECOIL SWITCH (fig. 166).

a. Removal.

(1) Throw the turret and stabilizer switches to the "OFF" position.

(2) Remove the two mounting bolts, holding switch to mounting bracket.

(3) Remove switch cover plate and disconnect wiring.

(4) Unscrew switch from shielded conduit fitting and remove switch.

MEDIUM ARMORED CAR T17E1

b. Installation.

(1) To install replacement switch, reverse the removal procedure.

(2) The recoil switch is normally closed and mounted so its plunger will be depressed by the breech of the gun. The plunger must not protrude more than $\frac{1}{16}$-inch when the gun is in battery position. Adjust by shifting the switch mounting bracket or the switch. If the switch mounting holes are not oversize, enlarge as necessary with rat tail file.

228. REPLACING GEAR BOX AND MOUNTING BRACKET (fig. 167).

a. Adjustments for the gear box are given in paragraph 222. If parts are damaged or worn, replace the entire unit as follows:

(1) Turn turret switch to "OFF" position.

(2) Press release lever and remove multi-prong plug from gyro control base (fig. 170).

(3) Remove gear box end cover (fig. 170).

(4) Loosen flexible shaft clamping screw and remove flexible shaft.

(5) Remove the gyro control from the worm wheel of the gear box.

(6) Disconnect the control link from the piston rod end mounting bracket.

(7) If the mounting bolt is to be replaced, remove the mounting bolt lock nut.

(8) Remove the locking wire, bolt, lock plate, and washer from the gyro control end of the mounting bolt.

(9) Turn the gear box to a position that will permit it to be slipped off the mounting bolt.

(10) Loosen the four mounting bolts (fig. 170) and remove the mounting bracket.

b. When installing the gear box and bracket, reverse the removal operations.

(1) Adjust worm and worm bracket.

(2) Check the operation of the complete stabilizer.

229. REPLACING WIRING AND SHIELDED CONDUIT (fig. 181).

a. Turn turret switch to "OFF" position.

b. Measure new piece of shielded conduit to length and tin at least $\frac{1}{2}$-inch on each side of cut using noncorrosive flux.

c. Cut with fine-toothed hack saw.

d. Tin inside of ferrule.

TM 9-741
229

GYROSTABILIZER

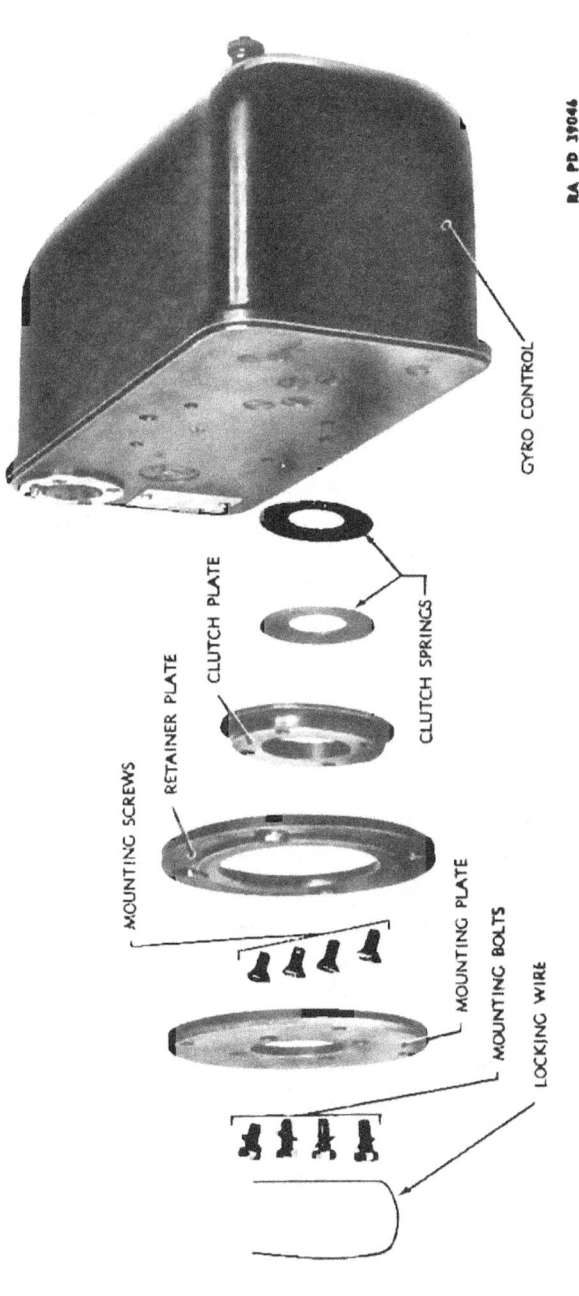

Figure 173 — Gyro Control Clutch Parts

335

e. Sweat solder cable into place. *If a torch is used,* use flame on fitting only, never directly on cable.

f. Cut new wires to length and run them through shielded cable.

g. When rewiring, follow the color scheme and connections shown in the wiring diagram (fig. 181).

230. REPLACING CLUTCH PARTS (fig. 173).

a. The clutch is not adjustable. To replace:

(1) Turn the turret switch to the "OFF" position.

(2) Remove gyro control.

(3) Remove locking wire from mounting plate bolts and remove mounting bolts and plate.

(4) Remove mounting screws from retainer plate and remove retainer plate.

(5) Remove clutch plate and clutch springs.

(6) Replace retainer plate, clutch plate, and the two clutch springs. Add one spring if only one is present.

(7) To install, reverse the removal procedure. Stake the mounting screws on retainer plate.

(8) Retest clutch, and check the operation of the complete stabilizer.

231. REPLACING FLEXIBLE SHAFT.

a. **Removing Flexible Shaft from the Flexible Shaft Gear Box.**

(1) Remove the locking wire from the gear box cover screws and shaft nut (fig. 174).

(2) Unscrew the shaft nut from the gear box.

(3) Pull the flexible shaft from its position in the gear box.

b. **Removing Flexible Shaft from Gyro Control Gear Box.**

(1) Remove the end cover from the gyro control gear box.

(2) Loosen the flexible shaft clamping screw (fig. 175) and pull the shaft from its position in the worm bracket.

c. **Installing Flexible Shaft.**

(1) To install a new shaft, reverse the removal operations.

(2) Be sure the couplings at the end of the flexible shaft mesh properly (fig. 175) before tightening the clamping screw and shaft nut.

232. REPLACING FLEXIBLE SHAFT GEAR BOX.

a. **Removing Flexible Shaft Gear Box.**

(1) Remove the flexible shaft (par. 231).

TM 9-741

GYROSTABILIZER

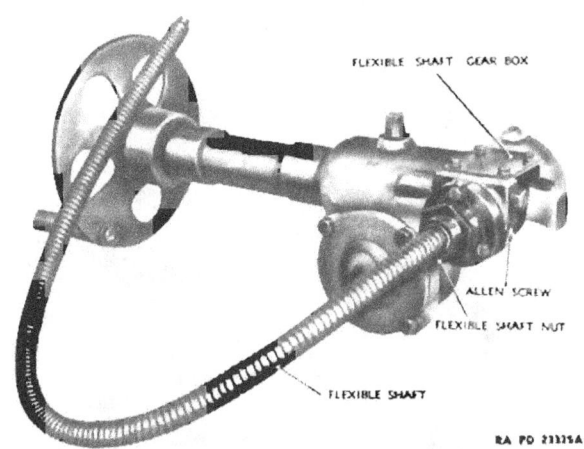

Figure 174 — Flexible Shaft Gear Box

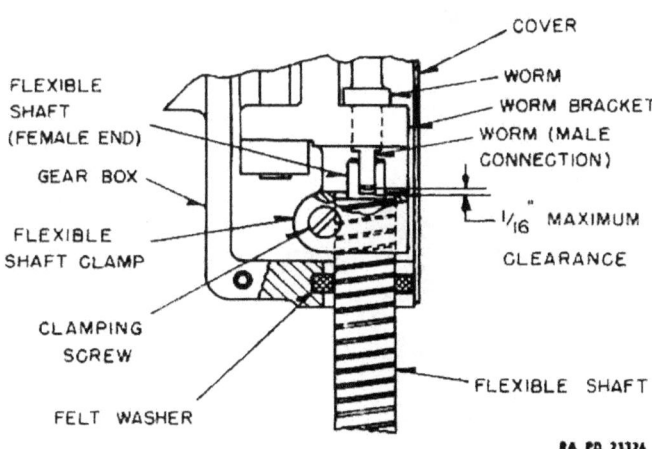

Figure 175 — Flexible Shaft Connection to Gear Box

Figure 176 — Miter Gear Installation

TM 9-741
232-233

GYROSTABILIZER

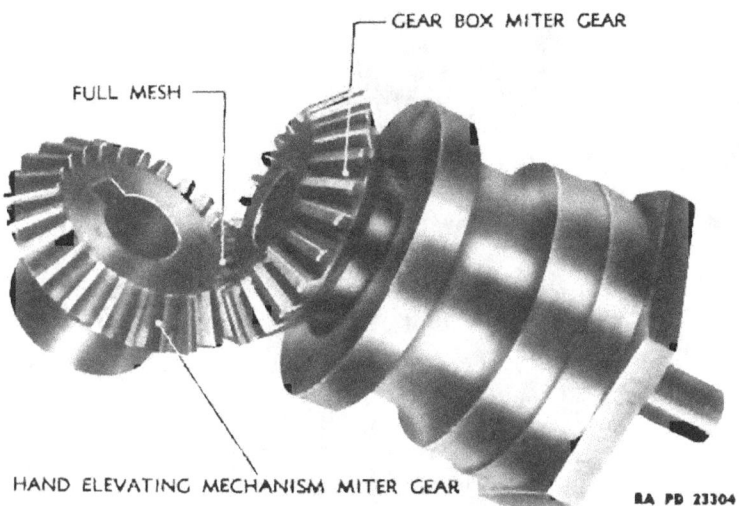

Figure 177 — Miter Gears in Full Mesh

(2) Unscrew the three bolts holding the gear box to the hand elevating mechanism (fig. 174).

(3) Remove the gear box.

(4) Remove the miter gear from the handwheel shaft (fig. 176).

b. Installing Flexible Shaft Gear Box.

(1) To install the gear box, reverse the removal procedure.

(2) Binding between the miter gears may be eliminated by installing shims between the gear box and hand elevating gear mechanism, or by loosening the Allen set screw and moving the miter gear assembly (fig. 177) in the gear box.

233. REPLACING CONTROL LINK.

a. Removing Control Link.

(1) Remove the gyro control (par. 223).

(2) Remove the gear box from the hand elevating mechanism (par. 228).

(3) Remove the worm wheel from the mounting yoke (fig. 178).

(4) Loosen and remove the bearing bolt from the mounting yoke.

(5) Remove the control link.

b. Installation of the control link is the reverse of removal.

GYROSTABILIZER

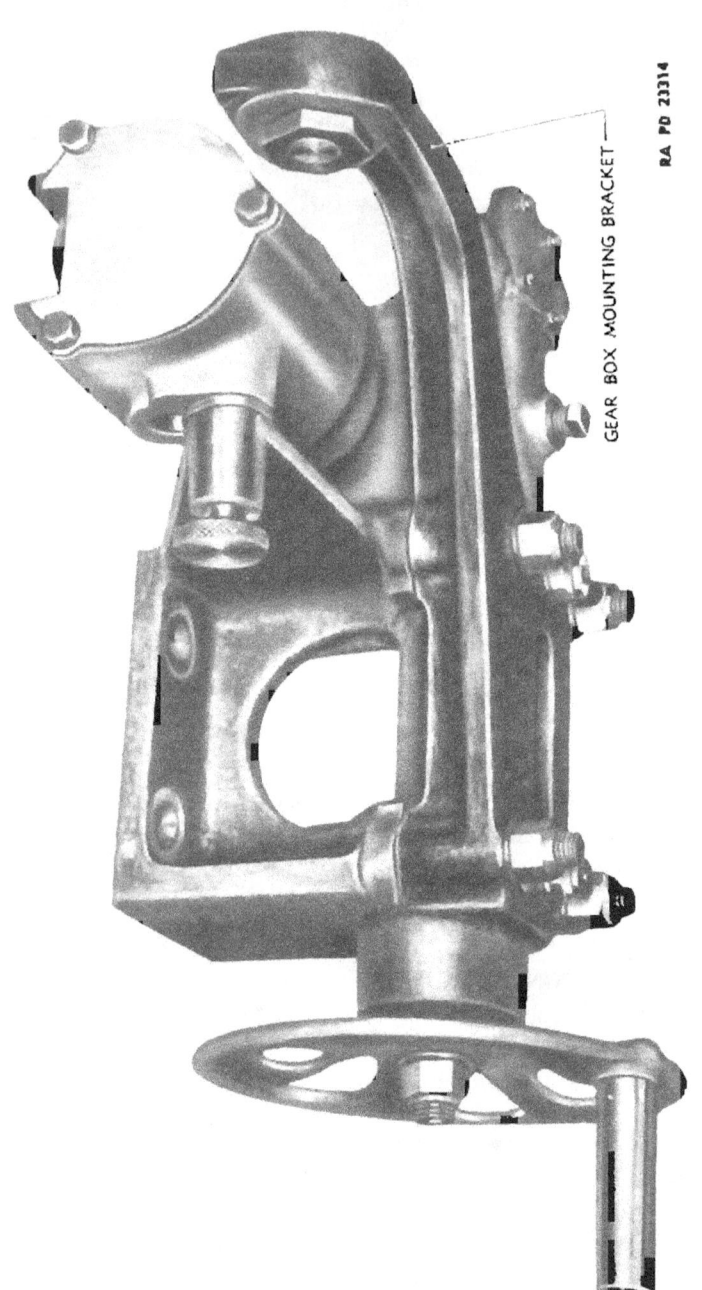

Figure 178 — Gear Box Mounting Bracket

GYROSTABILIZER

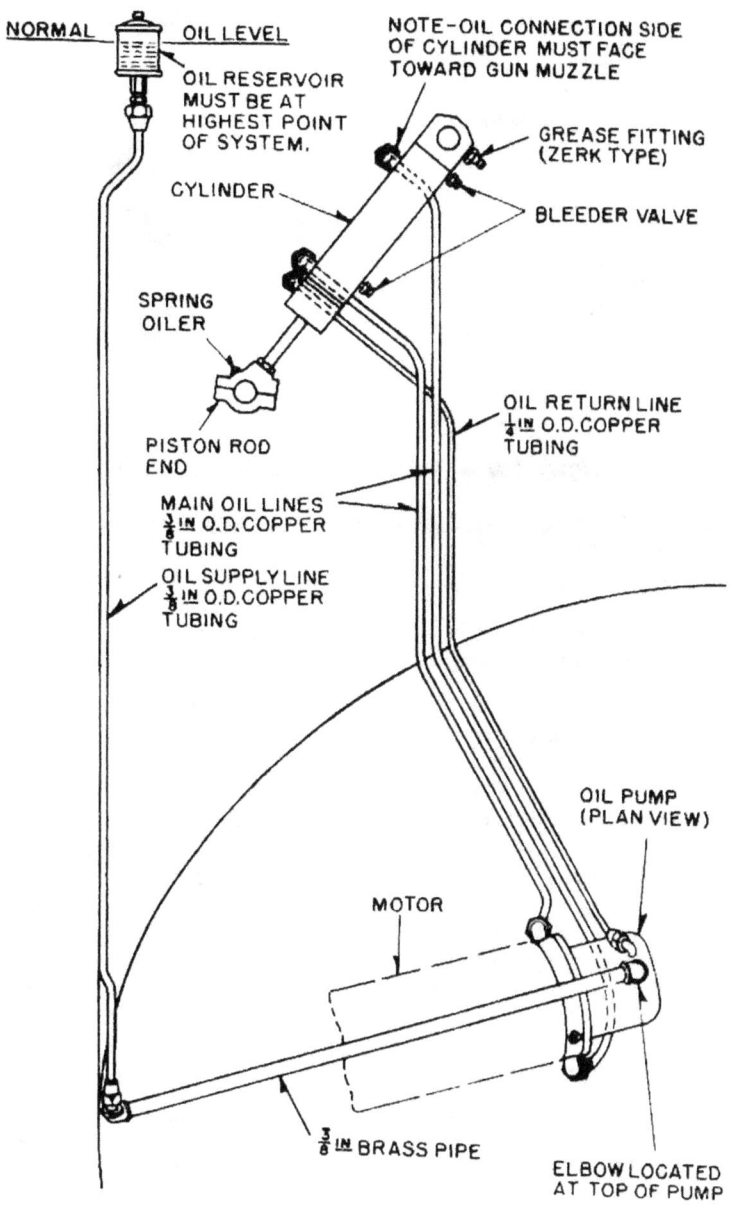

Figure 179 — Piping Diagram

TM 9-741
MEDIUM ARMORED CAR T17E1

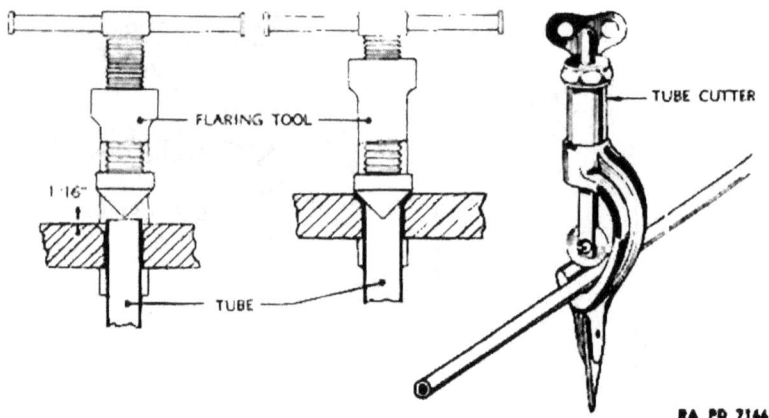

Figure 180 — Tube Cutter and Flaring Tool

234. REPAIRING AND REPLACING DAMAGED OIL LINES (fig. 179).

a. Turn turret switch to "OFF" position.

b. Disconnect the defective oil line and catch oil drainage from the affected part of the system. Replace damaged lines completely (fig. 179).

c. Use extreme care to keep dirt or other foreign material out of the hydraulic system. For this reason, when making a flare, cut the tubing with a tube cutter instead of a hack saw (fig. 180). Feed the cutting wheel a very small amount each time the tool is rotated 360 degrees until the tube is cut. If the cutter is fed too rapidly, the end of the tubing may be beveled inward. Slip the flare nut over the end of the tube and grip the tube in the flaring tool block, allowing the end to project approximately $\frac{1}{16}$-inch. Do not turn it down too hard as the flared end will be thinned out and become brittle. The flared end of the tube should be sufficiently large to form a good seat on the male part of the flare connection; yet it must not be so large that it will not clear the threads of the flare nut. When making the connections, draw the flare nuts up tightly, but if they are forced too tightly, the flare may be thinned out and consequently weakened.

d. Reconnect the flare nut to its proper connection.

e. Recharge the system with oil and check for air in the system.

f. Check the operation of the complete stabilizer.

TM 9-741

GYROSTABILIZER

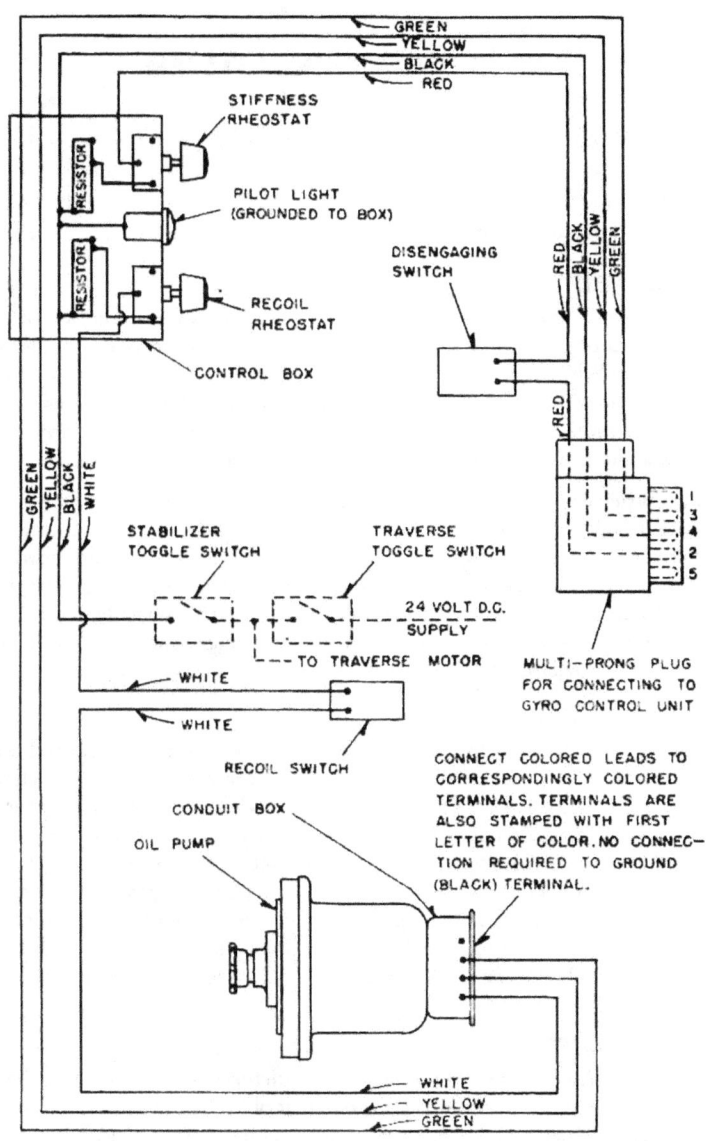

Figure 181 — Stabilizer Wiring Diagram

TM 9-741

MEDIUM ARMORED CAR T17E1

Section XXXIX

FIRE EXTINGUISHING SYSTEM

	Paragraph
Description	235
Operation	236
Inspection	237
Reset and recharge system	238

235. DESCRIPTION.

a. General (fig. 182). The fire extinguishing system consists of two steel cylinders containing 10 pounds each of Lux or carbon dioxide gas, a double-check tee, two remote release handles for each tank, three discharge nozzles in each engine compartment, and the necessary connecting tubing. In addition to this system, one 4-pound portable extinguisher is provided for small fires in or about the vehicle.

b. Principle of the System. Lux or carbon dioxide gas is not poisonous, but it is suffocating. It is noncorrosive, noninjurious to all substances, and is a nonconductor of electricity. When released from the cylinder, it resembles a cloud of steam and has a smothering effect on fires. When inhaled, it has a tingling effect on the nostrils; and if too much of the gas is inhaled, it will suffocate the person inhaling it. The gas is easily diffused and removed by ventilation.

236. OPERATION.

a. In the event of a fire, the extinguishing system can be turned on at three different places. One set of pull release handles for each cylinder is located at the left rear of turret on the outside of the hull, another set is above the driver's head, and the third set is on the control heads on the cylinders. To release the gas at either set of remote control handles, pull handle out as far as it will go. To release the gas at the control handle, remove the pull-out pin in the control head, and rotate the manual lever. Releasing one handle empties one of the cylinders, and if necessary, the other cylinder can be released by pulling the other handle.

237. INSPECTION.

a. Daily Inspection. The fire extinguishing system is used in an emergency in case of fire; therefore, it should be inspected at regular intervals to insure its operation when needed. Before starting off in vehicle, the operator should check to see that the red cap on the safety outlet of control valve is intact. If the red cap is not intact, the cylinder has been prematurely discharged due to high temperature and it must

TM 9-741
237

FIRE EXTINGUISHING SYSTEM

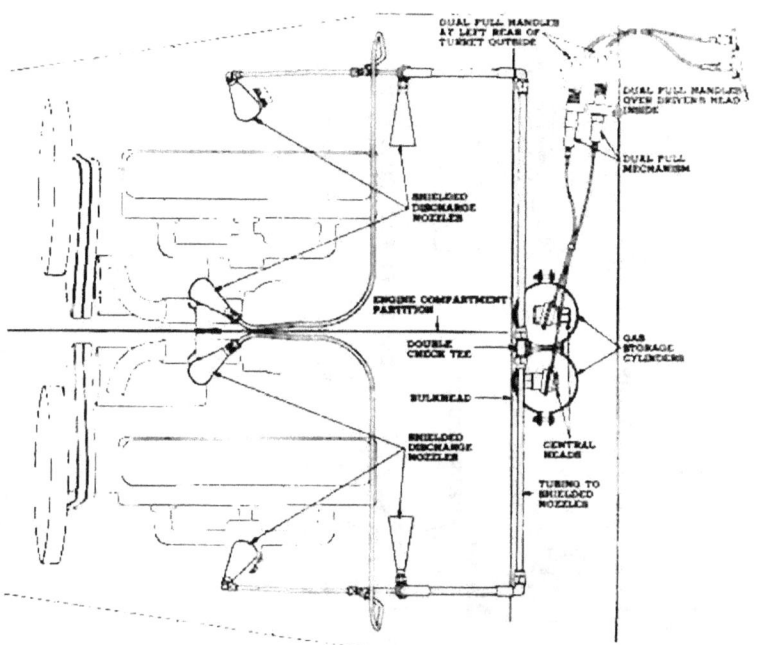

Figure 182 — Fire Extinguishing System

RA PD 32472

be recharged immediately. Also check the control head to see that pin is lined up with the arrow (fig. 183). If the pin is not lined up, the release handle(s) have been pulled, and the system will have to be reset and recharged (par. 238).

b. Periodic Inspection. Every 6 months, the system should be inspected in a routine manner as follows:

(1) Inspect entire system for any mechanical defects. Make certain that shielded nozzles are free of all foreign matter.

(2) Remove both cylinders as instructed in paragraph 238 and weigh each cylinder. Subtract from this weight the weight of the empty cylinder, which is stamped on the valve body. The weight of the empty cylinder includes the cylinder and cylinder valve but does not include the control head. If the resulting net weight of each cylinder is less than 9 pounds, the cylinder or cylinders must be recharged to 10 pounds or replaced with a cylinder that is charged with 10 pounds of gas.

(3) While both control heads are disassembled from cylinders, remove the cover exposing the cable set screws and check to see that

MEDIUM ARMORED CAR T17E1

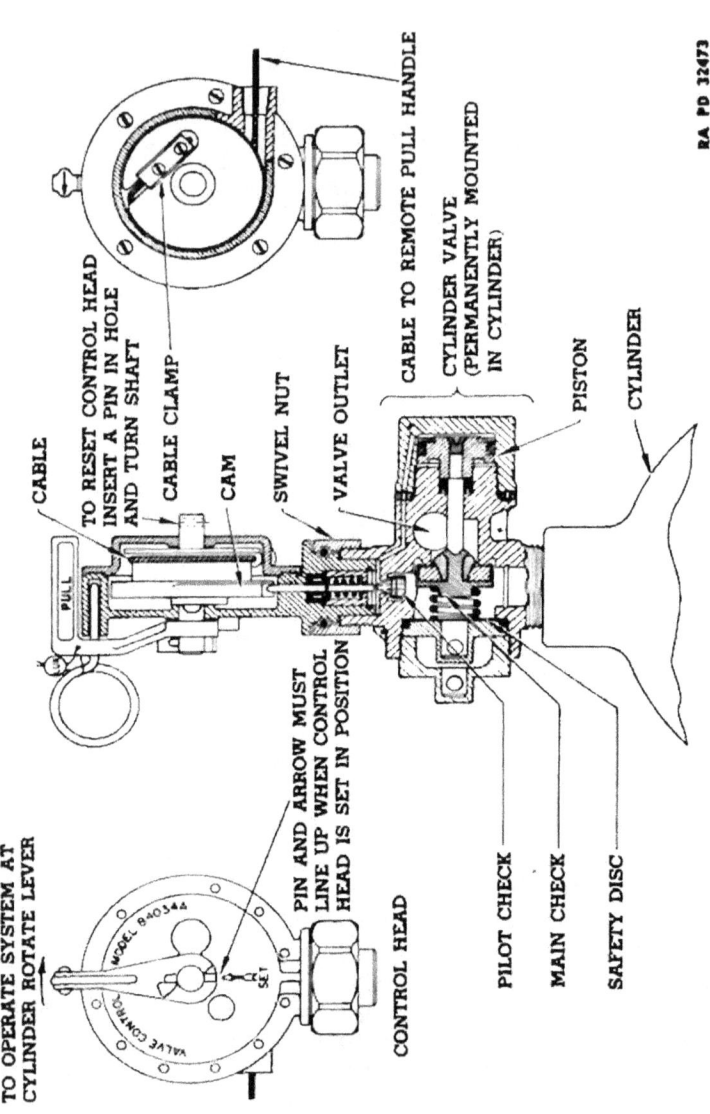

Figure 183 — Control Head and Cylinder Valve

FIRE EXTINGUISHING SYSTEM

these screws are tight. Install the cover and pull each remote control release handle to make sure that the control cam operates and that the cable does not bind. Reset the handles and control head and install the cylinders (par. 238).

238. RESET AND RECHARGE SYSTEM.

a. After using the extinguishing system, it is necessary to reset the system before it can be used again.

b. **Resetting Procedure.**

(1) REMOVE CYLINDERS. Remove the control head from the cylinder by turning the swivel nut (right-hand thread) off of cylinder valve and raise control head to clear the cylinder valve. Remove the connecting tube between the double-check tee and valve outlet. CAUTION: Never remove the cylinder with this connecting tube attached to cylinder valve outlet. Remove the cylinder clamps and remove the cylinder or cylinders.

(2) RESET CONTROLS. Push all remote control handles into original position. Remove cover from back of control head and using a narrow screwdriver, see that cable clamp screws are tight (fig. 183). Install the cover and insert a pin or nail in extended portion of shaft and turn the cam back until the pin in the front of shaft lines up with arrow below the pin (fig. 183). If the handle on the control valve was released, reset to upright position (fig. 183), and replace the pull-out pin and seal wire.

(3) INSTALL CYLINDER(S). Place a fully charged (10 pounds of gas) cylinder or cylinders in position in the rack and tighten the clamp nuts finger tight. Install the connecting tube between the double-check tee and valve outlet. Tighten clamp nuts securely. CAUTION: Before proceeding further, check control valve to make sure the valve has been reset. Then install the control head on the cylinder valve and tighten the swivel nut securely. Check shielded nozzles to see that they are not obstructed.

c. **Recharging Procedure.** Always remove the control head when handling or shipping cylinders. If carbon dioxide charging equipment is available, cylinders can be recharged as follows:

(1) CHARGE CYLINDER. Install charging adapter on cylinder valve and connect charging line to adapter. Open charging line valve and when cylinder is charged to 10 pounds of carbon dioxide gas, close line valve and disconnect charging line from adapter. Remove adapter. CAUTION: Do not charge the cylinder with more than 10 pounds of gas.

MEDIUM ARMORED CAR T17E1

(2) TEST CYLINDER. Fill cylinder valve chamber with water and watch for air bubbles. The bubbles may appear momentarily because of the trapped air. Blow off bubbles and watch to see that no more appear. If cylinder test shows no leaks, remove the water and fill in the weight on the record table on tag that is attached to cylinder.

TM 9-741
239

Section XL

STORAGE AND SHIPMENT

Paragraph

General instructions 239
Methods of securing Medium Armored Car T17E1 on freight
    cars 240

239. GENERAL INSTRUCTIONS.

a. General. The Medium Armored Car T17E1 will usually be shipped uncrated for domestic and overseas shipment. However, preparation of the armored cars will be different for domestic and overseas shipment (par. 239 b). Preparation for temporary storage (less than 60 days) will be the same as the preparation for domestic shipment. Preparation for indefinite storage (over 60 days) will be the same as the preparation for overseas shipment.

b. Preparation for Domestic Shipment and Temporary Storage.

(1) LUBRICATION. The vehicle should be completely lubricated before storage of same.

(2) FUEL IN TANKS. It will not be necessary to remove the fuel from the tanks nor to label these tanks under Interstate Commerce Commission Regulations.

(3) WATER IN RADIATOR. Radiator should be drained only when there is a possibility of freezing during storage or shipment. However, when the water is drained from the radiator, a conspicuous tag should be tied to the steering wheel of each vehicle, indicating that the radiator is empty.

(4) BATTERY. The battery should be disconnected by removing the positive battery cable and taping this cable and tying it away from the battery.

(5) UNPAINTED SURFACES. All unpainted and exposed surfaces must be treated with rust-preventive before the vehicle is stored or shipped. After cleaning the surface with solvents or a soap solution, all exterior surfaces will be treated with COMPOUND, rust-preventive, thin film. This preventive may be applied cold by spraying or brushing and will harden to a tough thin film. Surfaces from which it would be difficult to remove rust-preventives, such as the bore of a gun, will be treated with COMPOUND, rust-preventive, light. This compound is applied by brushing or slushing.

(6) TIRES. For domestic shipment the tires should be inflated to about 10 pounds per square inch above normal.

(7) INSPECTION. A systematic inspection should be made just before shipment or storage, and a list of all missing items or broken items that are not repaired should be attached to the steering wheel.

TM 9-741

MEDIUM ARMORED CAR T17E1

(8) SEALING. All ports will be closed and locked from the inside except the main port. This port will be sealed from the outside by means of an official seal. NOTE: The vehicles will be sealed after they have been loaded on the freight car and the brakes have been applied.

c. **Preparation for Overseas Shipment and Indefinite Storage.** All precautions given in paragraph 239 h above are included in the preparation for overseas shipment and indefinite storage. However, many additional precautions must be taken especially in overseas shipment (AR 850-18).

240. METHODS OF SECURING MEDIUM ARMORED CAR T17E1 ON FREIGHT CARS.

a. There are two approved methods of blocking armored cars on freight cars as described below.

(1) METHOD 1 (fig. 184).

(a) *Blocks B.* Eight blocks B should be located, one to the front and one to the rear of each wheel. The heel of the block should be nailed to the car floor with five 40-penny nails, and that portion of the block under the wheel should be toe-nailed to the car floor with two 40-penny nails.

(b) *Cleats C.* Two cleats C should be located against the outside face of each wheel. The lower cleat shall be nailed to the car floor with three 40-penny nails, and the top cleat to the cleats below with three 40-penny nails.

(c) *Strapping H.* Four strands, two wrappings, of No. 8 gage black annealed wire or 1½-inch flat steel strapping, indicated as H, figure 184, shall be passed through each of the four towing rings and fastened to the car floor by means of anchor plates if flat steel strapping is used or by running the wire around a block which is firmly nailed to the car floor if wire strapping is used. This strapping is not required when gondola cars are used.

(2) METHOD 2 (fig. 184). Method 2 is especially applicable to use in the field since no blocks are specially cut, and blocks F may be made from railroad ties or other available timbers.

(a) *Blocks F.* Four blocks F are placed, one to the front and one to the rear of each pair of wheels. These blocks should be at least 8 inches wider than the over-all width of the vehicle at the car floor.

(b) *Cleats E.* Sixteen cleats E shall be located, two against blocks F to the front and two cleats to the rear of each wheel (fig. 184).

(c) *Cleats G.* Four cleats G will be located, one against the outside of each wheel on the top of blocks F. These cleats will be nailed to each block F with two 40-penny nails.

TM 9-741
240

STORAGE AND SHIPMENT

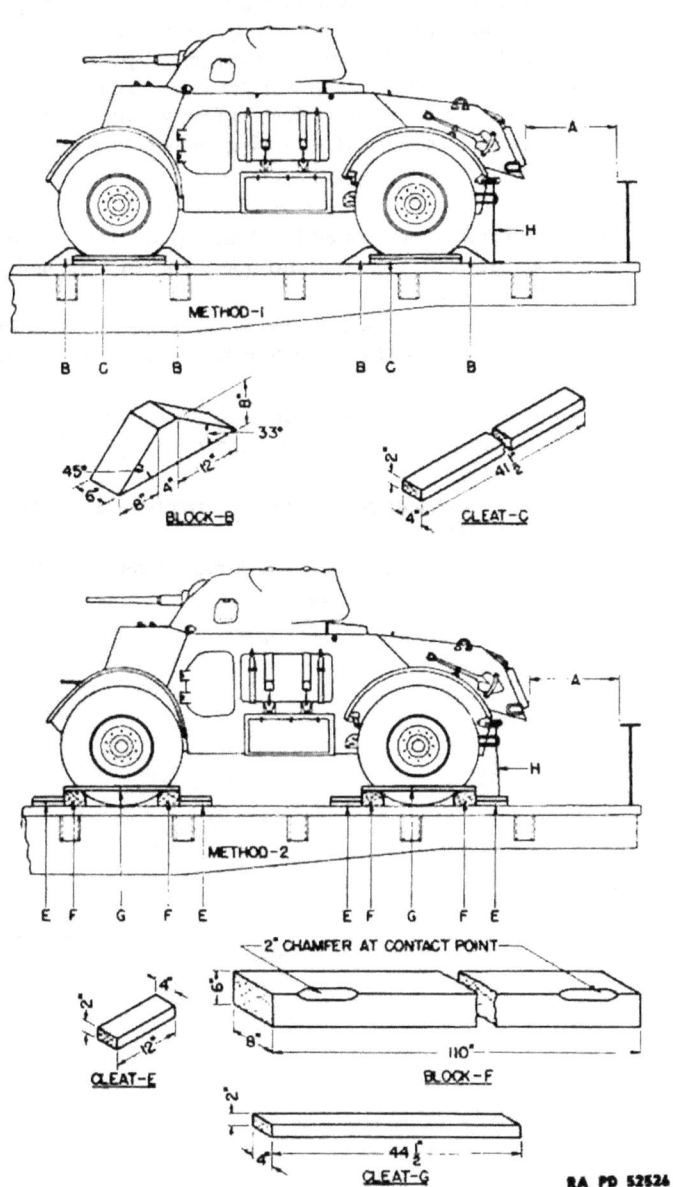

Figure 184 — Methods of Securing Armored Car T17E1 on Railroad Cars

MEDIUM ARMORED CAR T17E1

(*d*) *Strapping H.* Four strands, two wrappings, of No. 8 gage black annealed wire or 1½-inch flat steel strapping, indicated as H, figure 184, shall be passed through each of the four towing rings and fastened to the car floor by means of anchor plates if flat steel strapping is used or by running the wire around a block which is firmly nailed to the car floor if wire strapping is used. This strapping is not required when gondola cars are used.

(3) LOADING.

(*a*) *Inspection.* All railroad cars used for shipping armored cars must be inspected to make sure they have sound floors and that all nails or other projections have been removed.

(*b*) *Brake Wheel Clearance.* Each freight car must be loaded with a resulting brake wheel clearance of at least 6 inches in front, on top, and at each side of the brake wheel.

(*c*) *Distribution of Load.* The load must be so distributed that there will be as near as possible equal weight bearing on each truck of the railroad car.

(*d*) *Placarding.* Each railroad car containing armored cars must be placarded "DO NOT HUMP."

(*e*) *Brakes.* After loading and placing the armored cars, the brakes should be set.

Section XLI

REFERENCES

	Paragraph
Standard nomenclature lists	241
Explanatory publications	242

241. STANDARD NOMENCLATURE LISTS.

a. Ammunition.

Ammunition, fixed and semifixed, all types, for pack, light and medium field artillery	SNL R-1
Ammunition, revolver and automatic pistol and sub-machine guns	SNL T-2
Ammunition, rifle, carbine, and automatic gun	SNL T-1

b. Armament.

Gun, machine, cal. .30, Browning, M1919A2, M1919A4, fixed and flexible, and M1919A5, fixed, and ground mounts	SNL A-6
Gun, submachine, cal. .45, Thompson, M1928A1 and M1	SNL A-32
Gun, 37-mm, M5 and M6, and recoil mechanisms (tank)	SNL A-45
Mount, combat vehicles (combination gun mount M23)	SNL A-55

c. Car, armored, T17E1 (Chevrolet) .... SNL G-122

d. Cleaning, preserving and lubricating materials, recoil fluids, special oils, and similar items of issue . SNL K-1

Current Standard Nomenclature Lists are as tabulated here. An up-to-date list of SNL's is maintained in the "Ordnance Publications for Supply Index" OPSI

242. EXPLANATORY PUBLICATIONS.

a. Armament.

Browning machine gun, cal. .30, all types, U. S. machine gun cal. .22 and trainer, cal. .22	TM 9-1205
Browning machine gun, cal. .30, HB, M1919A4 (mounted in combat vehicles)	FM 23-50
Thompson submachine gun, cal. .45, M1928A1	FM 23-40
Thompson submachine gun, cal. .45, M1928A1	TM 9-1215
37-mm gun materiel (tank) M5 and M6	TM 9-1250
37-mm gun, tank, M6 (mounted in tanks)	FM 23-81

TM 9-741
242

MEDIUM ARMORED CAR T17E1

b. Cleaning, preserving, lubricating, and welding materials and similar items issued by the Ordnance Department .... TM 9-850

c. **Communications.**
Radio fundamentals .... TM 11-455
Radio set SCR-506.......................... TM 11-630
Radio set SCR-510.......................... TM 11-605
Radio sets SCR-508, SCR-528, and SCR-538.... TM 11-600
The radio operator .... TM 11-454

d. **Maintenance and Inspection.**
Echelon system of maintenance................ TM 10-525
Fire prevention, safety precautions, accidents..... TM 10-360
Motor transport inspections .... TM 10-545
Sheet metal work, body, fender, and radiator repairs TM 10-450

e. **Miscellaneous.**
Automotive brakes .... TM 10-565
Automotive electricity .... TM 10-580
Automotive lubrication .... TM 10-540
Camouflage .... FM 5-20
Defense against chemical attack................ FM 21-40
Electrical fundamentals .... TM 1-455
Fuels and carburetion........................ TM 10-550
Motor transport .... FM 25-10

f. **Stabilizers, All Types** (published as TM 9-1798A) TM 9-1739A

g. **Storage and Shipment.**
Storage of motor vehicle equipment............. AR 30-1055
Storage of motor vehicle equipment............. AR 850-18

354

INDEX

A

	Page No.
Acceleration troubles	154–157
Accelerator pedal	16
Accessories	
cold weather	67
electrical (See Electrical accessories)	
general discussion of	76
Adjustments (See also Maintenance)	
brake	208–209
parking	210
breaker points	122
carburetor	129–130
fan belts	139
hydraulic control rod	166–172
idling of engine	107
manual shift linkage (transmission)	174–177
recoil rheostat	320
servo bands during change of transmission oil	163–165
spark plug	120
stiffness rheostat	320
tires, tubes, wheels	240
toe-in of front wheels	199
valve clearance	89–91
wheel bearings	243–245
worm (bracket) end play	328–329
Air	
entering hydraulic system	172–173
leakage	239
removal from gyrostabilizer	324
Air cleaner	
checking in engine tune-up	105–106
description, replacement, and servicing	130
lubricant levels, maintaining	48–49
lubrication	40–42
venting system	128–129
Ammeters	18, 19
Ammunition (See also Armament)	
carried	26
cleaning gas exposed	68
data	9
Antifreeze, use of in cold weather operation	65–66
Armament	
data	9
gun mounts	
front machine gun mounting	30
37-mm gun and cal. .30 machine gun	30–32

	Page No.
gyrostabilizer	33
list and location	26
operation	30
Assembly	
fuel filter	134
steering connecting rod	236
(See also Reassembly)	
Automatic control, use of	48
Automotive materiel, special precautions for	70
Auxiliary blackout drive and marker lamp, description	259
Auxiliary blackout drive lamp sealed beam assembly replacement	262
Axles (See Front axle and Rear axle)	

B

	Page No.
Batteries and starting system	248–251
battery inspection and maintenance	248–249
filling	249
general	248–249
servicing battery terminals	249
testing	249
description	248, 249
maintenance	73
replacement	
battery	250
solenoid switch	251
starting motor	250–251
starting motor maintenance	249–250
trouble shooting	250
Battery and cables, checking in engine tune-up	104
Battery and electrical parts, checking in cold weather	66
Bearings, wheel	243–245
Blackout driving lamp switch	20
Blackout marker lamp(s)	
bulb replacement	262
description	257–259
replacement	261–262
Blackout tail and service tail and stop lamp	259
Blackout tail and stop lamp	259
Bleeding hydraulic system	173, 211–213
Body supports, data	10
Bolt, removal of to remove 37-mm gun	32

TM 9-741

TM 9-741

MEDIUM ARMORED CAR T17E1

B — Cont'd

	Page No.
Brake system	204–224
adjustment	
brake	208–209
parking	210
brake lines	221–224
brake shoe replacement	213–215
description	204
hydraulic system	206
bleeding	173, 211–213
hydrovac system	218–221
replacement	218–220
vacuum check valve	220–221
main cylinder	217–218
maintenance	73
operation	204–206
pedal	16
trouble shooting	207–208
wheel cylinder	215–216
Breaker point	
adjustment	122–123
replacement	122

C

Camouflage, paint as	54–55
Capacities, oil, table of	43
Carbon removing	93–94
Carburetor	
adjustment	129–130
checking in engine tune-up	106
choke lever	12–14
description	129
replacement	130
Care and preservation	52–55
cleaning the vehicle	52–53
paint(-ing)	
as camouflage	54–55
general information on	53
lubricating devices	55
metal surfaces	54
preparing for	53–54
removing	55
records	52
Care of equipment and tools	75
on vehicle	61
Caution plate	17
Charging gyrostabilizer with oil	322–324
Chemicals, liquid, removal of	68–70
Choke and throttle lever clamp bolts	17
Circuit breakers	278–280
fuel pump	279
generator cut-out relay	278–279

	Page No.
lighting switch	278
siren	279
steering gear motor	279
turret light switch and circuit breaker	280
turret traverse control motor relay switch	280
Cleaning:	
commutator	249
fuel filter	132
gas exposed materiel	68
gyrostablizer	326
propeller shafts	184
spark plug	120
vehicle	52–53
Clearance between gyro control gear and mounting bracket	331
Clutch parts, replacing	336
Coat contacting surfaces, lubrication	51
Coil replacement	121
Cold weather	
accessories	67
operation of vehicle	64–67
precautions	24
Communication	
data	10
equipment on vehicle	56
Commutators, cleaning	249
Compass	17
Compression check of engine	103
Condenser replacement	124
Condensers for radio suppression	275–276
Connecting rod, steering	235–236
Control box, replacing (gyrostabilizer)	332
Control link, replacing (gyrostabilizer)	339
Control of gun with stabilizer	318–319
Control valve assembly replacement	312
Controls (See also Generators and controls and Operation and controls)	
ammeters	18, 19
buttons	
fuel pump circuit breakers	20
head lamp release	16
lighting switch circuit breaker	20
starting	18
caution plate	17
choke and throttle lever clamp bolts	17
compass	17

INDEX

C — Cont'd

Controls (Cont'd)
- electrical sockets 20
- engine temperature indicator.... 20
- fire extinguishing system control handles 17
- fuel and oil pressure gages...... 20
- general information on 11
- hand fire extinguisher 17
- identification plate 18
- instrument panel lights 18
- levers
  - carburetor choke 12–14
  - engine selector 12
  - hand brake 15
  - hand throttle 14
  - Jettison fuel tank release..... 16
  - transfer case shift 11
  - front axle 15
  - transmission manual control... 11
- maintenance 73
- pedals
  - accelerator 16
  - brakes 16
  - hand brake assist 17
- periscopes 17
- siren foot control 17
- speedometer trip mileage reset.. 20
- switches
  - blackout driving lamp 20
  - ignition 18
  - lighting 18
  - master electric switch box.... 21
  - steering gear electric motor... 17
- transmission low pressure signal lights 21

Cooling system
- checking in engine tune-up..... 107
- description 135
- draining and refilling....44, 138–139
- fan belts 139
- inspection 135
- maintenance 72
- oil cooler 145–149
  - cleaning 146–147
  - description 145–146
  - installation 148–149
  - removal 147–148
- pressure filler cap 137–138
- radiator 140–145
  - core installation 143–145
  - description 140
  - removal 140–143

thermostat 139–140
trouble shooting 137
water pump 140
Coupling, fluid 151
Crankcase
- dilution 46
- lubrication 42
- oil, cold weather precautions.. 64–65
Crankcase ventilating system.. 101–102
- checking in engine tune-up..... 107
Cylinder block assembly 77–82
Cylinder head, components 77
Cylinder head or gasket replacement 99–100
- carbon removing 93–94
- cylinder head description 91
- left engine
  - installation 99–100
  - removal 98–99
- right engine
  - installation 96–98
  - removal 91–93
- rocker arm and shaft assemblies 94–96
Cylinders, numbering 77
(See specific names)

D

Data
- ammunition 9
- armament 9
- communication 10
- engine 9, 82
- fuel and oil 10
- performance 10
- protected vision 10
- seats, body supports, safety belts 10
- vehicle 9
Description
- batteries and starting system.... 248
- brake system 204
- cooling system 135
  - oil cooler 145–146
  - radiator 140
  - thermostat 139
  - water pump 140
- electrical wiring system 278–280
  - circuit breakers 278–280
  - general 278
  - switches 278
  - wiring 278

TM 9-741

MEDIUM ARMORED CAR T17E1

D — Cont'd

	Page No.
Description (Cont'd)	
engine heat indicator	268-269
engine ignition system	118
engine oil pressure gage	270
engines	71
cylinder head	91
oil filters	100
exhaust system	150
fire extinguishing system	344
front axle	194
fuel gage	266
fuel system	125
air cleaner	130
carburetor	129
fuel filter	132
fuel pump	131
venting system	128
generators and controls	252
hull	314
hydraulic throttle control	165
propeller shafts	179
radio suppression (system)	274
rear axle	200
shock absorbers	228-229
siren	272
springs	225-226
steering gear	230
motor	234-235
tire pump	245-246
tires	239
transfer case	188
transmission	152
transmission gear reduction assembly	178
turret and traversing system	307-310
turret and basket	307
turret traversing system	307-310
turret slip ring assembly	281
universal joints	179-181
vehicle	3-9
wheel bearings	239, 243
wheels	238-239
Design and construction of engines	77-82
Differentials, lubrication	43
Dilution, crankcase	46
Direction of rotation	77
Disassembly	
fuel filter	132
propeller shaft assembly service	
(gear case to transfer case)	183-184
(transfer case to axle)	186

	Page No.
Discharged battery	250
Distributor	
checking in engine tune-up	104-105
replacement	120-121
Dome lamp(s)	
bulb replacement	263
description	259
Domestic shipment and temporary storage, preparation for	349-350
Doors	
hull	315
vision	314
Drain plug locations	49
Drain plugs	
hull crew compartment	315-317
engine compartment	317
Draining:	
cooling system	44, 138
fuel filter	132
gear cases	44
Drive flange gasket, replacement (front axle)	196

E

Electric motor (See motor *under* Steering gear)	
Electrical accessories	272-273
maintenance	74
siren	
description	272
replacement	272
siren switch replacement	272
windshield wipers	
description	272-273
servicing wiper motors	273
wiper motor replacement	273
Electrical sockets	20
Electrical system wiring	278-306
description	278-280
engine terminal box and correction table	299-300
foot firing switch box	299
fuel pump and fuel gage terminal box and connection	304-306
instrument panel assemblies (See *under* Instrument panel)	
main light switch	283
connection table	284
master switch box	282-283
radio and phone terminal box and connection table	295-296

TM 9-741

INDEX

E — Cont'd

Electrical system wiring (Cont'd)
steering motor switch box connection table 292
tail lamp and fuel tank terminal box and connection table.. 300–301
tail lamp terminal box table
    left-hand 301–303
    right-hand 303–304
trouble shooting 282
turret
    control box and connection table 298–299
    motor relay switch box and connection table 298
    slip ring armature connection table 294–295
    slip ring assembly 281
    slip ring brush connection table 292–294
    slip ring terminal box table... 292
Engine compartment covers, use of 82
Engine heat indicator
    description 268–269
    maintenance 269–270
Engine ignition system 118–124
    description 118
    maintenance 72
    replacement (and adjustment):
        breaker point 122–123
        coil 121
        condenser 124
        distributor 120–121
        filter 121–122
        spark plug 120
        suppressor 122
    trouble shooting 118–120
Engine oil pressure gage
    description 270
    service operation 270–271
Engine selector
    lever 12
    shifting 23
Engine temperature indicator..... 20
Engine terminal box and connection table 295–300
Engine(s)
    crankcase ventilating system 101–102
    data 9, 82
    description, general 77
    design and construction...... 77–82
    ignition system (See Engine ignition system)

left engine
    cylinder head
        installation 99–100
        removal 98–99
    manifold gasket replacement 88–89
    valve clearance adjustment... 91
lubricant levels, maintaining.... 47
maintenance 72
oil filter service 100–101
replacement 107–117
    general discussion of 107
    installation 114–117
    removal 110–113
right engine
    cylinder head
        installation 96–98
    manifold gasket replacement 87–88
    removal 91–93
transmission, removal of from... 177
trouble shooting 82–86
tune-up 102–107
    air cleaner 105–106
    battery and cables........... 104
    carburetor 106
    compression check 103
    cooling system 107
    crankcase ventilator 107
    distributor 104–105
    general remarks on...... 102–103
    idling adjustment 107
    ignition timing 106–107
    road test 107
    spark plugs 103–104
    valve clearance adjustment.. 89–91
Equipment (See also Tools and equipment)
Equipment and tools on vehicle.. 56–61
    care of 61
    equipment, list of........... 56–61
        communication 56
        military 56–59
        miscellaneous 60–61
        service parts 60
        vehicular 59–60
    tools 61, 62
Exhaust pipe replacement........ 150
Exhaust system 150
    maintenance 72

TM 9-741

MEDIUM ARMORED CAR T17E1

F

	Page No.
Fan belts	139
V-type, use of	82
Filling batteries	249
Filter	
fuel	132–134
ignition system	121–122
radio suppression	274–275
(See also Oil filter(s))	
Fire extinguishing system	
control handles	17
description	344
general	346
principal of the system	346
inspection	344–347
daily	344–345
periodic	345–347
maintenance	74
operation	344
reset and recharge system	347–348
recharging procedure	347–348
resetting procedure	347
Fittings, lubrication	40
Flexible shaft, replacing (gyrostabilizer)	336
gear box	336–339
Fluid coupling	151
Fluids, hydraulic, brake	213
Foot firing switch box	299
Foot throttle (See hydraulic throttle control under Transmission)	
Front axle	194–199
adjusting toe-in of front wheels	199
description	194
lubricant levels, maintaining	48
maintenance	73
replacement:	
drive flange gasket	196
shaft	196–198
tie rod and tie rod yoke	198–199
trouble shooting	194–196
Front machine gun mounting	30
Front wheels, adjusting toe-in of	199
(See also Front axle)	
Fuel	
cold weather precautions	64
data	10
Fuel and oil pressure gages	20
Fuel filter	132–134
Fuel gage	
description	266
testing dash and tank units	266–268
unit replacement	268

	Page No.
Fuel pump	131–132
Fuel pump and fuel gage terminal box and connection	304–306
Fuel pump circuit breakers	20
Fuel pump condenser, radio suppression	275–276
Fuel system	
air cleaner	130
carburetor	129–130
description	125
fuel filter	132–134
fuel pump	131–132
fuel tanks	127–128
maintenance	72
trouble shooting	125–127
venting system	128–129

G

	Page No.
Gages, maintenance	74
(See also Instruments and gages)	
Gas, materiel affected by	68–70
automotive materiel, special precautions for	70
cleaning	68
decontamination	68–70
protective measures	68
Gear	
cases, draining	44
lubricants	66
ratio	152
steering (See Steering gear)	
troubles	158–161
(See also Transmission)	
Gear box and mounting bracket, replacing	334
Gear reduction assembly, transmission	178
Generator filters, radio suppression	274–275
Generators and controls	252–256
description	252
inspection and testing	252–253
maintenance	73
replacement	
generator	254–256
generator regulator unit	256
Guide, lubrication (See under Lubrication instructions)	
Gun and mount, service parts for	60
Gun, control of with stabilizer	318–319

INDEX

G — Cont'd

	Page No.
Gun, machine, cal. .30	30
Gun, 37-mm	30–32
Gyro control gear box adjustment (*See under* Gyrostabilizer)	
Gyro control, replacing	331
Gyrostabilizer	
adjusting:	
recoil rheostat	320
stiffness rheostat	320
charging with oil	322–324
checking stabilizer operation	321
cleanliness	326
control of gun	318–319
gyro control gear box adjustment	
clearance between gyro control gear and mounting bracket	331
controlling the mesh of the worm and worm wheel	329–331
worm (bracket) end play adjustment	328–329
lubrication	324
maintenance	74, 324
oil level	324
removing air from system	324
repairing and replacing damaged oil lines	342
replacement:	
clutch parts	336
control box	332
control link	339
flexible shaft	336
flexible shaft gear box	336–339
gear box and mounting bracket	334
gyro control	331
oil pump and/or motor	332
piston and cylinder	332–333
recoil switch	333–334
wiring and shielded conduit	334–336
starting the equipment	319–320
trouble shooting	321–322
"hunting"	322
trunnion friction and gear balance	326–328
use of	33

H

Hand brake assist pedal	17
Hand brake lever	15
Hand fire extinguisher	17
Hand throttle lever	14
Handwheel, elevating and depressing	32
Head lamp release button	16

TM 9-741

	Page No.
Head lamp(s)	
aiming	260–261
description	257
replacement	260
sealed beam assembly replacement	260
High temperature operation of vehicle	67
Hose, brake, replacement of	224
Hull	314–317
description	314
doors	315
drain plugs	315–317
periscope mounting	314–315
vision doors	314
"Hunting" in gun	322
Hydra-Matic transmission, maintaining lubricant levels	48
Hydraulic brake fluids	213
Hydraulic motor replacement	312–313
Hydraulic operation of turret and traversing system	310
Hydraulic pump replacement	313
Hydraulic system	
air entering, cause	172–173
bleeding	173, 211–213
general discussion of	206
Hydraulic throttle control (*See under* Transmission)	
Hydrovac brake system	218–221
replacement	218–220
vacuum check valve	220–221
Hydrovac cylinder, lubrication	44

I

Identification plate	18
Idling adjustment, checking in engine tune-up	107
Ignition coil filters, radio suppression	274
Ignition switch	18
Ignition system (*See* Engine ignition system)	
Ignition timing, checking in engine tune-up	106–107
Inspection	
at the halt	36
cooling system	135
fire extinguishing system	
daily	344–345
periodic	345–347
generator and controls	252

TM 9-741

MEDIUM ARMORED CAR T17E1

I — Cont'd

Page No.

Inspection (Cont'd)
operation
    inspection after 36–37
    inspection during 35–36
periodic
    1,000 miles 38
    6,000 miles 39
prestarting 34–35
propeller shafts 184
purpose of 34
Inspection lamp, description... 259–260
Instrument panel
  ammeter wiring assembly 286
  dome lamp wiring connector.... 286
  fuel pump, fuel gage and lamp
    wiring harness assembly...... 289
  heat indicator and oil gage wiring
    assembly 288
  ignition switch wiring assembly.. 287
    installation 265
  lights
    bulb replacement 263
    description 259
    operation 18
  neutral safety switch wiring connector 286
  removal of units 264–265
  stop light, and windshield wiper
    wiring 289
  supply assembly and connections 284–285
  units removal 264–265
Instruments and gages 264–271
  fuel gage 266–268
  engine heat indicator 268–270
  instrument panel 264–265
  maintenance 74
  (See also Controls)
Intervals of lubrication 40

J

Jettison fuel tank release lever.... 16

L

Lamps (See Lighting system)
Leaks (See Lubricant leaks)
Left side of engine.............. 77
Lighting switch 18
Lighting switch circuit breaker button 20
Lighting system
  description 257–260

Page No.

  inspection 259–260
  instrument panel lights 259
  maintenance 73–74
  replacement 260–263
  trouble shooting 260
Link, shock absorber, replacement
  of 229
Liquid chemicals, removal of.... 68–70
Lubricant leaks, cause and remedy
  front axle 194
  transfer case 188–189
Lubricants
  cold weather 66
  levels, maintaining 47–50
Lubricating devices, painting...... 55
Lubrication (See also Lubrication
  instructions)
  gyrostabilizer 324
  starting motor 249
  transmission gear reduction assembly 178
Lubrication instructions 40–51
  guide 40
    chassis 41
    turret 43
  lubrication information 45–51
    automatic control 46–47
    changing the oil 46
    coat contacting surfaces...... 51
    crankcase dilution 46
    drain plug major units....... 49
    engine oils 46
    general discussion of 45
    maintaining lubricant levels... 47
    oil can lubrication points... 49–51
    SAE viscosity numbers........ 45
  ordnance personnel duties...... 45
  references 40
  using arm, instructions for.... 40–45
    air cleaners 40–42
    crankcase 42
    draining cooling system and
      gear cases 44
    fittings 40
    hydrovac cylinder 44
    intervals 40
    oil can points 44
    oil filters 42
    transmission, differentials, and
      transfer case 42–44
    universal and slip joints...... 44

INDEX

M

	Page No.
Main cylinder	217–218
Main light switch	283
connection table	284
Maintenance	
definition of terms	71
engine heat indicator	269–270
oil filters	101
preventive (*See* Inspection)	
radio suppression	277
repair, allocation of	
batteries and starting system	73
brake system	73
cooling system	72
electrical accessories	74
engine ignition system	72
engines	72
exhaust system	72
fire extinguishing system	74
front and rear axle	73
fuel system	72
general	71
generators and controls	73
gyrostabilizer	74
instruments and gages	74
lighting system	73–74
propeller shafts and universal joints	72
radio suppression	74
springs, radius rods, and shock absorbers	73
steering gear	73
transfer case	72
transmission	72
turret and traversing system	74
wheels, tires, wheel bearings, tire pump	74
shock absorbers	228
stabilizer	324
tire pump	246
tires, tubes, wheels	240
Manifold gasket replacement	
left engine	88–89
right engine	87–88
Manual operation of turret traversing system	310
Manual shift control (*See under* Transmission)	
Master cylinder replacement	173–174
Master electric motor switch	21
Master switch box	282–283
Materiel affected by gas (*See* Gas, materiel affected by)	

	Page No.
Metal surfaces, painting	54
Military equipment on vehicle	56–59
Motor, steering gear (*See under* Steering gear)	
Mounts, gun (*See* gun mounts *under* Armament)	
Muffler replacement	150

O

	Page No.
Oil	
adjusting servo bands in changing transmission	163–165
checking and replenishing in engines	82
data	10
replacement of in air cleaner	130
table of capacities and temperatures	43
(*See also* Lubrication instructions)	
Oil can points, lubrication	94, 49–51
Oil cooler (*See under* Cooling system)	
Oil filter(s)	
description	100
lubricant leaks, maintaining	49
lubrication	42
service	100–101
Oil lever (gyrostabilizer)	324
Oil lines, damaged, repairing and replacing	342
Oil pan, use of	82
Oil pump and/or motor, replacing	332
Operating precautions, special, in cold weather	66–67
Operation (*See also* Operation and controls)	
armament	30
brake system	204–206
engine oil pressure gage	270–271
fire extinguishing system	344
hydraulic brake system	206
hydraulic throttle control	166
inspection after	36–37
inspection during	35–36
manual shift control (transmission)	174
shock absorber, checking	228
steering gear	230
transmission gear reduction assembly	178

TM 9-741

MEDIUM ARMORED CAR T17E1

O — Cont'd Page No.

Operation (Cont'd)
  turret and traversing system..... 310
    hydraulic 310
    manual 310
  under unusual conditions 64–67
    cold weather 64–67
    high temperature 67
  venting system 128
Operation and controls
  cold weather precautions 24
  inspection, prestarting 34–35
  instruments and controls (See Controls)
  operating the vehicle 22–24
  starting the engines 21–22
Organization spare parts 76
Organization tools and equipment.. 75
Overcooling in cooling system, cause and remedy 137
Overheating
  cooling system 137
  prevention of 67
Overseas shipment and indefinite storage, preparation for 350

P

Painting (See paint(-ing) under Care and preservation)
Parking brake
  adjustment 210
  location on transfer case........ 188
Parking vehicle on hills 152
Performance of vehicle, data on.... 10
Periodic inspection
  1,000 miles 38
  6,000 miles 39
Periscopes 17
  mounting 314–315
  vision device 32
Piston and cylinder, replacing. 332–333
Power tire pump
  installation 246
  removal 246–247
Pressure filler cap 137–138
Prestarting inspection 34–35
Preventive maintenance (See Inspection)
Propeller shaft (assembly)
  description 179
  gear case to transfer case
    installation 185

    removal 182–183
    service 183–185
  maintenance 72
  transfer case to axle
    installation 187
    removal 186
    service 186–187
  trouble shooting 181–182
Protected vision, data 10

R

Radiator (See under Cooling system)
Radio and phone terminal box and connection table 295–296
Radio interference, cause and remedy 118
Radio suppression (system)... 274–277
  description 274
  condensers 275–276
  filters 274–275
  general 274
  suppressor resistor 274
  wire shielding 277
  maintenance 74, 277
Radius rods 228
  maintenance 73
Rear axle
  description 200
  maintenance 73
  lubricant level 48
  shaft replacement 201–203
  trouble shooting 200–201
Reassembly
  propeller shaft assembly service
    (gear case to transfer case) 184–185
    (transfer case to axle)... 186–187
Recharging fire extinguishing system 347–348
Recoil switch, replacing 333–334
Records
  lubrication and servicing 45
  Ordnance Motor Book, use of... 52
  tools and equipment 75
Regimental tools and equipment... 75
Resetting fire extinguishing system 347
Resistor suppressors, function..... 274
Right side of engine 77
Road test of vehicle 38
  after engine tune-up 107
Rocker arm and shaft assemblies 94–96

364

INDEX

R — Cont'd

	Page No.
Rod, steering connecting	235–236
Rods, radius	228
Rotation, direction of	77

S

	Page No.
SAE viscosity numbers of oil	45
Safety belts, data	10
Seats, data	10
Service parts for gun and mount	60
Servicing (See also Lubrication instructions)	
air cleaner	130
battery terminals	249
engine oil pressure gage	270–271
steering gear	234
wiper motors	273
Servo bands, adjustment of in changing transmission oil	163–165
Shaft control, manual (See under Transmission)	
Shaft replacement	
front axle	196–198
rear axle	201–203
Shifting	
engine selector	23
transfer case	23, 188
transmissions	22–23
Shimmy of front wheel	195, 240
Shipment, preparation for	
domestic, and temporary storage	349–350
overseas, and indefinite storage	350
Shock absorbers	228–229
lubricant levels, maintaining	49
maintenance	73
Siren, description and replacement	272
Siren foot control	17
Siren switch replacement	272
Slave cylinder replacement	174
Slip joints, lubrication	44
Solenoid switch replacement	251
Spare parts and accessories	76
Spark plug	
adjustment, cleaning, and replacement	120
checking in engine tune-up	103–104
suppression replacement	122
Speedometer trip mileage reset	20
Springs	
description	225–226
maintenance	73
replacement	226–227

TM 9-741

	Page No.
Springs, brush, testing	250
Stabilizer (See Gyrostabilizer)	
Starting button	18
Starting	
engine(s)	21–22
towing vehicle to start	152
gyrostabilizer	319–320
vehicle	22
Starting motor	
maintenance	249–250
replacement	250–251
switch box connection table	282
Starting system	
description	248
(See also Batteries and starting system)	
Steering gear	230–237
description	230
lubricant levels, maintaining	49
maintenance	73
motor	234–235
condenser	275
description	234–235
removal and installation	235
switch	17
tests on	234
operation	230
service operations	234
steering connecting rod	235–236
trouble shooting	232–233
Steering, hard	194–195, 239
Storage and shipment	
general instructions	349
preparation for:	
domestic shipment and temporary storage	349–350
overseas shipment and indefinite storage	350
methods of securing vehicle on freight cars	350–352
Suppression radio (See Radio suppression (system))	
Suppressor, spark plug, replacement	122
Suppressors, resistor, function	274
Switches, wiring system, description	278

T

	Page No.
Tail and stop lamp (assembly)	
description	259
replacement	262
unit replacement	262

TM 9-741

MEDIUM ARMORED CAR T17E1

T — Cont'd

	Page No.
Tail lamp and fuel tank terminal box and connection table	300–301
Tail lamp terminal box and connection table	
left hand	301–302
right hand	303–304
Tank crew tools and equipment	75
Test(-ing)	
battery	249
brush springs	250
dash and tank units (fuel gage)	266–268
generator and controls	252–253
steering gear motor circuit	234
vehicle	38
after engine tune-up	107
Thermostat	139–140
Throttle control, hydraulic (See hydraulic throttle control under Transmission)	
Tie rod and yoke replacement (front axle)	198–199
Tire pump	245–247
description	245–246
maintenance	73, 246
power tire pump	
installation	246
removal	246–247
Tires (See Wheels, tires, wheel bearings, and tire pump)	
Toe-in of front wheels, adjustment	199
Tools and equipment	
organization	75
special tools	75
(See also Equipment and tools on vehicle)	
Towing vehicle to start engine	152
Transfer case	
description	188
installation	191–193
lubricant level	
checking	44
maintaining	48
lubrication	42–44
maintenance	72
removal	190–191
shifting	23
trouble shooting	188–189
Transfer case shift lever	11
front axle	15

	Page No.
Transmission	
description	152
gear reduction assembly, removal of from	178
hydra-matic transmission, removal	177
hydraulic throttle control	165–174
control rod adjustments	166–172
description	165
hydraulic system	172–173
master cylinder	173–174
operation	166
slave cylinder	174
lubrication	42–44
maintenance	72
manual shift control	
adjustment	174–177
operation	174
servo bands, adjustment	163–165
shifting	23
trouble shooting	152–163
(See also Transfer case)	
Transmission gear reduction assembly	178
Transmission level, checking	43–44
Transmission low pressure signal lights	21
Transmission manual control lever	11
Transmission oil, adjustment of servo bands in changing	163–165
Traversing gear assembly replacement	312
Traversing system (See Turret and traversing system)	
Trouble shooting	
batteries and starting system	250
brake system	207–208
cooling system	137
electrical system wiring	282
engine ignition system	118–120
engines	82–86
exhaust system	150
front axle	194–196
fuel system	125–127
generator and controls	253–254
gyrostabilizer	321–322
propeller shaft and universal joints	181–182
rear axle	200–201
steering gear	232–233
transfer case	188–189
transmission	152–163

INDEX

T—Cont'd Page No.

Trouble shooting (Cont'd)
  turret traversing system.... 310–311
  wheels, tires, wheel bearings, and tire pump 239–240

Trunnion friction and gun balance (gyrostabilizer) 326–328

Tune-up of engine (See tune-up under Engine(s))

Turret (See also Turret and traversing system)
  control switch box and connection table 298–299
  electric motor replacement...... 311
  handwheel lock, lubrication..... 44
  lubrication guide 43
  motor condensers, radio suppression 275
  traverse control lubricant levels, maintaining 49
  wiring tables (See under Electrical system wiring)

Turret and traversing system
  description 307–310
    turret and bracket 307
    turret traversing systems.. 307–310
  maintenance 74
  operation, hydraulic and manual. 310
  replacement 311–313
    control valve assembly 312
    hydraulic motor 312–313
    hydraulic pump 313
    traversing gear assembly..... 312
    turret electric motor 311
    trouble shooting 310–311

U

Universal joints
  description 179–181
  lubrication 44
  maintenance 72
  trouble shooting 181–182
  (See also Propeller shaft (assembly))

Unusual conditions, operation under (See under Operation)

Using arm, lubrication instructions for (See under Lubrication instructions)

V

Vacuum check valve (hydrovac system) 220–221

Valve clearance adjustment
  checking in engine tune-up.... 89–91
  left engine 91
  right engine 90–91

Vehicular equipment 59–60

Venting system 128–129

Viscosity numbers of oil.......... 45

Vision device 32

Vision doors, hull 314

V-type fan belts, use of.......... 82

W

Water pump 140

Wheel cylinder 215–216

Wheels, tires, wheel bearings, and tire pump
  description 238–239
  maintenance 73
  and adjustment 240
  replacement
    wheel and tire.......... 240–241
    tire 242–243
  tire pump 245–247
  trouble shooting 239–240
  wheel bearings 243–245

Windshield wiper motor condensers 276

Windshield wipers, description 272–273

Wiper motor(s)
  replacement 273
  servicing 273

Wire shielding, radio suppression.. 277

Wiring and shielded conduit, replacing 334–336

Wiring connection tables (See Electrical system wiring and Instrument panel)

Wiring, description 278

Wiring diagram, stabilizer 343

Worm and worm wheel, controlling the mesh of 329–331

Worm end play adjustment 329
  bracket 328–329

TM 9-741

MEDIUM ARMORED CAR T17E1

$$\begin{bmatrix} \text{A.G. 062.11 (12-4-42)} \\ \text{GRA WAO 11 Dec. 071737Z} \end{bmatrix}$$

BY ORDER OF THE SECRETARY OF WAR:

                                      G. C. MARSHALL,
                                            Chief of Staff.

OFFICIAL:

    J. A. ULIO,
        Major General,
            The Adjutant General.

DISTRIBUTION: X

    (For explanation of symbols, see **FM 21-6**)

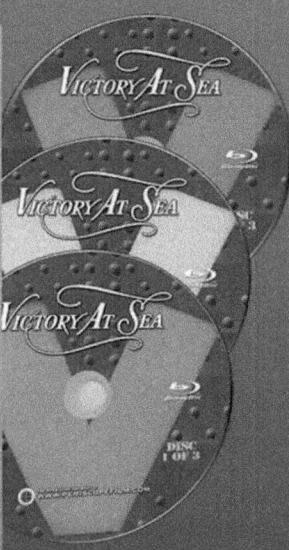

©2013 Periscope Film LLC
All Rights Reserved
ISBN#978-1-937684-40-2
www.PeriscopeFilm.com